QUANTUM COMPUTING

Scientific and Engineering Computation

William Gropp and Ewing Lusk, editors; Janusz Kowalik, founding editor

A complete list of the books in this series can be found at the back of this book.

QUANTUM COMPUTING

A Gentle Introduction

Eleanor Rieffel and Wolfgang Polak

The MIT Press
Cambridge, Massachusetts
London, England

First MIT Press paperback edition, 2014

©2011 Massachusetts Institute of Technology

For information about special quantity discounts, please email special_sales@mitpress.mit.edu

This book was set in Syntax and Times Roman by Westchester Book Group. Printed and bound in the United States of America.

Library of Congress Cataloging-in-Publication Data
Rieffel, Eleanor, 1965–
Quantum computing : a gentle introduction / Eleanor Rieffel and Wolfgang Polak.
 p. cm.—(Scientific and engineering computation)
Includes bibliographical references and index.
ISBN 978-0-262-01506-6 (hardcover : alk. paper)—978-0-262-52667-8 (pb.) 1. Quantum computers. 2. Quantum theory. I. Polak, Wolfgang, 1950– II. Title.
QA76.889.R54 2011
004.1—dc22

2010022682

10 9 8 7 6 5 4 3

Contents

Preface

Quantum computing is a beautiful combination of quantum physics, computer science, and information theory. The purpose of this book is to make this exciting research area accessible to a broad audience. In particular, we endeavor to help the reader bridge the conceptual and notational barriers that separate quantum computing from conventional computing.

The book is concerned with theory: what changes when the classical model underpinning conventional computing is replaced with a quantum one. It contains only a brief discussion of the ongoing efforts to build quantum computers, an active area which is still so young that it is impossible even for experts to predict which approaches will be most successful. While this book is about theory, it is important to ground the discussion of quantum computation in the physics that motivates it. For this reason, the text includes discussions of quantum physics and experiments that illustrate why the theory is defined the way it is.

We precisely define concepts used in quantum computation and emphasize subtle distinctions. This rigor is motivated in part by our experience working with members of the joint FXPAL[1]/PARC[2] reading group and with reviewing papers by authors new to the field. Mistakes commonly arise due to a lack of precision. For example, we take care to distinguish a quantum state from a vector that represents it. We make clear which notions are basis dependent (e.g., superposition) and which are not (e.g., entanglement), and emphasize the dependence of certain notions (e.g., entanglement) on a particular tensor decomposition. The distinction between tensor decompositions and direct sum decompositions, both used extensively in quantum mechanics, is discussed explicitly in both quantum mechanical and classical probabilistic settings. Definitions are carefully motivated. For example, instead of starting with axioms for density operators or mixed states, the definitions of these concepts are motivated by a discussion of what can be deduced about a subsystem from measurements of the subsystem alone.

One advantage of dealing only with theory, and not with the efforts to build quantum computers, is that the amount of quantum physics and supporting mathematics needed is reduced. We are able to develop all of the necessary quantum mechanics within the book; no previous exposure to quantum physics is required. We give careful and precise descriptions of fundamental concepts—such as quantum state spaces, quantum measurement, and entanglement—before covering the standard quantum algorithms and other quantum information processing tasks such as quantum key distribution and quantum teleportation.

The intent of this book is to make quantum computing accessible to a wide audience of computer scientists, engineers, mathematicians, and anyone with a general interest in the subject who knows sufficient mathematics. Basic concepts from college-level linear algebra such as vector spaces, linear transformations, eigenvalues, and eigenvectors are used throughout the book. A few sections require more mathematics; familiarity with group theory is required for sections 8.6.1 and 8.6.2, appendix B, and much of chapter 11. Group theory is reviewed in boxes, but readers who have never seen group theory should consult a book on the subject or skip those sections.

While we hope our book lives up to the *gentle* of its title, reading it will require effort. Many of the concepts are subtle and unintuitive, and much of the notation unfamiliar. Readers will need to spend time working with the concepts and notations to develop a level of fluency at each stage. For example, even readers with significant mathematical background may not have worked much with tensor products and may not be familiar with the relation of tensor product spaces to their component spaces. The early chapters of the book develop these notions carefully, since they are absolutely fundamental to quantum information processing. It is well worth the effort to master them, as well as the concise Dirac notation in which they are generally expressed, but mastery will require effort. The precise nature of these mathematical formalisms provides a means of working with quantum concepts before fully understanding them. Intuition for quantum mechanics and quantum information processing will develop from playing with the formal mathematics.

The book emphasizes features of quantum mechanics that give quantum computation its power and are responsible for its limitations. Neither the extent of the power of quantum computation nor its limitations have been fully understood. Research challenges remain not only in building quantum computers and developing novel algorithms and protocols, but also in answering fundamental questions as to the source of quantum computing's power and the reasons for its limitations. This book examines what is known about what quantum computers can and cannot do, and also explores what is known about why.

The focus on the reasons underlying quantum computing's effectiveness results in the inclusion of topics frequently left out of other expositions of the subject. For example, one theme of the book is the relationship of quantum information processing to probability. That many quantum algorithms are nonprobabilistic is emphasized. A section is devoted to modifications of Grover's original algorithm that preserve the speed-up but return a solution with certainty. On the other hand, the strong formal resemblance between quantum theory and probability theory is described in detail and distinctions are highlighted, illuminating, for example, how entanglement differs from correlation, and the difference between a superposition and a mixture.

As another example, while *quantum entanglement* is the most common explanation given for why quantum information processing works, multipartite entanglement remains poorly understood. Bipartite entanglement is much better understood but has limited use for understanding quantum computation. The book includes sections on multipartite entanglement, a topic often left out of introductory books, and discusses bipartite entanglement. Discussions of multipartite entanglement require examples, which made it natural to include a section on cluster states, the fundamental entanglement resource used for cluster state, or one-way, quantum computation. Cluster state quantum computation and adiabatic quantum computation, two alternatives to the standard circuit model, are briefly introduced and their strengths and applications discussed.

As a final example, while the conversion between general classical circuits and reversible classical circuits is a purely classical topic, it is the heart of the proof that anything a classical computer can do, a quantum computers can do with comparable efficiency. For this reason, the book includes a detailed account of this piece of classical, but nonstandard, computer science.

This is not a book about quantum mechanics. We treat quantum mechanics as an abstract mathematical theory and consider the physical aspects only to elucidate theoretical concepts. We do not discuss issues of interpretation of quantum mechanics; the occasional use of terms such as *quantum parallelism*, for example, is not to be construed as an endorsement of one or another particular interpretation.

Acknowledgments

We are enormously indebted to Michael B. Heaney and Paul McEvoy, both of whom read multiple versions of many of the chapters and provided valuable comments each time. It is largely due to their steadfast belief in this project that the book reached completion. The FXPAL/PARC reading group enabled us to discover which expository approaches worked and which did not. The group's comments, struggles, and insights spurred substantial improvements in the book. We are grateful to all of the members of that group, particularly Dirk Balfanz, Stephen Jackson, and Michael Plass. Many thanks to Tad Hogg and Marc Rieffel for their feedback on some of the most technical and notationally heavy sections. Thanks also go to Gene Golovchinsky for suggestions that clarified and streamlined the writing of an early draft, to Livia Polanyi for suggestions that positively impacted the flow and emphasis, to Garth Dales for comments on an early draft that improved our wording and use of notation, and to Denise Greaves for extensive editorial assistance. Many people provided valuable comments on drafts of the tutorial[3] that was the starting point for this book. Their comments improved this book as well as the tutorial. We gratefully acknowledge the support of FXPAL for part of this work. We are grateful to our friends, to our family, and especially to our spouses for their support throughout the years it took us to write this book.

Notes

1. FX Palo Alto Laboratory.
2. Palo Alto Research Center.
3. E. G. Rieffel and W. Polak. An introduction to quantum computing for non-physicists. *ACM Computing Surveys*, 32(3):300–335, 2000.

1 Introduction

In the last decades of the twentieth century, scientists sought to combine two of the century's most influential and revolutionary theories: information theory and quantum mechanics. Their success gave rise to a new view of computation and information. This new view, quantum information theory, changed forever how computation, information, and their connections with physics are thought about, and it inspired novel applications, including some wildly different algorithms and protocols. This view and the applications it spawned are the subject of this book.

Information theory, which includes the foundations of both computer science and communications, abstracted away the physical world so effectively that it became possible to talk about the major issues within computer science and communications, such as the efficiency of an algorithm or the robustness of a communication protocol, without understanding details of the physical devices used for the computation or the communication. This ability to ignore the underlying physics proved extremely powerful, and its success can be seen in the ubiquity of the computing and communications devices around us. The abstraction away from the physical had become such a part of the intellectual landscape that the assumptions behind it were almost forgotten. At its heart, until recently, information sciences have been firmly rooted in classical mechanics. For example, the Turing machine is a classical mechanical model that behaves according to purely classical mechanical principles.

Quantum mechanics has played an ever-increasing role in the development of new and more efficient computing devices. Quantum mechanics underlies the working of traditional, classical computers and communication devices, from the transistor through the laser to the latest hardware advances that increase the speed and power and decrease the size of computer and communications components. Until recently, the influence of quantum mechanics remained confined to the low-level implementation realm; it had no effect on how computation or communication was thought of or studied.

In the early 1980s, a few researchers realized that quantum mechanics had unanticipated implications for information processing. Charles Bennett and Gilles Brassard, building on ideas of Stephen Wiesner, showed how nonclassical properties of quantum measurement provided a provably secure mechanism for establishing a cryptographic key. Richard Feynman, Yuri Manin, and others recognized that certain quantum phenomena—phenomena associated with so-called entangled particles—could not be simulated efficiently by a Turing machine. This observation led to speculation that perhaps these quantum phenomena could be used to speed up computation in general. Such a program required rethinking the information theoretic model underlying computation, taking it out of the purely classical realm.

Quantum information processing, a field that includes quantum computing, quantum cryptography, quantum communications, and quantum games, explores the implications of using quantum mechanics instead of classical mechanics to model information and its processing. Quantum computing is not about changing the physical substrate on which computation is done from classical to quantum, but rather changing the notion of computation itself. The change starts at the most basic level: the fundamental unit of computation is no longer the bit, but rather the quantum bit or qubit. Placing computation on a quantum mechanical foundation led to the discovery of faster algorithms, novel cryptographic mechanisms, and improved communication protocols.

The phrase *quantum computing* does not parallel the phrases *DNA computing* or *optical computing*: these describe the substrate on which computation is done without changing the notion of computation. *Classical computers*, the ones we all have on our desks, make use of quantum mechanics, but they compute using bits, not qubits. For this reason, they are not considered quantum computers. A quantum or classical computer may or may not be an optical computer, depending on whether optical devices are used to carry out the computation. Whether the computer is quantum or classical depends on whether the information is represented and manipulated in a quantum or classical way. The phrase *quantum computing* is closer in character to *analog computing* because the computational model for analog computing differs from that of standard computing: a continuum of values, rather than only a discrete set, is allowed. While the phrases are parallel, the two models differ greatly in that analog computation does not support entanglement, a key resource for quantum computation, and measurements of a quantum computer's registers can yield only a small, discrete set of values. Furthermore, while a qubit can take on a continuum of values, in many ways a qubit resembles a bit, with its two discrete values, more than it does analog computation. For example, as we will see in section 4.3.1, only one bit's worth of information can be extracted from a qubit by measurement.

The field of quantum information processing developed slowly in the 1980s and early 1990s as a small group of researchers worked out a theory of quantum information and quantum information processing. David Deutsch developed a notion of a quantum mechanical Turing machine. Daniel Bernstein, Vijay Vazirani, and Andrew Yao improved upon his model and showed that a quantum Turing machine could simulate a classical Turing machine, and hence any classical computation, with at most a polynomial time slowdown. The standard quantum circuit model was then defined, which led to an understanding of quantum complexity in terms of a set of basic quantum transformations called quantum gates. These gates are theoretical constructs that may or may not have direct analogs in the physical components of an actual quantum computer.

In the early 1990s, researchers developed the first truly quantum algorithms. In spite of the probabilistic nature of quantum mechanics, the first quantum algorithms, for which superiority over classical algorithms could be proved, give the correct answer with certainty. They improve upon classical algorithms by solving in polynomial time with certainty a problem that can be solved in polynomial time only with high probability using classical techniques. Such a result is of no direct practical interest, since the impossibility of building a perfect machine reduces any practical machine running any algorithm to solving a problem only with high probability. But such results were of high theoretical interest, since they showed for the first time that quantum computation is theoretically more powerful than classical computation for certain computational problems.

These results caught the interest of various researchers, including Peter Shor, who in 1994 surprised the world with his polynomial-time quantum algorithm for factoring integers. This result provided a solution to a well-studied problem of practical interest. A classical polynomial-time solution had long been sought, to the point where the world felt sufficiently confident that no such solution existed that many security protocols, including the widely used RSA algorithm, base their security entirely on the computational difficulty of this problem. It is unknown whether an efficient classical solution exists, so Shor's result does not prove that quantum computers can solve a problem more efficiently than a classical computer. But even in the unlikely event that a polynomial-time classical algorithm is found for this problem, it would be an indication of the elegance and effectiveness of the quantum information theory point of view that a quantum algorithm, in spite of all the unintuitive aspects of quantum mechanics, was easier to find.

While Shor's result sparked a lot of interest in the field, doubts as to its practical significance remained. Quantum systems are notoriously fragile. Key properties, such as quantum entanglement, are easily disturbed by environmental influences that cause the quantum states to *decohere*. Properties of quantum mechanics, such as the impossibility of reliably copying an unknown quantum state, made it look unlikely that effective error-correction techniques for quantum computation could ever be found. For these reasons, it seemed unlikely that reliable quantum computers could be built.

Luckily, in spite of serious and widespread doubts as to whether quantum information processing could ever be practical, the theory itself proved so tantalizing that researchers continued to explore it. As a result, in 1996 Shor and Robert Calderbank, and independently Andrew Steane, saw a way to finesse the seemingly show-stopping problems of quantum mechanics to develop quantum error correction techniques. Today, quantum error correction is arguably the most mature area of quantum information processing.

How practical quantum computing and quantum information will turn out is still unknown. No fundamental physical principles are known that prohibit the building of large-scale and reliable quantum computers. Engineering issues, however, remain. As of this writing, laboratory experiments have demonstrated quantum computations with several quantum bits performing dozens of quantum operations. Myriad promising approaches are being explored by theorists and experimentalists around the world, but much uncertainty remains as to how, when, or even whether, a quantum computer capable of carrying out general quantum computations on hundreds of qubits will be built.

Quantum computational approaches improve upon classical methods for a number of special-ized tasks. The extent of quantum computing's applicability is still being determined. It does not provide efficient solutions to all problems; neither does it provide a universal way of circumvent-ing the slowing of Moore's law. Strong limitations on the power of quantum computation are known; for many problems, it has been proven that quantum computation provides no significant advantage over classical computation. Grover's algorithm, the other major algorithm of the mid-1990s, provides a small speedup for unstructured search algorithms. But it is also known that this small speedup is the most that quantum algorithms can attain. Grover's search algorithm applies to unstructured search. For other search problems, such as searching an ordered list, quantum computation provides no significant advantage over classical computation. Simulation of quan-tum systems is the other significant application of quantum computation known in the mid-1990s. Of interest in its own right, the simulation of increasingly larger quantum systems may provide a bootstrap that will ultimately lead to the building of a scalable quantum computer.

After Grover's algorithm, there was a hiatus of more than five years before a significantly new algorithm was discovered. During that time, other areas of quantum information processing, such as quantum error correction, advanced significantly. In the early 2000s, several new algorithms were discovered. Like Shor's algorithm, these algorithms solve specific problems with narrow, if important, applications. Novel approaches to constructing quantum algorithms also devel-oped. Investigations of quantum simulation from a quantum-information-processing point of view have led to improved classical techniques for simulating quantum systems, as well as novel quan-tum approaches. Similarly, the quantum-information-processing point of view has led to novel insights into classical computing, including new classical algorithms. Furthermore, alternatives to the standard circuit model of quantum computation have been developed that have led to new quan-tum algorithms, breakthroughs in building quantum computers, new approaches to robustness, and significant insights into the key elements of quantum computation.

However long it takes to build a scalable quantum computer and whatever the breadth of applications turns out to be, quantum information processing has changed forever the way in which quantum physics is understood. The quantum information processing view of quantum mechanics has done much to clarify the character of key aspects of quantum mechanics such as quantum measurement and entanglement. This advancement in knowledge has already had applications outside of quantum information processing to the creation of highly entangled states used for microlithography at scales below the wavelength limit and for extraordinarily accurate sensors. The precise practical consequences of this increased understanding of nature are hard to predict, but the unification of the two theories that had the most profound influence on the technological advances of the twentieth century can hardly fail to have profound effects on technological and intellectual developments throughout the twenty-first.

Part I of this book covers the basic building blocks of quantum information processing: quan-tum bits and quantum gates. Physical motivation for these building blocks is given and tied to the key quantum concepts of quantum measurement, quantum state transformations, and entangle-ment between quantum subsystems. Each of these concepts is explored in depth. Quantum key distribution, quantum teleportation, and quantum dense coding are introduced along the way. The final chapter of part I shows that anything that can be done on a classical computer can be done with comparable efficiency on a quantum computer.

Part II covers quantum algorithms. It begins with a description of some of the most common elements of quantum computation. Since the advantage of quantum computation over classical computation is all about efficiency, part II carefully defines notions of complexity. Part II also discusses known bounds on the power of quantum computation. A number of simple algorithms are described. Full chapters are devoted to Shor's algorithm and Grover's algorithm.

Part III explores entanglement and robust quantum computation. A discussion of quantum subsystems leads into discussions of quantifying entanglement and of decoherence, the environmental errors affecting a quantum system because it is really a part of a larger quantum system. The elegant and important topic of quantum error correction fills a chapter, followed by a chapter on techniques to achieve fault tolerance. The book finishes with brief descriptions and pointers to references for many quantum information processing topics the book could not cover in depth. These include further quantum algorithms and protocols, adiabatic, cluster state, holonomic, and topological quantum computing, and the impact quantum information processing has had on classical computer science and physics.

QUANTUM BUILDING BLOCKS

Quantum mechanics, that mysterious, confusing discipline, which none of us really understands, but which we know how to use.
—*Murray Gell-Mann* [126]

2 Single-Qubit Quantum Systems

Quantum bits are the fundamental units of information in quantum information processing in much the same way that bits are the fundamental units of information for classical processing. Just as there are many ways to realize classical bits physically (two voltage levels, lights on or off in an array, positions of toggle switches), there are many ways to realize quantum bits physically. As is done in classical computer science, we will concern ourselves only rarely with how the quantum bits are realized. For the sake of concretely illustrating quantum bits and their properties, however, section 2.1 looks at the behavior of polarized photons, one of many possible realizations of quantum bits.

Section 2.2 abstracts key properties from the polarized photon example of section 2.1 to give a precise definition of a quantum bit, or qubit, and a description of the behavior of quantum bits under measurement. Dirac's bra/ket notation, the standard notation used throughout quantum information processing as well as quantum mechanics, is introduced in this section. Section 2.4 describes the first application of quantum information processing: quantum key distribution. The chapter concludes with a detailed discussion of the state space of a single-qubit system.

2.1 The Quantum Mechanics of Photon Polarization

A simple experiment illustrates some of the nonintuitive behavior of quantum systems, behavior that is exploited to good effect in quantum algorithms and protocols. This experiment can be performed by the reader using only minimal equipment: a laser pointer and three polaroids (polarization filters), readily available from any camera supply store. The formalisms of quantum mechanics that describe this simple experiment lead directly to a description of the quantum bit, the fundamental unit of quantum information on which quantum information processing is done. The experiment not only gives a concrete realization of a quantum bit, but it also illustrates key properties of quantum measurement. We encourage you to obtain the equipment and perform the experiment yourself.

2.1.1 A Simple Experiment

Shine a beam of light on a projection screen. When polaroid A is placed between the light source and the screen, the intensity of the light reaching the screen is reduced. Let us suppose that the polarization of polaroid A is horizontal (figure 2.1).

Next, place polaroid C between polaroid A and the projection screen. If polaroid C is rotated so that its polarization is orthogonal (vertical) to the polarization of A, no light reaches the screen (figure 2.2).

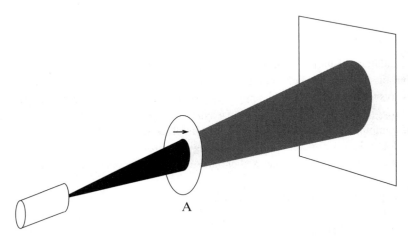

Figure 2.1
Single polaroid attenuates unpolarized light by 50 percent.

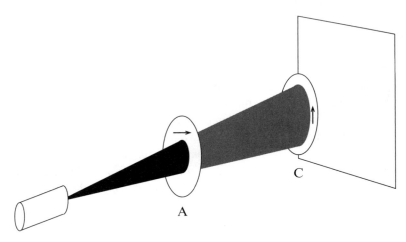

Figure 2.2
Two orthogonal polaroids block all photons.

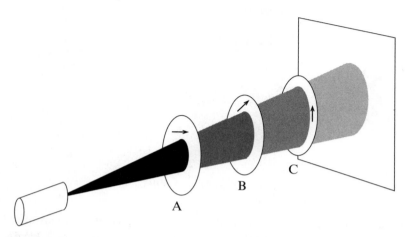

Figure 2.3
Inserting a third polaroid allows photons to pass.

Finally, place polaroid B between polaroids A and C. One might expect that adding another polaroid will not make any difference; if no light got through two polaroids, then surely no light will pass through three! Surprisingly, at most polarization angles of B, light shines on the screen. The intensity of this light will be maximal if the polarization of B is at 45 degrees to both A and C (figure 2.3).

Clearly the polaroids cannot be acting as simple sieves; otherwise, inserting polaroid B could not increase the number of photons reaching the screen.

2.1.2 A Quantum Explanation

For a bright beam of light, there is a classical explanation of the experiment in terms of waves. Versions of the experiment described here, using light so dim that only one photon at a time interacts with the polaroid, have been done with more sophisticated equipment. The results of these single photon experiments can be explained only using quantum mechanics; the classical wave explanation no longer works. Furthermore, it is not just light that behaves in this peculiar way. The quantum mechanical explanation of the experiment consists of two parts: a model of a photon's polarization state and a model of the interaction between a polaroid and a photon. The description of this experiment, and the definition of a qubit, use basic notions of linear algebra such as *vector*, *basis*, *orthonormal*, and *linear combination*. Linear algebra is used throughout the book; we briefly remind readers of the meanings of these concepts in section 2.2. Section 2.6 suggests some books on linear algebra.

Quantum mechanics models a photon's polarization state by a unit vector, a vector of length 1, pointing in the appropriate direction. We write $|\uparrow\rangle$ and $|\rightarrow\rangle$ for the unit vectors that represent vertical and horizontal polarization respectively. Think of $|v\rangle$ as a vector with some arbitrary label v. In quantum mechanics, the standard notation for a vector representing a quantum state

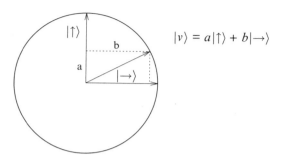

$$|v\rangle = a|\uparrow\rangle + b|\rightarrow\rangle$$

Figure 2.4
Measurement of state $|v\rangle = a|\uparrow\rangle + b|\rightarrow\rangle$ by a measuring device with preferred basis $\{|\uparrow\rangle, |\rightarrow\rangle\}$.

is $|v\rangle$, just as $\vec{v}$ or **v** are notations used for vectors in other settings. This notation is part of a more general notation, Dirac's notation, that will be explained in more detail in sections 2.2 and 4.1. An arbitrary polarization can be expressed as a linear combination $|v\rangle = a|\uparrow\rangle + b|\rightarrow\rangle$ of the two basis vectors $|\uparrow\rangle$ and $|\rightarrow\rangle$. For example, $|\nearrow\rangle = \frac{1}{\sqrt{2}}|\uparrow\rangle + \frac{1}{\sqrt{2}}|\rightarrow\rangle$ is a unit vector representing polarization of 45 degrees. The coefficients a and b in $|v\rangle = a|\uparrow\rangle + b|\rightarrow\rangle$ are called the *amplitudes* of $|v\rangle$ in the directions $|\uparrow\rangle$ and $|\rightarrow\rangle$ respectively (see figure 2.4). When a and b are both non-zero, $|v\rangle = a|\uparrow\rangle + b|\rightarrow\rangle$ is said to be a *superposition* of $|\uparrow\rangle$ and $|\rightarrow\rangle$.

Quantum mechanics models the interaction between a photon and a polaroid as follows. The polaroid has a preferred axis, its polarization. When a photon with polarization $|v\rangle = a|\uparrow\rangle + b|\rightarrow\rangle$ meets a polaroid with preferred axis $|\uparrow\rangle$, the photon will get through with probability $|a|^2$ and will be absorbed with probability $|b|^2$; the probability that a photon passes through the polaroid is the square of the magnitude of the amplitude of its polarization in the direction of the polaroid's preferred axis. The probability that the photon is absorbed by the polaroid is the square of the magnitude of the amplitude in the direction perpendicular to the polaroid's preferred axis. Furthermore, any photon that passes through the polaroid will now be polarized in the direction of the polaroid's preferred axis. The probabilistic nature of the interaction and the resulting change of state are features of all interactions between qubits and measuring devices, no matter what their physical realization.

In the experiment, any photons that pass through polaroid A will leave polarized in the direction of polaroid A's preferred axis, in this case horizontal, $|\rightarrow\rangle$. A horizontally polarized photon has no amplitude in the vertical direction, so it has no chance of passing through polaroid C, which was given a vertical orientation. For this reason, no light reaches the screen. Had polaroid C been in any other orientation, a horizontally polarized photon would have some amplitude in the direction of polaroid C's preferred axis, and some photons would reach the screen.

To understand what happens once polaroid B, with preferred axis $|\nearrow\rangle$, is inserted, it is helpful to write the horizontally polarized photon's polarization state $|\rightarrow\rangle$ as

$$|\rightarrow\rangle = \frac{1}{\sqrt{2}}|\nearrow\rangle - \frac{1}{\sqrt{2}}|\nwarrow\rangle.$$

Any photon that passes through polaroid A becomes horizontally polarized, so the amplitude of any such photon's state $|\rightarrow\rangle$ in the direction $|\nearrow\rangle$ is $\frac{1}{\sqrt{2}}$. Applying the quantum theory we just learned tells us that a horizontally polarized photon will pass through polaroid B with probability $\frac{1}{2} = |\frac{1}{\sqrt{2}}|^2$. Any photons that have passed through polaroid B now have polarization $|\nearrow\rangle$. When these photons hit polaroid C, they do have amplitude in the vertical direction, so some of them (half) will pass thorough polaroid C and hit the screen (see figure 2.3). In this way, quantum mechanics explains how more light can reach the screen when the third polaroid is added, and it provides a means to compute how much light will reach the screen.

In summary, the polarization state of a photon is modeled as a unit vector. Its interaction with a polaroid is probabilistic and depends on the amplitude of the photon's polarization in the direction of the polaroid's preferred axis. Either the photon will be absorbed or the photon will leave the polaroid with its polarization aligned with the polaroid's preferred axis.

2.2 Single Quantum Bits

The space of possible polarization states of a photon is an example of a *quantum bit*, or *qubit*. A qubit has a continuum of possible values: any state represented by a unit vector $a|\uparrow\rangle + b|\rightarrow\rangle$ is a legitimate qubit value. The amplitudes a and b can be complex numbers, even though complex amplitudes were not needed for the explanation of the experiment. (In the photon polarization case, the imaginary coefficients correspond to *circular polarization*.)

In general, the set of all possible states of a physical system is called the *state space* of the system. Any quantum mechanical system that can be modeled by a two-dimensional complex vector space can be viewed as a qubit. (There is redundancy in this representation in that any vector multiplied by a modulus one [unit length] complex number represents the same quantum state. We discuss this redundancy carefully in sections 2.5 and 3.1.) Such systems, called *two-state quantum systems*, include photon polarization, electron spin, and the ground state together with an excited state of an atom. The *two-state* label for these systems does not mean that the state space has only two states—it has infinitely many—but rather that all possible states can be represented as a linear combination, or superposition, of just two states. For a two-dimensional complex vector space to be viewed as a qubit, two linearly independent states, labeled $|0\rangle$ and $|1\rangle$, must be distinguished. For the theory of quantum information processing, all two-state systems, whether they be electron spin or energy levels of an atom, are equally good. From a practical point of view, it is as yet unclear which two-state systems will be most suitable for physical realizations of quantum information processing devices such as quantum computers; it is likely that a variety of physical representation of qubits will be used.

Dirac's bra / ket notation is used throughout quantum physics to represent quantum states and their transformations. In this section we introduce the part of Dirac's notation that is used for quantum states. Section 4.1 introduces Dirac's notation for quantum transformations. Familiarity and fluency with this notation will help greatly in understanding all subsequent material; we strongly encourage readers to work the exercises at the end of this chapter.

In Dirac's notation, a *ket* such as $|x\rangle$, where x is an arbitrary label, refers to a vector representing a state of a quantum system. A vector $|v\rangle$ is a *linear combination* of vectors $|s_1\rangle$, $|s_2\rangle$, ..., $|s_n\rangle$ if there exist complex numbers a_i such that $|v\rangle = a_1|s_1\rangle + a_2|s_2\rangle + \cdots + a_n|s_n\rangle$.

A set of vectors S *generates* a complex vector space V if every element $|v\rangle$ of V can be written as a complex linear combination of vectors in the set: every $|v\rangle \in V$ can be written as $|v\rangle = a_1|s_1\rangle + a_2|s_2\rangle + \cdots + a_n|s_n\rangle$ for some elements $|s_i\rangle \in S$ and complex numbers a_i. Given a set of vectors S, the subspace of all linear combinations of vectors in S is called the *span* of S and is denoted $\mathrm{span}(S)$. A set of vectors B for which every element of V can be written *uniquely* as a linear combination of vectors in B is called a *basis* for V. In a two-dimensional vector space, any two vectors that are not multiples of each other form a basis. In quantum mechanics, bases are usually required to be *orthonormal*, a property we explain shortly. The two distinguished states, $|0\rangle$ and $|1\rangle$, are also required to be orthonormal.

An *inner product* $\langle v_2|v_1\rangle$, or *dot product*, on a complex vector space V is a complex function defined on pairs of vectors $|v_1\rangle$ and $|v_2\rangle$ in V, satisfying

- $\langle v|v\rangle$ is non-negative real,
- $\langle v_2|v_1\rangle = \overline{\langle v_1|v_2\rangle}$, and
- $\langle v_1|(a|v_2\rangle + b|v_3\rangle) = a\langle v_1|v_2\rangle + b\langle v_1|v_3\rangle)$,

where $\overline{z}$ is the complex conjugate $\overline{z} = a - \mathbf{i}b$ of $z = a + \mathbf{i}b$.

Two vectors $|v_1\rangle$ and $|v_2\rangle$ are said to be *orthogonal* if $\langle v_1|v_2\rangle = 0$. A set of vectors is orthogonal if all of its members are orthogonal to each other. The *length*, or norm, of a vector $|v\rangle$ is $\||v\rangle\| = \sqrt{\langle v|v\rangle}$. Since all vectors $|x\rangle$ representing quantum states are of unit length, $\langle x|x\rangle = 1$ for any state vector $|x\rangle$. A set of vectors is said to be *orthonormal* if all of its elements are of length one and orthogonal to each other: a set of vectors $B = \{|\beta_1\rangle, |\beta_2\rangle, \ldots, |\beta_n\rangle\}$ is orthonormal if $\langle \beta_i|\beta_j\rangle = \delta_{ij}$ for all i, j, where

$$\delta_{ij} = \begin{cases} 1 & \text{if } i = j \\ 0 & \text{otherwise.} \end{cases}$$

In quantum mechanics we are mainly concerned with bases that are orthonormal, so whenever we say *basis* we mean *orthonormal basis* unless we say otherwise.

For the state space of a two-state system to represent a quantum bit, two orthonormal distinguished states, labeled $|0\rangle$ and $|1\rangle$, must be specified. Apart from the requirement that $|0\rangle$ and $|1\rangle$ be orthonormal, the states may be chosen arbitrarily. For instance, in the case of photon polarization, we may choose $|0\rangle$ and $|1\rangle$ to correspond to the states $|\uparrow\rangle$ and $|\rightarrow\rangle$, or to $|\nearrow\rangle$ and $|\nwarrow\rangle$. We follow the convention that $|0\rangle = |\uparrow\rangle$ and $|1\rangle = |\rightarrow\rangle$, which implies that $|\nearrow\rangle = \frac{1}{\sqrt{2}}(|0\rangle + |1\rangle)$ and $|\nwarrow\rangle = \frac{1}{\sqrt{2}}(|0\rangle - |1\rangle)$. In the case of electron spin, $|0\rangle$ and $|1\rangle$ could correspond to the spin-up and spin-down states, or spin-left and spin-right. When talking about qubits, and quantum information processing in general, a *standard basis* $\{|0\rangle, |1\rangle\}$ with respect to which all statements are made must be chosen in advance and remain fixed throughout the discussion. In quantum information processing, classical bit values of 0 and 1 will be encoded in the distinguished states $|0\rangle$ and $|1\rangle$. This encoding enables a direct comparison between bits and qubits: bits can take on only two values, 0 and 1, while qubits can take on not only the values $|0\rangle$ and $|1\rangle$ but also any superposition of these values, $a|0\rangle + b|1\rangle$, where a and b are complex numbers such that $|a|^2 + |b|^2 = 1$.

Vectors and linear transformations can be written using matrix notation once a basis has been specified. That is, if basis $\{|\beta_1\rangle, |\beta_2\rangle\}$ is specified, a ket $|v\rangle = a|\beta_1\rangle + b|\beta_2\rangle$ can be written $\begin{pmatrix} a \\ b \end{pmatrix}$; a ket $|v\rangle$ corresponds to a column vector v, where v is simply a label, a name for this vector. The conjugate transpose $v^\dagger$ of a vector

$$v = \begin{pmatrix} a_1 \\ \vdots \\ a_n \end{pmatrix} \quad \text{is} \quad v^\dagger = (\overline{a_1}, \ldots, \overline{a_n}) \ .$$

In Dirac's notation, the conjugate transpose of a ket $|v\rangle$ is called a *bra* and is written $\langle v|$, so

$$|v\rangle = \begin{pmatrix} a_1 \\ \vdots \\ a_n \end{pmatrix} \quad \text{and} \quad \langle v| = (\overline{a_1}, \ldots, \overline{a_n}) \ .$$

A bra $\langle v|$ corresponds to a row vector $v^\dagger$.

Given two complex vectors

$$|a\rangle = \begin{pmatrix} a_1 \\ \vdots \\ a_n \end{pmatrix} \quad \text{and} \quad |b\rangle = \begin{pmatrix} b_1 \\ \vdots \\ b_n \end{pmatrix},$$

the standard *inner product* $\langle a|b\rangle$ is defined to be the scalar obtained by multiplying the conjugate transpose $\langle a| = (\overline{a_1}, \ldots, \overline{a_n})$ with $|b\rangle$:

$$\langle a|b\rangle = \langle a||b\rangle = (\overline{a_1}, \ldots, \overline{a_n}) \begin{pmatrix} b_1 \\ \vdots \\ b_n \end{pmatrix} = \sum_{i=1}^{n} \overline{a_i} b_i.$$

When $\vec{a} = |a\rangle$ and $\vec{b} = |b\rangle$ are real vectors, this inner product is the same as the standard dot product on the n dimensional real vector space $\mathbf{R}^n$: $\langle a|b\rangle = a_1 b_1 + \cdots + a_n b_n = \vec{a} \cdot \vec{b}$. Dirac's choice of *bra* and *ket* arose as a play on words: an inner product $\langle a|b\rangle$ of a bra $\langle a|$ and a ket $|b\rangle$ is sometimes called a *bracket*. The following relations hold, where $v = a|0\rangle + b|1\rangle$: $\langle 0|0\rangle = 1$, $\langle 1|1\rangle = 1$, $\langle 1|0\rangle = \langle 0|1\rangle = 0$, $\langle 0|v\rangle = a$, and $\langle 1|v\rangle = b$.

In the standard basis, with ordering $\{|0\rangle, |1\rangle\}$, the basis elements $|0\rangle$ and $|1\rangle$ can be expressed as $\begin{pmatrix} 1 \\ 0 \end{pmatrix}$ and $\begin{pmatrix} 0 \\ 1 \end{pmatrix}$, and a complex linear combination $|v\rangle = a|0\rangle + b|1\rangle$ can be written $\begin{pmatrix} a \\ b \end{pmatrix}$.

This choice of basis and order of the basis vectors are mere convention. Representing $|0\rangle$ as $\begin{pmatrix} 1 \\ 0 \end{pmatrix}$ and $|1\rangle$ as $\begin{pmatrix} 0 \\ 1 \end{pmatrix}$ or representing $|0\rangle$ as $\frac{1}{\sqrt{2}}\begin{pmatrix} 1 \\ -1 \end{pmatrix}$ and $|1\rangle$ as $\frac{1}{\sqrt{2}}\begin{pmatrix} 1 \\ 1 \end{pmatrix}$ would be equally good as long as it is done consistently. Unless otherwise specified, all vectors and matrices in this book will be written with respect to the standard basis $\{|0\rangle, |1\rangle\}$ in this order.

A quantum state $|v\rangle$ is a *superposition* of basis elements $\{|\beta_1\rangle, |\beta_2\rangle\}$ if it is a nontrivial linear combination of $|\beta_1\rangle$ and $|\beta_2\rangle$, if $|v\rangle = a_1|\beta_1\rangle + a_2|\beta_2\rangle$ where a_1 and a_2 are non-zero. For the term *superposition* to be meaningful, a basis must be specified. In this book, if we say "superpostion" without explicitly specifying the basis, we implicitly mean with respect to the standard basis.

Initially the vector/matrix notation will be easier for many readers to use because it is familiar. Sometimes matrix notation is convenient for performing calculations, but it always requires the choice of a basis and an ordering of that basis. The bra/ket notation has the advantage of being independent of basis and the order of the basis elements. It is also more compact and suggests correct relationships, as we saw for the inner product, so once it becomes familiar, it is easier to read and faster to use.

Instead of qubits, physical systems with states modeled by three- or n-dimensional vector spaces could be used as fundamental units of computation. Three-valued units are called *qutrits*, and n-valued units are called *qudits*. Since qudits can be modeled using multiple qubits, a model of quantum information based on qudits has the same computational power as one based on qubits. For this reason we do not consider qudits further, just as in the classical case most people use a bit-based model of information.

We now have a mathematical model with which to describe quantum bits. In addition, we need a mathematical model for measuring devices and their interaction with quantum bits.

2.3 Single-Qubit Measurement

The interaction of a polaroid with a photon illustrates key properties of any interaction between a measuring device and a quantum system. The mathematical description of the experiment can be used to model all measurements of single qubits, whatever their physical instantiation. The measurement of more complicated systems retains many of the features of single-qubit measurement: the probabilistic outcomes and the effect measurement has on the state of the system. This section considers only measurements of single-qubit systems. Chapter 4 discusses measurements of more general quantum systems.

Quantum theory postulates that any device that measures a two-state quantum system must have two preferred states whose representative vectors, $\{|u\rangle, |u^\perp\rangle\}$, form an orthonormal basis for the associated vector space. Measurement of a state transforms the state into one of the measuring device's associated basis vectors $|u\rangle$ or $|u^\perp\rangle$. The probability that the state is measured as basis vector $|u\rangle$ is the square of the magnitude of the amplitude of the component of the state in the direction of the basis vector $|u\rangle$. For example, given a device for measuring the polarization of photons with associated basis $\{|u\rangle, |u^\perp\rangle\}$, the state $|v\rangle = a|u\rangle + b|u^\perp\rangle$ is measured as $|u\rangle$ with probability $|a|^2$ and as $|u^\perp\rangle$ with probability $|b|^2$.

This behavior of measurement is an axiom of quantum mechanics. It is not derivable from other physical principles; rather, it is derived from the empirical observation of experiments with measuring devices. If quantum mechanics is correct, all devices that measure single qubits must behave in this way; all have associated bases, and the measurement outcome is always one of the two basis vectors. For this reason, whenever anyone says "measure a qubit," they must specify with respect to which basis the measurement takes place. Throughout the book, if we say "measure a qubit" without further elaboration, we mean that the measurement is with respect to the standard basis $\{|0\rangle, |1\rangle\}$.

Measurement of a quantum state changes the state. If a state $|v\rangle = a|u\rangle + b|u^\perp\rangle$ is measured as $|u\rangle$, then the state $|v\rangle$ changes to $|u\rangle$. A second measurement with respect to the same basis will return $|u\rangle$ with probability 1. Thus, unless the original state happens to be one of the basis states, a single measurement will change that state, making it impossible to determine the original state from any sequence of measurements.

While the mathematics of measuring a qubit in the superposition state $a|0\rangle + b|1\rangle$ with respect to the standard basis is clear, measurement brings up questions as to the meaning of a superposition. To begin with, the notion of superposition is basis-dependent; all states are superpositions with respect to some bases and not with respect to others. For instance, $a|0\rangle + b|1\rangle$ is a superposition with respect to the basis $\{|0\rangle, |1\rangle\}$ but not with respect to $\{a|0\rangle + b|1\rangle, \overline{b}|0\rangle - \overline{a}|1\rangle\}$.

Also, because the result of measuring a superposition is probabilistic, some people are tempted to think of the state $|v\rangle = a|0\rangle + b|1\rangle$ as a probabilistic mixture of $|0\rangle$ and $|1\rangle$. It is not. In particular, it is not true that the state is really either $|0\rangle$ or $|1\rangle$ and that we just do not happen to know which. Rather, $|v\rangle$ is a definite state, which, when measured in certain bases, gives deterministic results, while in others it gives random results: a photon with polarization $|\nearrow\rangle = \frac{1}{\sqrt{2}}(|\uparrow\rangle + |\rightarrow\rangle)$ behaves deterministically when measured with respect to the Hadamard basis $\{|\nearrow\rangle, |\nwarrow\rangle\}$, but it gives random results when measured with respect to the standard basis $\{|\uparrow\rangle, |\rightarrow\rangle\}$. It is okay to think of a superposition $|v\rangle = a|0\rangle + b|1\rangle$ as in some sense being in both state $|0\rangle$ and state $|1\rangle$ at the same time, as long as that statement is not taken too literally: states that are combinations of $|0\rangle$ and $|1\rangle$ in similar proportions but with different amplitudes, such as $\frac{1}{\sqrt{2}}(|0\rangle + |1\rangle)$, $\frac{1}{\sqrt{2}}(|0\rangle - |1\rangle)$ and $\frac{1}{\sqrt{2}}(|0\rangle + i|1\rangle)$, represent distinct states that behave differently in many situations.

Given that qubits can take on any one of infinitely many states, one might hope that a single qubit could store lots of classical information. However, the properties of quantum measurement severely restrict the amount of information that can be extracted from a qubit. Information about a quantum bit can be obtained only by measurement, and any measurement results in one of only two states, the two basis states associated with the measuring device; thus, a single measurement yields at most a single classical bit of information. Because measurement changes the state, one cannot make two measurements on the original state of a qubit. Furthermore, section 5.1.1 shows that an unknown quantum state cannot be cloned, which means it is not possible to measure a qubit's state in two ways, even indirectly by copying the qubit's state and measuring the copy.

Thus, even though a quantum bit can be in infinitely many different superposition states, it is possible to extract only a single classical bit's worth of information from a single quantum bit.

2.4 A Quantum Key Distribution Protocol

The quantum theory introduced so far is sufficient to describe a first application of quantum information processing: a key distribution protocol that relies on quantum effects for its security and for which there is no classical analog.

Keys—binary strings or numbers chosen randomly from a sufficiently large set—provide the security for most cryptographic protocols, from encryption to authentication to secret sharing. For this reason, the establishment of keys between the parties who wish to communicate is of fundamental importance in cryptography. Two general classes of keys exist: symmetric keys and public-private key pairs. Both types are used widely, often in conjunction, in a wide variety of practical settings, from secure e-commerce transactions to private communication over public networks.

Public-private key pairs consist of a public key, knowable by all, and a corresponding private key whose secrecy must be carefully guarded by the owner. Symmetric keys consist of a single key (or a pair of keys easily computable from one another) that are known to all of the legitimate parties and no one else. In the symmetric key case, multiple parties are responsible for guarding the security of the key.

Quantum key distribution protocols establish a symmetric key between two parties, who are generally known in the cryptographic community as Alice and Bob. Quantum key distribution protocols can be used securely anywhere classical key agreement protocols such as Diffie-Hellman can be used. They perform the same task; however, the security of quantum key distribution rests on fundamental properties of quantum mechanics, whereas classical key agreement protocols rely on the computational intractability of a certain problem. For example, while Diffie-Hellman remains secure against all known classical attacks, the problem on which it is based, the discrete logarithm problem, is tractable on a quantum computer. Section 8.6.1 discusses Shor's quantum algorithm for the discrete log problem.

The earliest quantum key distribution protocol is known as BB84 after its inventors, Charles Bennett and Gilles Brassard, and the year of the invention. The aim of the BB84 protocol is to establish a secret key, a random sequence of bit values 0 and 1, known only to the two parties, Alice and Bob, who may use this key to support a cryptographic task such as exchanging secret messages or detecting tampering. The BB84 protocol enables Alice and Bob to be sure that if they detect no problems while attempting to establish a key, then with high probability it is secret. The protocol does not guarantee, however, that they will succeed in establishing a private key.

Suppose Alice and Bob are connected by two public channels: an ordinary bidirectional classical channel and a unidirectional quantum channel. The quantum channel allows Alice to send a sequence of single qubits to Bob; in our case we suppose the qubits are encoded in the polarization states of individual photons. Both channels can be observed by an eavesdropper Eve. This situation

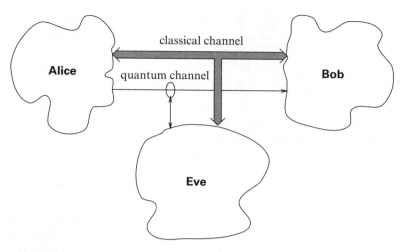

Figure 2.5
Alice and Bob wish to agree on a common key not known to Eve.

is illustrated in figure 2.5. To begin the process of establishing a private key, Alice uses quantum
or classical means to generate a random sequence of classical bit values. As we will see, a random
subset of this sequence will be the final private key. Alice then randomly encodes each bit of
this sequence in the polarization state of a photon by randomly choosing for each bit one of the
following two agreed-upon bases in which to encode it: the standard basis,

$0 \rightarrow |\uparrow\rangle$

$1 \rightarrow |\rightarrow\rangle$,

or the Hadamard basis,

$0 \rightarrow |\nearrow\rangle = \frac{1}{\sqrt{2}}(|\uparrow\rangle + |\rightarrow\rangle))$
$1 \rightarrow |\nwarrow\rangle = \frac{1}{\sqrt{2}}(|\uparrow\rangle - |\rightarrow\rangle))$.

She sends this sequence of photons to Bob through the quantum channel.

Bob measures the state of each photon he receives by randomly picking either basis. Over
the classical channel, Alice and Bob check that Bob has received a photon for every one Alice
has sent, and only then do Alice and Bob tell each other the bases they used for encoding and
decoding (measuring) each bit. When the choice of bases agree, Bob's measured bit value agrees
with the bit value that Alice sent. When they chose different bases, the chance that Bob's bit
matches Alice's is only 50 percent. Without revealing the bit values themselves, which would
also reveal the values to Eve, there is no way for Alice and Bob to figure out which of these bit
values agree and which do not. So they simply discard all the bits on which their choice of bases
differed. An average of 50 percent of all bits transmitted remain. Then, depending on the level of

assurance they require, Alice and Bob compare a certain number of bit values to check that no eavesdropping has occurred. These bits will also be discarded, and only the remaining bits will be used as their private key.

We describe one sort of attack that Eve can make and how quantum aspects of this protocol guard against it. On the classical channel, Alice and Bob discuss only the choice of bases and not the bit values themselves, so Eve cannot gain any information about the key from listening to the classical channel alone. To gain information, Eve must intercept the photons transmitted by Alice through the quantum channel. Eve must send photons to Bob before knowing the choice of bases made by Alice and Bob, because they compare bases only after Bob has confirmed receipt of the photons. If she sends different photons to Bob, Alice and Bob will detect that something is wrong when they compare bit values, but if she sends the original photons to Bob without doing anything, she gains no information.

To gain information, Eve makes a measurement before sending the photons to Bob. Instead of using a polaroid to measure, she can use a calcite crystal and a photon detector; a beam of light passing through a calcite crystal is split into two spatially separated beams, one polarized in the direction of the crystal's optic axis and the other polarized in the direction perpendicular to the optic axis. A photon detector placed in one of the beams performs a quantum measurement: the probability with which a photon ends up in one of the beams can be calculated just as described in section 2.3.

Since Alice has not yet told Bob her sequence of bases, Eve does not know in which basis to measure each bit. If she randomly measures the bits, she will measure using the wrong basis approximately half of the time. (Exercise 2.10 examines the case in which Eve does not even know which two bases to choose from.) When she uses the wrong basis to measure, the measurement changes the polarization of the photon before it is resent to Bob. This change in the polarization means that, even if Bob measures the photon in the same basis as Alice used to encode the bit, he will get the correct bit value only half the time.

Overall, for each of the qubits Alice and Bob retain, if the qubit was measured by Eve before she sent it to Bob, there will be a 25 percent chance that Bob measures a different bit value than the one Alice sent. Thus, this attack on the quantum channel is bound to introduce a high error rate that Alice and Bob detect by comparing a sufficient number of bits over the classical channel. If these bits agree, they can confidently use the remaining bits as their private key. So, not only is it likely that 25 percent of Eve's version of the key is incorrect, but the fact that someone is eavesdropping can be detected by Alice and Bob. Thus Alice and Bob run little risk of establishing a compromised key; either they succeed in creating a private key or they detect that eavesdropping has taken place.

Eve does not know in which basis to measure the qubits, a property crucial to the security of this protocol, because Alice and Bob share information about which bases they used only after Bob has received the photons; if Eve knew in which basis to measure the photons, her measurements would not change the state, and she could obtain the bit values without Bob and Alice noticing anything suspicious. A seemingly easy way for Eve to overcome this obstacle is for her to copy the qubit, keeping a copy for herself while sending the original on to Bob. Then she can measure her copy later after learning the correct basis from listening in on the classical channel. Such a protocol is defeated by an important property of quantum information. As we will show in section 5.1.1, the no-cloning principle of quantum mechanics means that it is impossible to reliably copy quantum information unless a basis in which it is encoded is known; all quantum copying machines are basis dependent. Copying with the wrong machine not only does not produce an accurate copy, but it also changes the original in much the same way measuring in the wrong basis does. So Bob and Alice would detect attempts to copy with high probability.

The security of this protocol, like other pure key distribution protocols such as Diffie-Hellman, is vulnerable to a *man-in-the-middle attack* in which Eve impersonates Bob to Alice and impersonates Alice to Bob. To guard against such an attack, Alice and Bob need to combine it with an authentication protocol, be it recognizing each other's voices or a more mathematical authentication protocol.

More sophisticated versions of this protocol exist that support quantum key distribution through noisy channels and stronger guarantees about the amount of information Eve can gain. In the noisy case, Eve is able to gain some information initially, but techniques of quantum error correction and privacy amplification can reduce the amount of information Eve gains to arbitrarily low levels as well as compensate for the noise in the channels.

2.5 The State Space of a Single-Qubit System

The *state space* of a classical or quantum physical system is the set of all possible states of the system. Depending on which properties of the system are under consideration, a state of the system consists of any combination of the positions, momenta, polarizations, spins, energy, and so on of the particles in the system. When we are considering only polarization states of a single photon, the state space is all possible polarizations. More generally, the state space for a single qubit, no matter how it is realized, is the set of possible qubit values,

$$\{a|0\rangle + b|1\rangle\},$$

where $|a|^2 + |b|^2 = 1$ and $a|0\rangle + b|1\rangle$ and $a'|0\rangle + b'|1\rangle$ are considered the same qubit value if $a|0\rangle + b|1\rangle = c(a'|0\rangle + b'|1\rangle)$ for some modulus one complex number c.

2.5.1 Relative Phases versus Global Phases

That the same quantum state is represented by more than one vector means that there is a critical distinction between the complex vector space in which we write our qubit values and the quantum state space itself. We have reduced the ambiguity by requiring that vectors representing quantum states be unit vectors, but some ambiguity remains: unit vectors equivalent up to multiplication by a complex number of modulus one represent the same state. The multiple by which two vectors representing the same quantum state differ is called the *global phase* and has no physical meaning. We use the equivalence relation $|v\rangle \sim |v'\rangle$ to indicate that $|v\rangle = c|v'\rangle$ for some complex global phase $c = e^{i\phi}$. The space in which two two-dimensional complex vectors are considered equivalent if they are multiples of each other is called *complex projective space* of dimension one.

This *quotient space*, a space obtained by identifying sets of equivalent vectors with a single point in the space, is expressed with the compact notation used for quotient spaces:

$$\mathbf{CP}^1 = \{a|0\rangle + b|1\rangle\}/ \sim .$$

So the quantum state space for a single-qubit system is in one-to-one correspondence with the points of the complex projective space $\mathbf{CP}^1$. We will make no further use of $\mathbf{CP}^1$ in this book, but it is used in the quantum information processing literature.

Because the linearity of vector spaces makes them easier to work with than projective spaces (we know how to add vectors and there is no corresponding way of adding points in projective spaces), we generally perform all calculations in the vector space corresponding to the quantum state space. The multiplicity of representations of a single quantum state in this vector space representation, however, is a common source of confusion for newcomers to the field.

A physically important quantity is the *relative phase* of a single-qubit state $a|0\rangle + b|1\rangle$. The relative phase (in the standard basis) of a superposition $a|0\rangle + b|1\rangle$ is a measure of the angle in the complex plane between the two complex numbers a and b. More precisely, the relative phase is the modulus one complex number $e^{i\phi}$ satisfying $a/b = e^{i\phi}|a|/|b|$. Two superpositions $a|0\rangle + b|1\rangle$ and $a'|0\rangle + b'|1\rangle$ whose amplitudes have the same magnitudes but that differ in a relative phase represent different states.

The physically meaningful relative phase and the physically meaningless global phase should not be confused. While multiplication with a unit constant does not change a quantum state vector, relative phases in a superposition do represent distinct quantum states: even though $|v_1\rangle \sim e^{i\phi}|v_1\rangle$, the vectors $\frac{1}{\sqrt{2}}(e^{i\phi}|v_1\rangle + |v_2\rangle)$ and $\frac{1}{\sqrt{2}}(|v_1\rangle + |v_2\rangle)$ do *not* represent the same state. We must always be cognizant of the $\sim$ equivalence when we interpret the results of our computations as quantum states.

A few single-qubit states will be referred to often enough that we give them special labels:

$$|+\rangle = 1/\sqrt{2}(|0\rangle + |1\rangle) \tag{2.1}$$

$$|-\rangle = 1/\sqrt{2}(|0\rangle - |1\rangle) \tag{2.2}$$

$$|i\rangle = 1/\sqrt{2}(|0\rangle + i|1\rangle) \tag{2.3}$$

$$|-i\rangle = 1/\sqrt{2}(|0\rangle - i|1\rangle). \tag{2.4}$$

The basis $\{|+\rangle, |-\rangle\}$ is referred to as the Hadamard basis. We sometimes use the notation $\{|\nwarrow\rangle, |\nearrow\rangle\}$ for the Hadamard basis when discussing photon polarization.

Some authors omit normalization factors, allowing vectors of any length to represent a state where two vectors represent the same state if they differ by any complex factor. We will explicitly write the normalization factors, both because then the amplitudes have a more direct relation to the measurement probabilities and because keeping track of the normalization factor provides a check that helps avoid errors.

2.5.2   Geometric Views of the State Space of a Single Qubit

While we primarily use vectors to represent quantum states, it is helpful to have models of the single-qubit state space in which there is a one-to-one correspondence between states and points in the space. We give two related but different geometric models with this property. The second of these, the Bloch sphere model, will be used in section 5.4.1 to illustrate single-qubit quantum transformations, and in chapter 10 it will be generalized to aid in the discussion of single-qubit subsystems. These models are just different ways of looking at complex projective space of dimension 1. As we will see, complex projective space of dimension 1 can be viewed as a sphere. First we show that it can be viewed as the extended complex plane, the complex plane $\mathbf{C}$ together with an additional point traditionally labeled ∞.

Extended Complex Plane $\mathbf{C} \cup \{\infty\}$ A correspondence between the set of all complex numbers and single-qubit states is given by

$$a|0\rangle + b|1\rangle \mapsto b/a = \alpha$$

and its inverse

$$\alpha \mapsto \frac{1}{\sqrt{1+|\alpha|^2}}|0\rangle + \frac{\alpha}{\sqrt{1+|\alpha|^2}}|1\rangle.$$

The preceding mapping is not defined for the state with $a = 0$ and $b = 1$. To make this correspondence one-to-one we need to add a single point, which we label ∞, to the complex plane and define $\infty \leftrightarrow |1\rangle$. For example, we have

$$|0\rangle \;\mapsto\; 0$$

$$|1\rangle \;\mapsto\; \infty$$

$$|+\rangle \;\mapsto\; +1$$

$$|-\rangle \;\mapsto\; -1$$

$$|\mathbf{i}\rangle \;\mapsto\; \mathbf{i}$$

$$|-\mathbf{i}\rangle \;\mapsto\; -\mathbf{i}.$$

We now describe another useful model, related to but different from the previous one.

Bloch Sphere   Starting with the previous representation, we can map each state, represented by the complex number $\alpha = s + \mathbf{i}t$, onto the unit sphere in three real dimensions, the points $(x, y, z) \in \mathbf{R}^3$ satisfying $|x|^2 + |y|^2 + |z|^2 = 1$, via the standard *stereographic projection*

$$(s, t) \mapsto \left(\frac{2s}{|\alpha|^2 + 1}, \frac{2t}{|\alpha|^2 + 1}, \frac{1 - |\alpha|^2}{|\alpha|^2 + 1} \right),$$

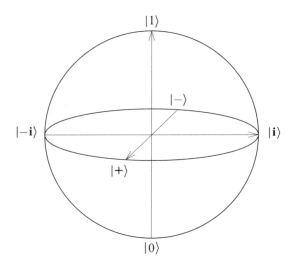

Figure 2.6
Location of certain single-qubit states on the surface of the Bloch sphere.

further requiring that $\infty \mapsto (0, 0, -1)$. Figure 2.6 illustrates the following correspondences:

$|0\rangle \;\;\mapsto (0, 0, 1)$

$|1\rangle \;\;\mapsto (0, 0, -1)$

$|+\rangle \mapsto (1, 0, 0)$

$|-\rangle \mapsto (-1, 0, 0)$

$|\mathbf{i}\rangle \;\;\mapsto (0, 1, 0)$

$|-\mathbf{i}\rangle \mapsto (0, -1, 0).$

We have given three representations of the quantum state space for a single-qubit system.

1. Vectors written in ket notation: $a|0\rangle + b|1\rangle$ with complex coefficients a and b, subject to $|a|^2 + |b|^2 = 1$, where a and b are unique up to a unit complex factor. Because of this factor, the global phase, this representation is not one-to-one.

2. Extended complex plane: a single complex number $\alpha \in \mathbf{C}$ or ∞. This representation is one-to-one.

3. Bloch sphere: points (x, y, z) on the unit sphere. This representation is also one-to-one.

As we will see in section 10.1, the points in the interior of the sphere also have meaning for quantum information processing. For historical reasons, the entire ball, including the interior, is called the Bloch sphere, instead of just the states on the surface, which truly form a sphere. For this reason, we refer to the state space of a single-qubit system as the surface of the Bloch sphere (figure 2.6).

One of the advantages of the Bloch sphere representation is that it is easy to read off all possible bases from the model; orthogonal states correspond to antipodal points of the Bloch sphere. In particular, every diameter of the Bloch sphere corresponds to a basis for the single-qubit state space.

The illustration we gave in figure 2.4 differs from the Bloch sphere representation of single-qubit quantum states in that the angles are half that of those in the Bloch sphere representation: in particular, the angle between two states in figure 2.4 has the usual relation to the inner product, whereas in the Bloch sphere representation the angle is twice that of the angle in the inner product formula.

2.5.3 Comments on General Quantum State Spaces

The states of all quantum systems satisfy certain properties that are encapsulated by a linear differential equation called the Schrödinger wave equation. For this reason, solutions to the Schrödinger equation are called wave functions, so all quantum states have representations as wave functions. For the theory of quantum information processing, we do not need to concern ourselves with properties specific to any of the various possible physical realizations of quantum bits, so we do not need to look at the details of specific wave function solutions; we can simply view wave functions as abstract vectors which we will denote by kets such as $|\rightarrow\rangle$ or $|0\rangle$.

Since the Schrödinger equation is linear and homogeneous, the addition of two solutions to the Schrödinger equation or a constant multiple of a solution of the Schrödinger equation are also solutions to the Schrödinger equation. Thus, the set of solutions to the Schrödinger equation for any quantum system is a complex vector space. Furthermore, the set of solutions has a natural inner product. For the theoretical aspects of quantum information processing, considering only finite dimensional vector spaces usually suffices. We simply mention that, in the infinite dimensional case, the space of solutions satisfies the conditions needed to form a *Hilbert space*. Hilbert spaces are frequently mentioned in the literature, since they are the most general case, but in most papers on quantum information processing, the Hilbert spaces discussed are finite-dimensional, in which case they are nothing more or less than finite-dimensional complex vector spaces equipped with inner product. We discuss the state spaces of multiple-qubit systems in chapter 3. Just as in the single-qubit case, there is redundancy in this model. In fact, there is greater redundancy in the vector space representation of larger quantum systems, which leads to a significantly more complicated geometry.

2.6 References

The early essays of Feynman and Manin can be found in [119, 120, 121] and [202, 203] respectively. The bra/ket notation was first introduced by Dirac in 1958 [103]. It is found in most quantum mechanics textbooks and is used in virtually all papers on quantum computing.

More information about linear algebra, in particular proofs of facts stated here, can be found in any linear algebra text, including Strang's *Linear Algebra and Its Applications* [265] and Hoffman and Kunze's *Linear Algebra* [152], or in a book on mathematics for physicists such as Bamberg and Sternberg's *A Course in Mathematics for Students of Physics* [30].

The BB84 quantum key distribution protocol was developed by Charles Bennett and Gilles Brassard [42, 43, 45] building on work of Stephen Wiesner [284]. A related protocol was shown to be unconditionally secure by Lo and Chau [198]. Their proof was later simplified by Shor and Preskill [255] and extended to BB84. Another proof was given by Mayers [206]. The BB84 protocol was first demonstrated experimentally by Bennett et al. in 1992 over 30 cm of free space [37]. Since then, several groups have demonstrated this protocol and other quantum key distribution protocols over 100 km of fiber optic cable. Bienfang et al. [51] demonstrated quantum key distribution over 23 km of free space at night, and Hughes et al. have achieved distances of 10 km through free space in daylight [156]. See the ARDA roadmap [157], the QIPC strategic report [295], and Gisin et al. [130] for detailed overviews of implementation efforts and the challenges involved. The companies id Quantique, MagiQ, and SmartQuantum currently sell quantum cryptographic systems implementing the BB84 protocol. Other quantum key distribution protocols exist. Exercise 2.11 develops the B92 protocol, and section 3.4 describes Ekert's entanglement-based quantum key distribution protocol.

While we explain all quantum mechanics needed for the topics covered in this book, the reader may be interested in books on quantum mechanics. Countless books on quantum mechanics are available. Greenstein and Zajonc [140] give a readable high-level exposition of quantum mechanics, including descriptions of many experiments. The third volume of the *Feynman Lectures on Physics* [122] is accessible to a large audience. A classical explanation of the polarization experiment is given in the first volume. Shankar's textbook [247] defines much more of the notation and mathematics required for performing calculations than do the previously mentioned books, and it is quite readable as well. Other textbooks, such as Liboff [194], may be more appropriate for readers with a physics background.

2.7 Exercises

Exercise 2.1. Let the direction $|v\rangle$ of polaroid B's preferred axis be given as a function of θ, $|v\rangle = \cos\theta|\rightarrow\rangle + \sin\theta|\uparrow\rangle$, and suppose that the polaroids A and C remain horizontally and vertically polarized as in the experiment of Section 2.1.1. What fraction of photons reach the screen? Assume that each photon generated by the laser pointer has random polarization.

Exercise 2.2. Which pairs of expressions for quantum states represent the same state? For those pairs that represent different states, describe a measurement for which the probabilities of the two outcomes differ for the two states and give these probabilities.

a. $|0\rangle$ and $-|0\rangle$

b. $|1\rangle$ and $\mathbf{i}|1\rangle$

c. $\frac{1}{\sqrt{2}}(|0\rangle + |1\rangle)$ and $\frac{1}{\sqrt{2}}(-|0\rangle + i|1\rangle)$

d. $\frac{1}{\sqrt{2}}(|0\rangle + |1\rangle)$ and $\frac{1}{\sqrt{2}}(|0\rangle - |1\rangle)$

e. $\frac{1}{\sqrt{2}}(|0\rangle - |1\rangle)$ and $\frac{1}{\sqrt{2}}(|1\rangle - |0\rangle)$

f. $\frac{1}{\sqrt{2}}(|0\rangle + i|1\rangle)$ and $\frac{1}{\sqrt{2}}(i|1\rangle - |0\rangle)$

g. $\frac{1}{\sqrt{2}}(|+\rangle + |-\rangle)$ and $|0\rangle$

h. $\frac{1}{\sqrt{2}}(|i\rangle - |-i\rangle)$ and $|1\rangle$

i. $\frac{1}{\sqrt{2}}(|i\rangle + |-i\rangle)$ and $\frac{1}{\sqrt{2}}(|-\rangle + |+\rangle)$

j. $\frac{1}{\sqrt{2}}(|0\rangle + e^{i\pi/4}|1\rangle)$ and $\frac{1}{\sqrt{2}}(e^{-i\pi/4}|0\rangle + |1\rangle)$

Exercise 2.3. Which states are superpositions with respect to the standard basis, and which are not? For each state that is a superposition, give a basis with respect to which it is not a superposition.

a. $|+\rangle$

b. $\frac{1}{\sqrt{2}}(|+\rangle + |-\rangle)$

c. $\frac{1}{\sqrt{2}}(|+\rangle - |-\rangle)$

d. $\frac{\sqrt{3}}{2}|+\rangle - \frac{1}{2}|-\rangle)$

e. $\frac{1}{\sqrt{2}}(|i\rangle - |-i\rangle)$

f. $\frac{1}{\sqrt{2}}(|0\rangle - |1\rangle)$

Exercise 2.4. Which of the states in 2.3 are superpositions with respect to the Hadamard basis, and which are not?

Exercise 2.5. Give the set of all values of θ for which the following pairs of states are equivalent.

a. $|1\rangle$ and $\frac{1}{\sqrt{2}}(|+\rangle + e^{i\theta}|-\rangle)$

b. $\frac{1}{\sqrt{2}}(|i\rangle + e^{i\theta}|-i\rangle)$ and $\frac{1}{\sqrt{2}}(|-i\rangle + e^{-i\theta}|i\rangle)$

c. $\frac{1}{2}|0\rangle - \frac{\sqrt{3}}{2}|1\rangle$ and $e^{i\theta}\left(\frac{1}{2}|0\rangle - \frac{\sqrt{3}}{2}|1\rangle\right)$

Exercise 2.6. For each pair consisting of a state and a measurement basis, describe the possible measurement outcomes and give the probability for each outcome.

a. $\frac{\sqrt{3}}{2}|0\rangle - \frac{1}{2}|1\rangle$, $\{|0\rangle, |1\rangle\}$

b. $\frac{\sqrt{3}}{2}|1\rangle - \frac{1}{2}|0\rangle, \{|0\rangle, |1\rangle\}$

c. $|-\mathbf{i}\rangle, \{|0\rangle, |1\rangle\}$

d. $|0\rangle, \{|+\rangle, |-\rangle\}$

e. $\frac{1}{\sqrt{2}}(|0\rangle - |1\rangle), \{|\mathbf{i}\rangle, |-\mathbf{i}\rangle\}$

f. $|1\rangle, \{|\mathbf{i}\rangle, |-\mathbf{i}\rangle\}$

g. $|+\rangle, \{\frac{1}{2}|0\rangle + \frac{\sqrt{3}}{2}|1\rangle, \frac{\sqrt{3}}{2}|0\rangle - \frac{1}{2}|1\rangle\}$

Exercise 2.7. For each of the following states, describe all orthonormal bases that include that state.

a. $\frac{1}{\sqrt{2}}(|0\rangle + \mathbf{i}|1\rangle)$

b. $\frac{1+\mathbf{i}}{2}|0\rangle - \frac{1-\mathbf{i}}{2}|1\rangle$

c. $\frac{1}{\sqrt{2}}(|0\rangle + e^{\mathbf{i}\pi/6}|1\rangle)$

d. $\frac{1}{2}|+\rangle - \frac{\mathbf{i}\sqrt{3}}{2}|-\rangle$

Exercise 2.8. Alice is confused. She understands that $|1\rangle$ and $-|1\rangle$ represent the same state. But she does not understand why that does not imply that $\frac{1}{\sqrt{2}}(|0\rangle + |1\rangle)$ and $\frac{1}{\sqrt{2}}(|0\rangle - |1\rangle)$ would be the same state. Can you help her out?

Exercise 2.9. In the BB84 protocol, how many bits do Alice and Bob need to compare to have a 90 percent chance of detecting Eve's presence?

Exercise 2.10. Analyze Eve's success in eavesdropping on the BB84 protocol if she does not even know which two bases to choose from and so chooses a basis at random at each step.

a. On average, what percentage of bit values of the final key will Eve know for sure after listening to Alice and Bob's conversation on the public channel?

b. On average, what percentage of bits in her string are correct?

c. How many bits do Alice and Bob need to compare to have a 90 percent chance of detecting Eve's presence?

Exercise 2.11. *B92 quantum key distribution protocol.* In 1992 Bennett proposed the following quantum key distribution protocol. Instead of encoding each bit in either the standard basis or the Hadamard basis as is done in the BB84 protocol, Alice encodes her random string x as follows

$$0 \mapsto |0\rangle$$
$$1 \mapsto |+\rangle = \frac{1}{\sqrt{2}}(|0\rangle + |1\rangle)$$

and sends them to Bob. Bob generates a random bit string y. If $y_i = 0$ he measures the i th qubit in the Hadamard basis $\{|+\rangle, |-\rangle\}$, if $y_i = 1$ he measures in the standard basis $\{|0\rangle, |1\rangle\}$. In this protocol, instead of telling Alice over the public classical channel which basis he used to measure

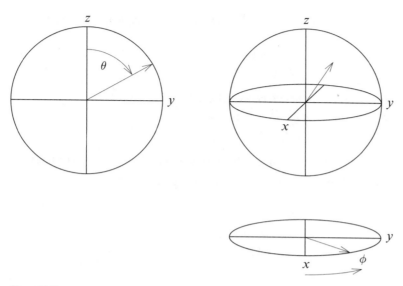

Figure 2.7
Bloch sphere representation of single-qubit quantum states.

each qubit, he tells her the results of his measurements. If his measurement resulted in $|+\rangle$ or $|0\rangle$ Bob sends 0; if his measurement indicates the state is $|1\rangle$ or $|-\rangle$, he sends 1. Alice and Bob discard all bits from strings x and y for which Bob's bit value from measurement yielded 0, obtaining strings x' and y'. Alice uses x' as the secret key and Bob uses y'. Then, depending on the security level they desire, they compare a number of bits to detect tampering. They discard these check bits from their key.

a. Show that if Bob receives exactly the states Alice sends, then the strings x' and y' are identical strings.

b. Why didn't Alice and Bob decide to keep the bits of x and y for which Bob's bit value from measurement was 0?

c. What if an eavesdropper Eve measures each bit in either the standard basis or the Hadamard basis to obtain a bit string z and forwards the measured qubits to Bob? On average, how many bits of Alice and Bob's key does she know for sure after listening in on the public classical? If Alice and Bob compare s bit values of their strings x' and y', how likely are they to detect Eve's presence?

Exercise 2.12. *Bloch Sphere: Spherical coordinates:*

a. Show that the surface of the Bloch sphere can be parametrized in terms of two real-valued parameters, the angles θ and ϕ illustrated in figure 2.7. Make sure your parametrization is in one-to-one correspondence with points on the sphere, and therefore single-qubit quantum states, in the range $\theta \in [0, \pi]$ and $\phi \in [0, 2\pi]$ except for the points corresponding to $|0\rangle$ and $|1\rangle$.

b. What are θ and ϕ for each of the states $|+\rangle$, $|-\rangle$, $|\mathbf{i}\rangle$, and $|-\mathbf{i}\rangle$?

Exercise 2.13. Relate the four parametrizations of the state space of a single qubit to each other: Give formulas for

a. vectors in ket notation

b. elements of the extended complex plane

c. spherical coordinates for the Bloch sphere (see exercise 2.12)

in terms of the x, y, and z coordinates of the Bloch sphere.

Exercise 2.14.

a. Show that antipodal points on the surface of the Block sphere represent orthogonal states.

b. Show that any two orthogonal states correspond to antipodal points.

3 Multiple-Qubit Systems

The first glimpse into why encoding information in quantum states might support more efficient computation comes when examining systems of more than one qubit. Unlike classical systems, the state space of a quantum system grows exponentially with the number of particles. Thus, when we encode computational information in quantum states of a system of n particles, there are vastly more possible computation states available than when classical states are used to encode the information. The extent to which these large state spaces corresponding to small amounts of physical space can be used to speed up computation will be the subject of much of the rest of this book.

The enormous difference in dimension between classical and quantum state spaces is due to a difference in the way the spaces combine. Imagine a macroscopic physical system consisting of several components. The state of this classical system can be completely characterized by describing the state of each of its component pieces separately. A surprising and unintuitive aspect of quantum systems is that often the state of a system cannot be described in terms of the states of its component pieces. States that cannot be so described are called *entangled states*. Entangled states are a critical ingredient of quantum computation.

Entangled states are a uniquely quantum phenomenon; they have no classical counterpart. Most states in a multiple-qubit system are entangled states; they are what fills the vast quantum state spaces. The impossibility of efficiently simulating the behavior of entangled states on classical computers suggested to Feynman, Manin, and others that it might be possible to use these quantum behaviors to compute more efficiently, leading to the development of the field of quantum computation.

The first few sections of this chapter will be fairly abstract as we develop the mathematical formalism to discuss multiple-qubit systems. We will try to make this material more concrete by including many examples. Section 3.1 formally describes the difference between the way quantum and classical state spaces combine, the difference between the *direct sum* of two or more vector spaces and the *tensor product* of a set of vector spaces. Section 3.1 then explores some of the implications of this difference, including the exponential increase in the dimension of a quantum state space with the number of particles. Section 3.2 formally defines entangled states and begins to describe their uniquely quantum behavior. As a first illustration of the usefulness of this behavior, section 3.4 discusses a second quantum key distribution scheme.

3.1 Quantum State Spaces

In classical physics, the possible states of a system of n objects, whose individual states can be described by a vector in a two-dimensional vector space, can be described by vectors in a vector space of $2n$ dimensions. Classical state spaces combine through the direct sum. However, the combined state space of n quantum systems, each with states modeled by two-dimensional vectors, is much larger. The vector spaces associated with the quantum systems combine through the tensor product, resulting in a vector space of 2^n dimensions. We begin by reviewing the formal definition of a direct sum as well as of the tensor product in order to compare the two and the difference in size between the resulting spaces.

3.1.1 Direct Sums of Vector Spaces

The *direct sum* $V \oplus W$ of two vector spaces V and W with bases $A = \{|\alpha_1\rangle, |\alpha_2\rangle, \ldots, |\alpha_n\rangle\}$ and $B = \{|\beta_1\rangle, |\beta_2\rangle, \ldots, |\beta_m\rangle\}$ respectively is the vector space with basis $A \cup B = \{|\alpha_1\rangle, |\alpha_2\rangle, \ldots, |\alpha_n\rangle, |\beta_1\rangle, |\beta_2\rangle, \ldots, |\beta_m\rangle\}$. The order of the basis is arbitrary. Every element $|x\rangle \in V \oplus W$ can be written as $|x\rangle = |v\rangle \oplus |w\rangle$ for some $|v\rangle \in V$ and $|w\rangle \in W$. For V and W of dimension n and m respectively, $V \oplus W$ has dimension $n + m$:

$$\dim(V \oplus W) = \dim(V) + \dim(W).$$

Addition and scalar multiplication are defined by performing the operation on the two component vector spaces separately and adding the results. When V and W are inner product spaces, the standard inner product on $V \oplus W$ is given by

$$((\langle v_2| \oplus \langle w_2|)(|v_1\rangle \oplus |w_1\rangle)) = \langle v_2|v_1\rangle + \langle w_2|w_1\rangle.$$

The vector spaces V and W embed in $V \oplus W$ in the obvious canonical way, and the images are orthogonal under the standard inner product.

Suppose that the state of each of three classical objects O_1, O_2, and O_3 is fully described by two parameters, the position x_i and the momentum p_i. Then the state of the system can be described by the direct sum of the states of the individual objects:

$$\begin{pmatrix} x_1 \\ p_1 \end{pmatrix} \oplus \begin{pmatrix} x_2 \\ p_2 \end{pmatrix} \oplus \begin{pmatrix} x_3 \\ p_3 \end{pmatrix} = \begin{pmatrix} x_1 \\ p_1 \\ x_2 \\ p_2 \\ x_3 \\ p_3 \end{pmatrix}.$$

More generally, the state space of n such classical objects has dimension $2n$. Thus the size of the state space grows linearly with the number of objects.

3.1.2 Tensor Products of Vector Spaces

The *tensor product* $V \otimes W$ of two vector spaces V and W with bases $A = \{|\alpha_1\rangle, |\alpha_2\rangle, \ldots, |\alpha_n\rangle\}$ and $B = \{|\beta_1\rangle, |\beta_2\rangle, \ldots, |\beta_m\rangle\}$ respectively is an nm-dimensional vector space with a basis consisting of the nm elements of the form $|\alpha_i\rangle \otimes |\beta_j\rangle$ where $\otimes$ is the tensor product, an abstract binary operator that satisfies the following relations:

$$(|v_1\rangle + |v_2\rangle) \otimes |w\rangle = |v_1\rangle \otimes |w\rangle + |v_2\rangle \otimes |w\rangle$$

$$|v\rangle \otimes (|w_1\rangle + |w_2\rangle) = |v\rangle \otimes |w_1\rangle + |v\rangle \otimes |w_2\rangle$$

$$(a|v\rangle) \otimes |w\rangle = |v\rangle \otimes (a|w\rangle) = a(|v\rangle \otimes |w\rangle).$$

Taking $k = \min(n, m)$, all elements of $V \otimes W$ have form

$$|v_1\rangle \otimes |w_1\rangle + |v_2\rangle \otimes |w_2\rangle + \cdots + |v_k\rangle \otimes |w_k\rangle$$

for some $v_i \in V$ and $w_i \in W$. Due to the relations defining the tensor product, such a representation is not unique. Furthermore, while all elements of $V \otimes W$ can be written

$$\alpha_1(|\alpha_1\rangle \otimes |\beta_1\rangle) + \alpha_2(|\alpha_2\rangle \otimes |\beta_1\rangle) + \cdots + \alpha_{nm}(|\alpha_n\rangle \otimes |\beta_m\rangle),$$

most elements of $V \otimes W$ cannot be written as $|v\rangle \otimes |w\rangle$, where $v \in V$ and $w \in W$. It is common to write $|v\rangle|w\rangle$ for $|v\rangle \otimes |w\rangle$.

Example 3.1.1 Let V and W be two-dimensional vector spaces with orthonormal bases $A = \{|\alpha_1\rangle, |\alpha_2\rangle\}$ and $B = \{|\beta_1\rangle, |\beta_2\rangle\}$ respectively. Let $|v\rangle = a_1|\alpha_1\rangle + a_2|\alpha_2\rangle$ and $|w\rangle = b_1|\beta_1\rangle + b_2|\beta_2\rangle$ be elements of V and W. Then

$$|v\rangle \otimes |w\rangle = a_1b_1|\alpha_1\rangle \otimes |\beta_1\rangle + a_1b_2|\alpha_1\rangle \otimes |\beta_2\rangle + a_2b_1|\alpha_2\rangle \otimes |\beta_1\rangle + a_2b_2|\alpha_2\rangle \otimes |\beta_2\rangle.$$

If V and W are vector spaces corresponding to a qubit, each with standard basis $\{|0\rangle, |1\rangle\}$, then $V \otimes W$ has $\{|0\rangle \otimes |0\rangle, |0\rangle \otimes |1\rangle, |1\rangle \otimes |0\rangle, |1\rangle \otimes |1\rangle\}$ as basis. The tensor product of two single-qubit states $a_1|0\rangle + b_1|1\rangle$ and $a_2|0\rangle + b_2|1\rangle$ is $a_1a_2|0\rangle \otimes |0\rangle + a_1b_2|0\rangle \otimes |1\rangle + a_2b_1|1\rangle \otimes |0\rangle + a_2b_2|1\rangle \otimes |1\rangle$.

To write examples in the more familiar matrix notation for vectors, we must choose an ordering for the basis of the tensor product space. For example, we can choose the dictionary ordering $\{|\alpha_1\rangle|\beta_1\rangle, |\alpha_1\rangle|\beta_2\rangle, |\alpha_2\rangle|\beta_1\rangle, |\alpha_2\rangle|\beta_2\rangle\}$.

Example 3.1.2 With the dictionary ordering of the basis for the tensor product space, the tensor product of the unit vectors with matrix representation $|v\rangle = \frac{1}{\sqrt{5}}(1, -2)^\dagger$ and $|w\rangle = \frac{1}{\sqrt{10}}(-1, 3)^\dagger$ is the unit vector $|v\rangle \otimes |w\rangle = \frac{1}{5\sqrt{2}}(-1, 3, 2, -6)^\dagger$.

If V and W are inner product spaces, then $V \otimes W$ can be given an inner product by taking the product of the inner products on V and W; the inner product of $|v_1\rangle \otimes |w_1\rangle$ and $|v_2\rangle \otimes |w_2\rangle$ is given by

$$(\langle v_2| \otimes \langle w_2|) \cdot (|v_1\rangle \otimes |w_1\rangle) = \langle v_2|v_1\rangle\langle w_2|w_1\rangle,$$

The tensor product of two unit vectors is a unit vector, and given orthonormal bases $\{|\alpha_i\rangle\}$ for V and $\{|\beta_i\rangle\}$ for W, the basis $\{|\alpha_i\rangle \otimes |\beta_j\rangle\}$ for $V \otimes W$ is also orthonormal. The tensor product $V \otimes W$ has dimension $\dim(V) \times \dim(W)$, so the tensor product of n two-dimensional vector spaces has 2^n dimensions.

Most elements $|w\rangle \in V \otimes W$ cannot be written as the tensor product of a vector in V and a vector in W (though they are all linear combinations of such elements). This observation is of crucial importance to quantum computation. States of $V \otimes W$ that cannot be written as the tensor product of a vector in V and a vector in W are called *entangled* states. As we will see, for most quantum states of an n-qubit system, in particular for all entangled states, it is not meaningful to talk about the state of a single qubit of the system.

A tensor product structure also underlies probability theory. While the tensor product structure there is rarely mentioned, a common source of confusion is a tendency to try to impose a direct sum structure on what is actually a tensor product structure. Readers may find it useful to read section A.1, which discusses the tensor product structure inherent in probability theory, which illustrates the use of tensor product in another, more familiar, context. Readers may also wish to do exercises A.1 through A.4.

3.1.3 The State Space of an n-Qubit System

Given two quantum systems with states represented by unit vectors in V and W respectively, the possible states of the joint quantum system are represented by unit vectors in the vector space $V \otimes W$. For $0 \leq i < n$, let V_i be the vector space, with basis $\{|0\rangle_i, |1\rangle_i\}$, corresponding to a single qubit. The standard basis for the vector space $V_{n-1} \otimes \cdots \otimes V_1 \otimes V_0$ for an n-qubit system consists of the 2^n vectors

$$\{|0\rangle_{n-1} \otimes \cdots \otimes |0\rangle_1 \otimes |0\rangle_0,$$

$$|0\rangle_{n-1} \otimes \cdots \otimes |0\rangle_1 \otimes |1\rangle_0,$$

$$|0\rangle_{n-1} \otimes \cdots \otimes |1\rangle_1 \otimes |0\rangle_0,$$

$$\vdots,$$

$$|1\rangle_{n-1} \otimes \cdots \otimes |1\rangle_1 \otimes |1\rangle_0\}.$$

The subscripts are often dropped, since the corresponding qubit is clear from position. The convention that adjacency of kets means the tensor product enables us to write this basis more compactly:

$$\{|0\rangle \cdots |0\rangle |0\rangle,$$

$$|0\rangle \cdots |0\rangle |1\rangle,$$

$$|0\rangle \cdots |1\rangle |0\rangle,$$

$$\vdots,$$

$$|1\rangle \cdots |1\rangle |1\rangle\}.$$

Since the tensor product space corresponding to an n-qubit system occurs so frequently throughout quantum information processing, an even more compact and readable notation uses $|b_{n-1} \ldots b_0\rangle$ to represent $|b_{n-1}\rangle \otimes \cdots \otimes |b_0\rangle$. In this notation the standard basis for an n-qubit system can be written

$$\{|0 \cdots 00\rangle, |0 \cdots 01\rangle, |0 \cdots 10\rangle, \ldots, |1 \cdots 11\rangle\}.$$

Finally, since decimal notation is more compact than binary notation, we will represent the state $|b_{n-1} \ldots b_0\rangle$ more compactly as $|x\rangle$, where b_i are the digits of the binary representation for the decimal number x. In this notation, the standard basis for an n-qubit system is written

$$\{|0\rangle, |1\rangle, |2\rangle, \ldots, |2^n - 1\rangle\}.$$

The standard basis for a two-qubit system can be written as

$$\{|00\rangle, |01\rangle, |10\rangle, |11\rangle\} = \{|0\rangle, |1\rangle, |2\rangle, |3\rangle\},$$

and the standard basis for a three-qubit system can be written as

$$\{|000\rangle, |001\rangle, |010\rangle, |011\rangle, |100\rangle, |101\rangle, |110\rangle, |111\rangle\}$$

$$= \{|0\rangle, |1\rangle, |2\rangle, |3\rangle, |4\rangle, |5\rangle, |6\rangle, |7\rangle\}.$$

Since the notation $|3\rangle$ corresponds to two different quantum states in these two bases, one a two-qubit state, the other a three-qubit state, in order for such notation to be unambiguous, the number of qubits must be clear from context.

We often revert to a less compact notation when we wish to set apart certain sets of qubits, to indicate separate registers of a quantum computer, or to indicate qubits controlled by different people. If Alice controls the first two qubits and Bob the last three, we may write a state as $\frac{1}{\sqrt{2}}(|00\rangle|101\rangle + |10\rangle|011\rangle)$, or even as $\frac{1}{\sqrt{2}}(|00\rangle_A|101\rangle_B + |10\rangle_A|011\rangle_B)$, where the subscripts indicate which qubits Alice controls and which qubits Bob controls.

Example 3.1.3 The superpositions

$$\frac{1}{\sqrt{2}}|0\rangle + \frac{1}{\sqrt{2}}|7\rangle = \frac{1}{\sqrt{2}}|000\rangle + \frac{1}{\sqrt{2}}|111\rangle$$

and

$$\frac{1}{2}(|1\rangle + |2\rangle + |4\rangle + |7\rangle) = \frac{1}{2}(|001\rangle + |010\rangle + |100\rangle + |111\rangle)$$

represent possible states of a three-qubit system.

To use matrix notation for state vectors of an n-qubit system, the order of basis vectors must be established. Unless specified otherwise, basis vectors labeled with numbers are assumed to be sorted numerically. Using this convention, the two qubit state

$$\frac{1}{2}|00\rangle + \frac{\mathbf{i}}{2}|01\rangle + \frac{1}{\sqrt{2}}|11\rangle = \frac{1}{2}|0\rangle + \frac{\mathbf{i}}{2}|1\rangle + \frac{1}{\sqrt{2}}|3\rangle$$

will have matrix representation

$$\begin{pmatrix} \frac{1}{2} \\ \frac{i}{2} \\ 0 \\ \frac{1}{\sqrt{2}} \end{pmatrix}.$$

We use the standard basis predominantly, but we use other bases from time to time. For example, the following basis, the Bell basis for a two-qubit system, $\{|\Phi^+\rangle, |\Phi^-\rangle, |\Psi^+\rangle, |\Psi^-\rangle\}$, where

$$\begin{aligned} |\Phi^+\rangle &= 1/\sqrt{2}(|00\rangle + |11\rangle) \\ |\Phi^-\rangle &= 1/\sqrt{2}(|00\rangle - |11\rangle) \\ |\Psi^+\rangle &= 1/\sqrt{2}(|01\rangle + |10\rangle) \\ |\Psi^-\rangle &= 1/\sqrt{2}(|01\rangle - |10\rangle), \end{aligned} \qquad (3.1)$$

is important for various applications of quantum information processing including quantum teleportation. As in the single-qubit case, a state $|v\rangle$ is a superposition with respect to a set of orthonormal states $\{|\beta_1\rangle, \dots, |\beta_i\rangle\}$ if it is a linear combination of these states, $|v\rangle = a_1|\beta_1\rangle + \dots + a_i|\beta_i\rangle$, and at least two of the a_i are non-zero. When no set of orthonormal states is specified, we will mean that the superposition is with respect to the standard basis.

Any unit vector of the 2^n-dimensional state space represents a possible state of an n-qubit system, but just as in the single-qubit case there is redundancy. In the multiple-qubit case, not only do vectors that are multiples of each other refer to the same quantum state, but properties of the tensor product also mean that phase factors distribute over tensor products; the same phase factor in different qubits of a tensor product represent the same state:

$$|v\rangle \otimes (e^{i\phi}|w\rangle) = e^{i\phi}(|v\rangle \otimes |w\rangle) = (e^{i\phi}|v\rangle) \otimes |w\rangle.$$

Phase factors in individual qubits of a single term of a superposition can always be factored out into a single coefficient for that term.

Example 3.1.4 $\frac{1}{\sqrt{2}}(|0\rangle + |1\rangle) \otimes \frac{1}{\sqrt{2}}(|0\rangle + |1\rangle) = \frac{1}{2}(|00\rangle + |01\rangle + |10\rangle + |11\rangle)$

Example 3.1.5 $(\frac{1}{2}|0\rangle + \frac{\sqrt{3}}{2}|1\rangle) \otimes (\frac{1}{\sqrt{2}}|0\rangle + \frac{i}{\sqrt{2}}|1\rangle) = \frac{1}{2\sqrt{2}}|00\rangle + \frac{i}{2\sqrt{2}}|01\rangle + \frac{\sqrt{3}}{2\sqrt{2}}|10\rangle + \frac{i\sqrt{3}}{2\sqrt{2}}|11\rangle)$

Just as in the single-qubit case, vectors that differ only in a global phase represent the same quantum state. If we write every quantum state as

$$a_0|0\ldots00\rangle + a_1|0\ldots01\rangle + \cdots + a_{2^n-1}|1\ldots11\rangle$$

and require the first non-zero a_i to be real and non-negative, then every quantum state has a unique representation. Since this representation uniquely represents quantum states, the quantum state space of an n-qubit system has $2^n - 1$ complex dimensions. For any complex vector space of dimension N, the space in which vectors that are multiples of each other are considered equivalent is called *complex projective space* of dimension $N - 1$. So the space of distinct quantum states of an n-qubit system is a complex projective space of dimension $2^n - 1$.

Just as in the single-qubit case, we must be careful not to confuse the vector space in which we write our computations with the quantum state space itself. Again, we must be careful to avoid confusion between the relative phases between terms in the superposition, of critical importance in quantum mechanics, and the global phase which has no physical meaning. Using the notation of section 2.5.1, we write $|v\rangle \sim |w\rangle$ when two vectors $|v\rangle$ and $|w\rangle$ differ only by a global phase and thus represent the same quantum state. For example, even though $|00\rangle \sim e^{i\phi}|00\rangle$, the vectors $|v\rangle = \frac{1}{\sqrt{2}}(e^{i\phi}|00\rangle + |11\rangle)$ and $|w\rangle = \frac{1}{\sqrt{2}}(|00\rangle + |11\rangle)$ represent different quantum states, which behave differently in many situations:

$$\frac{1}{\sqrt{2}}(e^{i\phi}|00\rangle + |11\rangle) \not\sim \frac{1}{\sqrt{2}}(|00\rangle + |11\rangle).$$

However,

$$\frac{1}{\sqrt{2}}(e^{i\phi}|00\rangle + e^{i\phi}|11\rangle) \sim \frac{e^{i\phi}}{\sqrt{2}}(|00\rangle + |11\rangle) \sim \frac{1}{\sqrt{2}}(|00\rangle + |11\rangle).$$

Quantum mechanical calculations are usually performed in the vector space rather than in the projective space because linearity makes vector spaces easier to work with. But we must always be aware of the $\sim$ equivalence when we interpret the results of our calculations as quantum states. Further confusions arise when states are written in different bases. Recall from section 2.5.1 that $|+\rangle = \frac{1}{\sqrt{2}}(|0\rangle + |1\rangle)$ and $|-\rangle = \frac{1}{\sqrt{2}}(|0\rangle - |1\rangle)$. The expression $\frac{1}{\sqrt{2}}(|+\rangle + |-\rangle)$ is a different way of writing $|0\rangle$, and $\frac{1}{\sqrt{2}}(|0\rangle|0\rangle + |1\rangle|1\rangle)$ and $\frac{1}{\sqrt{2}}(|+\rangle|+\rangle + |-\rangle|-\rangle)$ are simply different expressions for the same vector.

Fluency with properties of tensor products, and with the notation just presented, will be crucial for understanding the rest of the book. The reader is strongly encouraged to work exercises 3.1 through 3.9 at this point to begin to develop that fluency.

3.2 Entangled States

As we saw in section 2.5.2, a single-qubit state can be specified by a single complex number so any tensor product of n individual single-qubit states can be specified by n complex numbers. But in the last section, we saw that it takes $2^n - 1$ complex numbers to describe states of an n-qubit system. Since $2^n \gg n$, the vast majority of n-qubit states cannot be described in terms of the state of n separate single-qubit systems. States that cannot be written as the tensor product of n single-qubit states are called *entangled* states. Thus the vast majority of quantum states are entangled.

Example 3.2.1 The elements of the Bell basis (Equation 3.1) are entangled. For instance, the Bell state $|\Phi^+\rangle = \frac{1}{\sqrt{2}}(|00\rangle + |11\rangle)$ cannot be described in terms of the state of each of its component qubits separately. This state cannot be decomposed, because it is impossible to find a_1, a_2, b_1, b_2 such that

$$(a_1|0\rangle + b_1|1\rangle) \otimes (a_2|0\rangle + b_2|1\rangle) = \frac{1}{\sqrt{2}}(|00\rangle + |11\rangle),$$

since

$$(a_1|0\rangle + b_1|1\rangle) \otimes (a_2|0\rangle + b_2|1\rangle) = a_1a_2|00\rangle + a_1b_2|01\rangle + b_1a_2|10\rangle + b_1b_2|11\rangle$$

and $a_1b_2 = 0$ implies that either $a_1a_2 = 0$ or $b_1b_2 = 0$. Two particles in the Bell state $|\Phi^+\rangle$ are called an EPR pair for reasons that will become apparent in section 4.4.

Example 3.2.2 Other examples of two-qubit entangled states include

$$|\Psi^+\rangle = \frac{1}{\sqrt{2}}(|01\rangle + |10\rangle),$$

$$\frac{1}{\sqrt{2}}(|00\rangle - i|11\rangle),$$

$$\frac{i}{10}|00\rangle + \frac{\sqrt{99}}{10}|11\rangle),$$

and

$$\frac{7}{10}|00\rangle + \frac{1}{10}|01\rangle + \frac{1}{10}|10\rangle + \frac{7}{10}|11\rangle).$$

The four entangled states

$$|\Phi^+\rangle = \frac{1}{\sqrt{2}}(|00\rangle + |11\rangle)$$

$$|\Phi^-\rangle = \frac{1}{\sqrt{2}}(|00\rangle - |11\rangle)$$

and

$$|\Psi^+\rangle = \frac{1}{\sqrt{2}}(|01\rangle + |10\rangle)$$

$$|\Psi^-\rangle = \frac{1}{\sqrt{2}}(|01\rangle - |10\rangle)$$

of equation 3.1 are called *Bell states*. Bell states are of fundamental importance to quantum information processing. For example, section 5.3 exhibits their use for quantum teleportation and dense coding. Section 10.2.1 shows that these states are maximally entangled.

Strictly speaking, entanglement is always with respect to a specified tensor product decomposition of the state space. More formally, given a state $|\psi\rangle$ of some quantum system with associated vector space V and a tensor decomposition of V, $V = V_1 \otimes \cdots \otimes V_n$, the state $|\psi\rangle$ is *separable*, or *unentangled*, with respect to that decomposition if it can be written as

$$|\psi\rangle = |v_1\rangle \otimes \cdots \otimes |v_n\rangle,$$

where $|v_i\rangle$ is contained in V_i. Otherwise, $|\psi\rangle$ is *entangled* with respect to this decomposition.

Unless we specify a different decomposition, when we say an n-qubit state is entangled, we mean it is entangled with respect to the tensor product decomposition of the vector space V associated to the n-qubit system into the n two-dimensional vector spaces $V_{n-1}, \ldots V_0$ associated with each of the individual qubits. For such statements to have meaning, it must be specified or clear from context which of the many possible tensor decompositions of V into two-dimensional spaces corresponds with the set of qubits under consideration.

It is vital to remember that entanglement is not an absolute property of a quantum state, but depends on the particular decomposition of the system into subsystems under consideration; states entangled with respect to the single-qubit decomposition may be unentangled with respect to other decompositions into subsystems. In particular, when discussing entanglement in quantum computation, we will be interested in entanglement with respect to a decomposition into registers, subsystems consisting of multiple qubits, as well as entanglement with respect to the decomposition into individual qubits. The following example demonstrates how a state can be entangled with respect to one decomposition and not with respect to another.

Example 3.2.3 *Multiple meanings of entanglement.* We say that the four-qubit state

$$|\psi\rangle = \frac{1}{2}(|00\rangle + |11\rangle + |22\rangle + |33\rangle) = \frac{1}{2}(|0000\rangle + |0101\rangle + |1010\rangle + |1111\rangle)$$

is entangled, since it cannot be expressed as the tensor product of four single-qubit states. That the entanglement is with respect to the decomposition into single qubits is implicit in this statement. There are other decompositions with respect to which this state is unentangled. For example, $|\psi\rangle$ can be expressed as the product of two two-qubit states:

$$|\psi\rangle = \frac{1}{2}(|0\rangle_1|0\rangle_2|0\rangle_3|0\rangle_4 + |0\rangle_1|1\rangle_2|0\rangle_3|1\rangle_4 + |1\rangle_1|0\rangle_2|1\rangle_3|0\rangle_4 + |1\rangle_1|1\rangle_2|1\rangle_3|1\rangle_4$$

$$= \frac{1}{\sqrt{2}}(|0\rangle_1|0\rangle_3 + |1\rangle_1|1\rangle_3) \otimes \frac{1}{\sqrt{2}}(|0\rangle_2|0\rangle_4 + |1\rangle_2|1\rangle_4),$$

where the subscripts indicate which qubit we are talking about. So $|\psi\rangle$ is not entangled with respect to the system decomposition consisting of a subsystem of the first and third qubit and a subsystem consisting of the second and fourth qubit. On the other hand, the reader can check that $|\psi\rangle$ is entangled with respect to the decomposition into the two two-qubit systems consisting of the first and second qubits and the third and fourth qubits.

It is important to recognize that the notion of entanglement is not basis dependent, even though it depends on the tensor decomposition under consideration; there is no reference, explicit or implicit, to a basis in the definition of entanglement. Certain bases may be more or less convenient to work with, depending for instance on how much they reflect the tensor decomposition under consideration, but that choice does not affect what states are considered entangled.

In section 2.3, we puzzled over the meaning of quantum superpositions. We now extend the remarks we made on the meaning of superpositions in section 2.3 to the multiple-qubit case. As in the single-qubit case, most n-qubit states are superpositions, nontrivial linear combinations of basis vectors. As always, the notion of superposition is basis-dependent; all states are superpositions with respect to some bases, and not superpositions with respect to other bases. For multiple qubits, the answer to the question of what superpositions mean is more involved than in the single-qubit case.

The common way of talking about superpositions in terms of the system being in two states "at the same time" is even more suspect in the multiple-qubit case. This way of thinking fails to distinguish between states like $\frac{1}{\sqrt{2}}(|00\rangle + |11\rangle)$ and $\frac{1}{\sqrt{2}}(|00\rangle + \mathbf{i}|11\rangle)$ that differ only by a relative phase and behave differently under a variety of circumstances. Furthermore, which states a system is viewed as "being in at the same time" is basis-dependent; the expressions $\frac{1}{\sqrt{2}}(|00\rangle + |11\rangle)$ and $\frac{1}{\sqrt{2}}(|+\rangle|+\rangle + |-\rangle|-\rangle)$ represent the same state but have different interpretations, one as being in the states $|00\rangle$ and $|11\rangle$ at the same time, and the other as being in the states $|++\rangle$ and $|--\rangle$ at the same time, in spite of being the same state and thus behaving in precisely the same way under all circumstances. This example underscores that quantum superpositions are not probabilistic mixtures.

Sections 3.4 and 4.4 will illustrate how the basis dependence of this interpretation obscures an essential part of the quantum nature of these states, an aspect that becomes apparent only when such states are considered in different bases. Nevertheless, as long as one is aware that this description should not be taken too literally, it can be helpful at first to think of superpositions as being in multiple states at once. Over the course of this chapter and the next, you will begin to develop more of a feel for the workings of these states.

Not only is entanglement between qubits key to the exponential size of quantum state spaces of multiple-qubit systems, but, as we will see in sections 3.4, 5.3.1, and 5.3.2, particles in an entangled state can also be used to aid communication of both classical and quantum information. Furthermore, the quantum algorithms of part II exploit entanglement to speed up computation. The way entangled states behave when measured is one of the central mysteries of quantum mechanics, as well as a source of power for quantum information processing. Entanglement and quantum measurement are two of the uniquely quantum properties that are exploited in quantum information processing.

3.3 Basics of Multi-Qubit Measurement

The experiment of section 2.1.2 illustrates how measurement of a single qubit is probabilistic and transforms the quantum state into a state compatible with the measuring device. A similar statement is true for measurements of multiple-qubit systems, except that the set of possible measurements and measurement outcomes is significantly richer than in the single-qubit case. The next paragraph develops some mathematical formalism to handle the general case.

Let V be the $N = 2^n$ dimensional vector space associated with an n-qubit system. Any device that measures this system has an associated direct sum decomposition into orthogonal subspaces

$$V = S_1 \oplus \cdots \oplus S_k$$

for some $k \leq N$. The number k corresponds to the maximum number of possible measurement outcomes for a state measured with that particular device. This number varies from device to device, even between devices measuring the same system. That any device has an associated direct sum decomposition is a direct generalization of the single-qubit case. Every device measuring a single-qubit system has an associated orthonormal basis $\{|v_1\rangle, |v_2\rangle\}$ for the vector space V associated with the single-qubit system; the vectors $|v_i\rangle$ each generate a one-dimensional subspace S_i (consisting of all multiples $a|v_i\rangle$ where a is a complex number), and $V = S_1 \oplus S_2$. Furthermore, the only nontrivial decompositions of the vector space V are into two one-dimensional subspaces, and any choice of unit length vectors, one from each of the subspaces, yields an orthonormal basis.

When a measuring device with associated direct sum decomposition $V = S_1 \oplus \cdots \oplus S_k$ interacts with an n-qubit system in state $|\psi\rangle$, the interaction changes the state to one entirely contained within one of the subspaces, and chooses the subspace with probability equal to the square of the absolute value of the amplitude of the component of $|\psi\rangle$ in that subspace. More formally, the state $|\psi\rangle$ has a unique direct sum decomposition $|\psi\rangle = a_1|\psi_1\rangle \oplus \cdots \oplus a_k|\psi_k\rangle$, where $|\psi_i\rangle$ is a unit vector in S_i and a_i is real and non-negative. When $|\psi\rangle$ is measured, the state $|\psi_i\rangle$ is obtained with probability $|a_i|^2$. That any measuring device has an associated direct sum decomposition, and that the interaction can be modeled in this way, is an axiom of quantum mechanics. It is not possible to prove that every device behaves in this way, but so far it has provided an excellent model that predicts the outcome of experiments with high accuracy.

Example 3.3.1 *Single-qubit measurement in the standard basis.* Let V be the vector space associated with a single-qubit system. A device that measures a qubit in the standard basis has, by definition, the associated direct sum decomposition $V = S_1 \oplus S_2$, where S_1 is generated by $|0\rangle$ and S_2 is generated by $|1\rangle$. An arbitrary state $|\psi\rangle = a|0\rangle + b|1\rangle$ measured by such a device will be $|0\rangle$ with probability $|a|^2$, the amplitude of $|\psi\rangle$ in the subspace S_1, and $|1\rangle$ with probability $|b|^2$.

Example 3.3.2 *Single-qubit measurement in the Hadamard basis.* A device that measures a single qubit in the Hadamard basis

$$\{|+\rangle = \frac{1}{\sqrt{2}}(|0\rangle + |1\rangle), |-\rangle = \frac{1}{\sqrt{2}}(|0\rangle - |1\rangle)\}$$

has associated subspace decomposition $V = S_+ \oplus S_-$, where S_+ is generated by $|+\rangle$ and S_- is generated by $|-\rangle$. A state $|\psi\rangle = a|0\rangle + b|1\rangle$ can be rewritten as $|\psi\rangle = \frac{a+b}{\sqrt{2}}|+\rangle + \frac{a-b}{\sqrt{2}}|-\rangle$, so the probability that $|\psi\rangle$ is measured as $|+\rangle$ will be $|\frac{a+b}{\sqrt{2}}|^2$ and $|-\rangle$ will be $|\frac{a-b}{\sqrt{2}}|^2$.

The next two examples describe measurements of two-qubit states that are used in the entanglement-based quantum key distribution protocol described in section 3.4. Chapter 4 explores measurement of multiple-qubit systems in more detail and builds up the standard notational shorthand for describing quantum measurements.

Example 3.3.3 *Measurement of the first qubit of a two-qubit state in the standard basis.* Let V be the vector space associated with a two-qubit system. A device that measures the first qubit in the standard basis has associated subspace decomposition $V = S_1 \oplus S_2$ where $S_1 = |0\rangle \otimes V_2$, the two-dimensional subspace spanned by $\{|00\rangle, |01\rangle\}$, and $S_2 = |1\rangle \otimes V_2$, which is spanned by $\{|10\rangle, |11\rangle\}$. To see what happens when such a device measures an arbitrary two-qubit state $|\psi\rangle = a_{00}|00\rangle + a_{01}|01\rangle + a_{10}|10\rangle + a_{11}|11\rangle$, we write $|\psi\rangle = c_1|\psi_1\rangle + c_2|\psi_2\rangle$ where $|\psi_1\rangle = 1/c_1(a_{00}|00\rangle + a_{01}|01\rangle) \in S_1$ and $|\psi_2\rangle = 1/c_2(a_{10}|10\rangle + a_{11}|11\rangle) \in S_2$, with $c_1 = \sqrt{|a_{00}|^2 + |a_{01}|^2}$ and $c_2 = \sqrt{|a_{10}|^2 + |a_{11}|^2}$ as the normalization factors. Measurement of $|\psi\rangle$ with this device results in the state $|\psi_1\rangle$ with probability $|c_1|^2 = |a_{00}|^2 + |a_{01}|^2$ and the state $|\psi_2\rangle$ with probability $|c_2|^2 = |a_{10}|^2 + |a_{11}|^2$. In particular, when the Bell state $|\Phi^+\rangle = \frac{1}{\sqrt{2}}(|00\rangle + |11\rangle)$ is measured, we obtain $|00\rangle$ and $|11\rangle$ with equal probability.

Example 3.3.4 *Measurement of the first qubit of a two-qubit state in the Hadamard basis.* A device that measures the first qubit of a two-qubit system with respect to the Hadamard basis $\{|+\rangle, |-\rangle\}$ has an associated direct sum decomposition $V = S_1' \oplus S_2'$, where $S_1' = |+\rangle \otimes V_2$, the two-dimensional subspace spanned by $\{|+\rangle|0\rangle, |+\rangle|1\rangle\}$, and $S_2' = |-\rangle \otimes V_2$. We write $|\psi\rangle = a_{00}|00\rangle + a_{01}|01\rangle + a_{10}|10\rangle + a_{11}|11\rangle$ as $|\psi\rangle = a_1'|\psi_1'\rangle + a_2'|\psi_2'\rangle$, where

$$|\psi_1'\rangle = c_1' \left(\frac{a_{00} + a_{10}}{\sqrt{2}}|+\rangle|0\rangle + \frac{a_{01} + a_{11}}{\sqrt{2}}|+\rangle|1\rangle \right)$$

and

$$|\psi_2'\rangle = c_2' \left(\frac{a_{00} - a_{10}}{\sqrt{2}}|-\rangle|0\rangle + \frac{a_{01} - a_{11}}{\sqrt{2}}|-\rangle|1\rangle \right).$$

We leave it to the reader to calculate c_1' and c_2' and the probabilities for the two outcomes, and to show that such a measurement on the state $|\Phi^+\rangle = \frac{1}{\sqrt{2}}(|00\rangle + |11\rangle)$ yields $|+\rangle|+\rangle$ and $|-\rangle|-\rangle$ with equal probability.

3.4 Quantum Key Distribution Using Entangled States

In 1991, Artur Ekert developed a quantum key distribution scheme that makes use of special properties of entangled states. The Ekert 91 protocol resembles the BB84 protocol of section 2.4 in some ways. In his protocol, Alice and Bob establish a shared key by separately performing random measurements on their halves of an EPR pair and then comparing which bases they used over a classical channel.

Because Alice and Bob do not exchange quantum states during the protocol, and an eavesdropper Eve cannot learn anything useful by listening in on the classical exchange alone, Eve's only chance to obtain information about the key is for her to interact with the purported EPR pair as it is being created or transmitted in the setup for the protocol. For this reason it is easier to prove the security of protocols based on entangled states. Such proofs have then been modified to prove the security of other QKD protocols like BB84. As with BB84, we describe only the protocol; tools developed in later chapters are needed to describe many of Eve's possible attacks and to give a proof of security. Exercise 3.15 analyzes the limited effectiveness of some simple attacks Eve could make.

The protocol begins with the creation of a sequence of pairs of qubits, all in the entangled state $|\Phi^+\rangle = \frac{1}{\sqrt{2}}(|00\rangle + |11\rangle)$. Alice receives the first qubit of each pair, while Bob receives the second. When they wish to create a secret key, for each qubit they both independently and randomly choose either the standard basis $\{|0\rangle, |1\rangle\}$ or the Hadamard basis $\{|+\rangle, |-\rangle\}$ in which to measure, just as in the BB84 protocol. After they have made their measurements, they compare bases and discard those bits for which their bases differ.

If Alice measures the first qubit in the standard basis and obtains $|0\rangle$, then the entire state becomes $|00\rangle$. If Bob now measures in the standard basis, he obtains the result $|0\rangle$ with certainty. If instead he measures in the Hadamard basis $\{|+\rangle, |-\rangle\}$, he obtains $|+\rangle$ and $|-\rangle$ with equal probability, since $|00\rangle = |0\rangle(\frac{1}{\sqrt{2}}(|+\rangle + |-\rangle))$. Just as in the BB84 protocol, he interprets the states $|+\rangle$ and $|-\rangle$ as corresponding to the classical bit values 0 and 1 respectively; thus when he measures in the basis $\{|+\rangle|-\rangle\}$ and Alice measures in the standard basis, he obtains the same bit value as Alice only half the time. The behavior is similar when Alice's measurement indicates her qubit is in state $|1\rangle$. If instead Alice measures in the Hadamard basis and obtains the result that her qubit is in the state $|+\rangle$, the whole state becomes $|+\rangle|+\rangle$. If Bob now measures in the Hadamard basis, he obtains $|+\rangle$ with certainty, whereas if he measures in the standard basis he obtains $|0\rangle$ and $|1\rangle$ with equal probability. Since they always get the same bit value if they measure in the same basis, the protocol results in a shared random key, as long as the initial pairs were EPR pairs. The security of the scheme relies on adding steps to the protocol we have just described that enable Alice and Bob to test the fidelity of their EPR pairs. We are not yet in a position to describe such tests. The tests Ekert suggested are based on Bell's inequalities (section 4.4.3). Other, more efficient tests have been devised.

This protocol has the intriguing property that in theory Alice and Bob can prepare shared keys as they need them, never needing to store keys for any length of time. In practice, to prepare keys on an as-needed basis in this way, Alice and Bob would need to be able to store their EPR pairs so that they are not corrupted during that time. The capability of long-term reliable storage of entangled states does not exist at present.

3.5 References

In the early 1980s, Richard Feynman and Yuri Manin separately recognized that certain quantum phenomena associated with entangled particles could not be simulated efficiently on standard computers. Turning this observation around caused them to speculate whether these quantum phenomena could be used to speed up computation in general. Their early musings on quantum computation can be found in [121], [150], [202], and [203].

More extensive treatments of the tensor product can be found in Arno Bohm's *Quantum Mechanics* [53], Paul Bamberg and Shlomo Sternberg's *A Course in Mathematics for Students of Physics* [30], and Thomas Hungerford's *Algebra* [158].

Ekert's key distribution protocol based on EPR pairs, originally proposed in [111], has been demonstrated in the laboratory [163, 294]. Gisin et al. [130] provide a detailed survey of work on quantum key distribution including Ekert's algorithm.

3.6 Exercises

Exercise 3.1. Let V be a vector space with basis $\{(1, 0, 0), (0, 1, 0), (0, 0, 1)\}$. Give two different bases for $V \otimes V$.

Exercise 3.2. Show by example that a linear combination of entangled states is not necessarily entangled.

Exercise 3.3. Show that the state

$$|W_n\rangle = \frac{1}{\sqrt{n}}(|0\ldots001\rangle + |0\ldots010\rangle + |0\ldots100\rangle + \cdots + |1\ldots000\rangle)$$

is entangled, with respect to the decomposition into the n qubits, for every $n > 1$.

Exercise 3.4. Show that the state

$$|GHZ_n\rangle = \frac{1}{\sqrt{2}}(|00\ldots0\rangle + |11\ldots1\rangle)$$

is entangled, with respect to the decomposition into the n qubits, for every $n > 1$.

Exercise 3.5. Is the state $\frac{1}{\sqrt{2}}(|0\rangle|+\rangle + |1\rangle|-\rangle)$ entangled?

Exercise 3.6. If someone asks you whether the state $|+\rangle$ is entangled, what will you say?

Exercise 3.7. Write the following states in terms of the Bell basis.

a. $|00\rangle$

b. $|+\rangle|-\rangle$

c. $\frac{1}{\sqrt{3}}(|00\rangle + |01\rangle + |10\rangle)$

Exercise 3.8.

a. Show that $\frac{1}{\sqrt{2}}(|0\rangle|0\rangle + |1\rangle|1\rangle)$ and $\frac{1}{\sqrt{2}}(|+\rangle|+\rangle + |-\rangle|-\rangle)$ refer to the same quantum state.

b. Show that $\frac{1}{\sqrt{2}}(|0\rangle|0\rangle - |1\rangle|1\rangle)$ refers to the same state as $\frac{1}{\sqrt{2}}(|\mathbf{i}\rangle|\mathbf{i}\rangle + |-\mathbf{i}\rangle|-\mathbf{i}\rangle)$.

Exercise 3.9.

a. Show that any n-qubit quantum state can be represented by a vector of the form

$$a_0|0\ldots00\rangle + a_1|0\ldots01\rangle + \cdots + a_{2^n-1}|1\ldots11\rangle$$

where the first non-zero a_i is real and non-negative.

b. Show that this representation is unique in the sense that any two different vectors of this form represent different quantum states.

Exercise 3.10. Show that for any orthonormal basis $B = \{|\beta_1\rangle, |\beta_2\rangle, \ldots, |\beta_n\rangle\}$ and vectors $|v\rangle = a_1|\beta_1\rangle + a_2|\beta_2\rangle + \cdots + a_n|\beta_n\rangle$ and $|w\rangle = c_1|\beta_1\rangle + c_2|\beta_2\rangle + \cdots + c_n|\beta_n\rangle$

a. the inner product $\langle w|v\rangle$ of $|v\rangle$ and $|w\rangle$ is $\bar{c}_1 a_1 + \bar{c}_2 a_2 + \cdots + \bar{c}_n a_n$, and

b. the length squared of $|v\rangle$ is $\||v\rangle\|^2 = \langle v|v\rangle = |a_1|^2 + |a_2|^2 + \cdots + |a_n|^2$.

Write all steps in Dirac's bra/ket notation.

Exercise 3.11. Let $|\psi\rangle$ be an n-qubit state. Show that the sum of the distances from $|\psi\rangle$ to the standard basis vectors $|j\rangle$ is bounded below by a positive constant that depends only on n,

$$\sum_j ||\psi\rangle - |j\rangle| \geq C,$$

where $|\vec{v}|$ indicates the length of the enclosed vector. Specify such a constant C in terms of n.

Exercise 3.12. Give an example of a two-qubit state that is a superposition with respect to the standard basis but that is not entangled.

Exercise 3.13.

a. Show that the four-qubit state $|\psi\rangle = \frac{1}{2}(|00\rangle + |11\rangle + |22\rangle + |33\rangle)$ of example 3.2.3 is entangled with respect to the decomposition into two two-qubit subsystems consisting of the first and second qubits and the third and fourth qubits.

b. For the four decompositions into two subsystems consisting of one and three qubits, say whether $|\psi\rangle$ is entangled or unentangled with respect to each of these decompositions.

Exercise 3.14.

a. For the standard basis, the Hadamard basis, and the basis $B = \{\frac{1}{\sqrt{2}}(|0\rangle + \mathbf{i}|1\rangle), \frac{1}{\sqrt{2}}|0\rangle - \mathbf{i}|1\rangle)\}$, determine the probability of each outcome when the second qubit of a two-qubit system in the state $|00\rangle$ is measured in each of the bases.

b. Determine the probability of each outcome when the second qubit of the state $|00\rangle$ is first measured in the Hadamard basis and then in the basis B of part a).

c. Determine the probability of each outcome when the second qubit of the state $|00\rangle$ is first measured in the Hadamard basis and then in the standard basis.

Exercise 3.15. This exercise analyzes the effectiveness of some simple attacks an eavesdropper Eve could make on Ekert's entangled state based QKD protocol.

a. Say Eve can measure Bob's half of each of the EPR pairs before it reaches him. Say she always measures in the standard basis. Describe a method by which Alice and Bob can determine that there is only a 2^{-s} chance that this sort of interference by Eve has gone undetected. What happens if Eve instead measures each qubit randomly in either the standard basis of the Hadamard basis? What happens if she uniformly at random chooses a basis from all possible bases?

b. Say Eve can pose as the entity sending the purported EPR pairs. Say instead of sending EPR pairs she sends a random mixture of qubit pairs in the states $|00\rangle$, $|11\rangle$, $|+\rangle|+\rangle$, and $|-\rangle|-\rangle$. After Alice and Bob perform the protocol of section 3.4, on how many bits on average do their purported shared secret keys agree? On average, how many of these bits does Eve know?

4 Measurement of Multiple-Qubit States

The nonclassical behavior of quantum measurement is critical to quantum information processing applications. This chapter develops the standard formalism used for measurement of multiple-qubit systems, and uses this formalism to describe the highly nonclassical behavior of entangled states under measurement. In particular, it discusses the *EPR paradox* and *Bell's theorem*, which illustrate the nonclassical nature of these states. Section 4.1 extends the Dirac bra/ket notation to linear transformations. It will be used in this chapter to describe measurements, and in chapter 5 to describe quantum transformations acting on quantum systems. Section 4.2 slowly introduces some of the notation and standard formalism for quantum measurement. Section 4.3 uses this material to give a full description of the standard formalism. Both sections contain a myriad of examples. The chapter concludes with a detailed discussion in section 4.4 of the behavior under measurement of the most famous of entangled states, *EPR pairs*.

4.1 Dirac's Bra/Ket Notation for Linear Transformations

Dirac's bra/ket notation provides a convenient way of specifying linear transformations on quantum states. Recall from section 2.2 that the conjugate transpose of the vector denoted by ket $|\psi\rangle$ is denoted by bra $\langle\psi|$, and the inner product of vectors $|\psi\rangle$ and $|\phi\rangle$ is given by $\langle\psi|\phi\rangle$. The notation $|x\rangle\langle y|$ represents the outer product of the vectors $|x\rangle$ and $|y\rangle$. Matrix multiplication is associative, and scalars commute with everything, so relations such as the following hold:

$$(|a\rangle\langle b|)|c\rangle = |a\rangle(\langle b||c\rangle)$$

$$= (\langle b|c\rangle)|a\rangle.$$

Let V be a vector space associated with a single-qubit system. The matrix for the operator $|0\rangle\langle 0|$ with respect to the standard basis in the standard order $\{|0\rangle, |1\rangle\}$ is

$$|0\rangle\langle 0| = \begin{pmatrix} 1 \\ 0 \end{pmatrix} \begin{pmatrix} 1 & 0 \end{pmatrix} = \begin{pmatrix} 1 & 0 \\ 0 & 0 \end{pmatrix}.$$

The notation $|0\rangle\langle 1|$ represents the linear transformation that maps $|1\rangle$ to $|0\rangle$ and $|0\rangle$ to the null vector, a relationship suggested by the notation:

$$(|0\rangle\langle 1|)|1\rangle = |0\rangle(\langle 1|1\rangle) = |0\rangle(1) = |0\rangle,$$

$$(|0\rangle\langle 1|)|0\rangle = |0\rangle(\langle 1|0\rangle) = |0\rangle(0) = 0.$$

Similarly

$$|1\rangle\langle 0| = \begin{pmatrix} 0 & 0 \\ 1 & 0 \end{pmatrix},$$

$$|1\rangle\langle 1| = \begin{pmatrix} 0 & 0 \\ 0 & 1 \end{pmatrix}.$$

Thus, all two-dimensional linear transformations on V can be written in Dirac's notation:

$$\begin{pmatrix} a & b \\ c & d \end{pmatrix} = a|0\rangle\langle 0| + b|0\rangle\langle 1| + c|1\rangle\langle 0| + d|1\rangle\langle 1|.$$

Example 4.1.1 The linear transformation that exchanges $|0\rangle$ and $|1\rangle$ is given by

$$X = |0\rangle\langle 1| + |1\rangle\langle 0|.$$

We will also use notation

$$X : \quad |0\rangle \mapsto |1\rangle$$
$$|1\rangle \mapsto |0\rangle,$$

which specifies a linear transformation in terms of its effect on the basis vectors. The transformation $X = |0\rangle\langle 1| + |1\rangle\langle 0|$ can also be represented by the matrix

$$\begin{pmatrix} 0 & 1 \\ 1 & 0 \end{pmatrix}$$

with respect to the standard basis.

Example 4.1.2 The transformation that exchanges the basis vectors $|00\rangle$ and $|10\rangle$ and leaves the others alone is written $|10\rangle\langle 00| + |00\rangle\langle 10| + |11\rangle\langle 11| + |01\rangle\langle 01|$ and has matrix representation

$$\begin{pmatrix} 0 & 0 & 1 & 0 \\ 0 & 1 & 0 & 0 \\ 1 & 0 & 0 & 0 \\ 0 & 0 & 0 & 1 \end{pmatrix}$$

in the standard basis.

An operator on an n-qubit system that maps the basis vector $|j\rangle$ to $|i\rangle$ and all other standard basis elements to 0 can be written

$$O = |i\rangle\langle j|$$

in the standard basis; the matrix for O has a single non-zero entry 1 in the ij^{th} place. A general operator O with entries a_{ij} in the standard basis can be written

$$O = \sum_i \sum_j a_{ij} |i\rangle\langle j|.$$

Similarly, the ij^{th} entry of the matrix for O in the standard basis is given by $\langle i|O|j\rangle$.

As an example of working with this notation, we write out the result of applying operator O to a vector $|\psi\rangle = \sum_k b_k |k\rangle$:

$$O|\psi\rangle = \left(\sum_i \sum_j a_{ij} |i\rangle\langle j|\right)\left(\sum_k b_k |k\rangle\right) = \sum_i \sum_j \sum_k a_{ij} b_k |i\rangle\langle j||k\rangle$$

$$= \sum_i \sum_j a_{ij} b_j |i\rangle.$$

More generally, if $\{|\beta_i\rangle\}$ is a basis for an N-dimensional vector space V, then an operator $O : V \to V$ can be written as

$$\sum_{i=1}^{N} \sum_{j=1}^{N} b_{ij} |\beta_i\rangle\langle \beta_j|$$

with respect to this basis. In particular, the matrix for O with respect to basis $\{|\beta_i\rangle\}$ has entries $O_{ij} = b_{ij}$.

Initially the vector/matrix notation may be easier for the reader to comprehend because it is more familiar, and sometimes this notation is convenient for performing calculations. But it requires choosing a basis and an ordering of that basis. The bra/ket notation is independent of the basis and the order of the basis elements. It is also more compact, and it suggests correct relationships, as we saw for the outer product, so that once it becomes familiar, it is easier to read.

4.2 Projection Operators for Measurement

Section 2.3 described measurement of a single qubit in terms of projection onto a basis vector associated with the measurement device. This notion generalizes to measurement in multiple-qubit systems. For any subspace S of V, the subspace $S^\perp$ consists of all vectors that are perpendicular to all vectors in S. The subspaces S and $S^\perp$ satisfy $V = S \oplus S^\perp$; thus, any vector $|v\rangle \in V$ can be written uniquely as the sum of a vector $\vec{s}_1 \in S$ and a vector $\vec{s}_2 \in S^\perp$. For any S, the *projection operator* P_S is the linear operator $P_S : V \to S$ that sends $|v\rangle \mapsto \vec{s}_1$ where $|v\rangle = \vec{s}_1 + \vec{s}_2$ with

$\vec{s}_1 \in S_1$ and $\vec{s}_2 \in S_2$. We use the notation $\vec{s}_i$ because $\vec{s}_1$ and $\vec{s}_2$ are generally not unit vectors. The operator $|\psi\rangle\langle\psi|$ is the projection operator onto the subspace spanned by $|\psi\rangle$. Projection operators are sometimes called *projectors* for short. For any direct sum decomposition of $V = S_1 \oplus \cdots \oplus S_k$ into orthogonal subspaces S_i there are k related projection operators $P_i : V \to S_i$ where $P_i|v\rangle = \vec{s}_i$ where $|v\rangle = \vec{s}_1 + \cdots + \vec{s}_k$ with $s_i \in S_i$. In this terminology, a measuring device with associated decomposition $V = S_1 \oplus \cdots \oplus S_k$ acting on a state $|\psi\rangle$ results in the state

$$|\phi\rangle = \frac{P_i|\psi\rangle}{|P_i|\psi\rangle|}$$

with probability $|P_i|\psi\rangle|^2$.

Example 4.2.1 The projector $|0\rangle\langle 0|$ acts on a single-qubit state $|\psi\rangle$ and obtains the component of $|\psi\rangle$ in the subspace generated by $|0\rangle$. Let $|\psi\rangle = a|0\rangle + b|1\rangle$. Then $(|0\rangle\langle 0|)|\psi\rangle = a\langle 0|0\rangle|0\rangle + b\langle 0|1\rangle|0\rangle = a|0\rangle$.

The projector $|1\rangle|0\rangle\langle 1|\langle 0|$ acts on two-qubit states. Let

$$|\phi\rangle = a_{00}|00\rangle + a_{01}|01\rangle + a_{10}|10\rangle + a_{11}|11\rangle.$$

Then

$$(|1\rangle|0\rangle\langle 1|\langle 0|)\ |\phi\rangle = a_{10}|1\rangle|0\rangle.$$

Let P_S be the projection operator from an n-dimensional vector space V onto an s-dimensional subspace S with basis $\{|\alpha_0\rangle, \ldots, |\alpha_{s-1}\rangle\}$. Then

$$P_S = \sum_{i=1}^{s-1} |\alpha_i\rangle\langle\alpha_i| = |\alpha_0\rangle\langle\alpha_0| + \cdots + |\alpha_{s-1}\rangle\langle\alpha_{s-1}|.$$

Example 4.2.2 Let $|\psi\rangle = a_{00}|00\rangle + a_{01}|01\rangle + a_{10}|10\rangle + a_{11}|11\rangle$ represent a state of a two-qubit system with associated vector space V. Let S_1 be the subspace spanned by $|00\rangle, |01\rangle$. The operator $P_S = |00\rangle\langle 00| + |01\rangle\langle 01|$ is the projection operator that sends $|\psi\rangle$ to the (non-normalized) vector $a_{00}|00\rangle + a_{01}|01\rangle$.

Let V and W be two vector spaces with inner product. The *adjoint operator* or *conjugate transpose* $O^\dagger : V \to W$ of an operator $O : W \to V$ is defined to be the operator that satisfies the following inner product relation. For any $\vec{v} \in V$ and $O\vec{w} \in W$, the inner product between $O^\dagger\vec{v}$ and $\vec{w}$ is the same as the inner product between $\vec{v}$ and $O\vec{w}$:

$$O^\dagger\vec{v} \cdot \vec{w} = \vec{v} \cdot O\vec{w}.$$

The matrix for the adjoint operator $O^\dagger$ of O is obtained by taking the complex conjugate of all entries and then the transpose of the matrix for O, where we are assuming consistent use of bases for V and W. Recall from section 2.2 that $\langle x|$ is the conjugate transpose of $|x\rangle$. The reader can check that $(A|x\rangle)^\dagger = \langle x|A^\dagger$. In bra/ket notation, the relation between the inner product of $O^\dagger|x\rangle$ and $|w\rangle$ and the inner product of $|x\rangle$ and $O|w\rangle$ is reflected in the notation:

$$(\langle x|O)|w\rangle = \langle x|(O|w\rangle) = \langle x|O|w\rangle.$$

The definition of a projection operator P implies that applying a projection operator many times in succession has the same effect as just applying it once: $PP = P$. Furthermore, any projection operator is its own adjoint: $P = P^\dagger$. Thus

$$|P|v\rangle|^2 = (\langle v|P^\dagger)(P|v\rangle) = \langle v|P|v\rangle$$

for any projection operator P and all $|v\rangle \in V$.

To solidify our understanding of projection operators and Dirac's notation, let us describe single-qubit measurement in the standard basis in terms of this formalism.

Example 4.2.3 *Formal treatment of single-qubit measurement in the standard basis.* Let V be the vector space associated with a single-qubit system. The direct sum decomposition for V associated with measurement in the standard basis is $V = S \oplus S'$, where S is the subspace generated by $|0\rangle$ and S' is the subspace generated by $|1\rangle$. The related projection operators are $P : V \to S$ and $P' : V \to S'$, where $P = |0\rangle\langle 0|$ and $P' = |1\rangle\langle 1|$. Measurement of the state $|\psi\rangle = a|0\rangle + b|1\rangle$ results in the state $\frac{P|\psi\rangle}{|P|\psi\rangle|}$ with probability $|P|\psi\rangle|^2$. Since

$$P|\psi\rangle = (|0\rangle\langle 0|)|\psi\rangle = |0\rangle\langle 0|\psi\rangle = a|0\rangle$$

and

$$|P|\psi\rangle|^2 = \langle\psi|P|\psi\rangle = \langle\psi|(|0\rangle\langle 0|)|\psi\rangle = \langle\psi|0\rangle\langle 0|\psi\rangle = \bar{a}a = |a|^2,$$

the result of the measurement is $\frac{a|0\rangle}{|a|}$ with probability $|a|^2$. Since by section 2.5 an overall phase factor is physically meaningless, the state represented by $|0\rangle$ has been obtained with probability $|a|^2$. A similar calculation shows that the state represented by $|1\rangle$ is obtained with probability $|b|^2$.

Before giving examples of more interesting measurements, we describe measurement of a two-qubit state with respect to the full decomposition associated with the standard basis.

Example 4.2.4 *Measuring a two-qubit state with respect to the full standard basis decomposition.* Let V be the vector space associated with a two-qubit system and $|\phi\rangle = a_{00}|00\rangle + a_{01}|01\rangle + a_{10}|10\rangle + a_{11}|11\rangle$ an arbitrary two-qubit state. Consider a measurement with decomposition $V = S_{00} \oplus S_{01} \oplus S_{10} \oplus S_{11}$, where S_{ij} is the one-dimensional complex subspace spanned by $|ij\rangle$. The related projection operators $P_{ij} : V \to S_{ij}$ are $P_{00} = |00\rangle\langle 00|$, $P_{01} = |01\rangle\langle 01|$, $P_{10} = |10\rangle\langle 10|$, and $P_{11} = |11\rangle\langle 11|$. The state after measurement will be $\frac{P_{ij}|\psi\rangle}{|P_{ij}|\psi\rangle|}$ with probability

$|P_{ij}|\psi\rangle|^2$. Recall from sections 2.5.1 and 3.1.3 that two unit vectors $|v\rangle$ and $|w\rangle$ represent the same quantum state if $|v\rangle = e^{i\theta}|w\rangle$ for some θ, and that $|v\rangle \sim |w\rangle$ indicates that $|v\rangle$ and $|w\rangle$ represent the same quantum state. The state after measurement is either

$$\frac{P_{00}|\psi\rangle}{|P_{00}|\psi\rangle|} = \frac{a_{00}|00\rangle}{|a_{00}|} \sim |00\rangle$$

with probability $\langle\psi|P_{00}|\psi\rangle = |a_{00}|^2$, or $|01\rangle$ with probability $|a_{01}|^2$, or $|10\rangle$ with probability $|a_{10}|^2$, or $|11\rangle$ with probability $|a_{11}|^2$.

To develop fluency with this material, the reader may now want to rewrite, using this notation, the examples of section 3.3.

More interesting are measurements that give information about the relations between qubit values without giving any information about the qubit values themselves. For example, we can measure two qubits for bit equality without determining the actual value of the bits. Such measurements will be used heavily in quantum error correction schemes.

Example 4.2.5 *Measuring a two-qubit state for bit equality in the standard basis.* Let V be the vector space associated with a two-qubit system. Consider a measurement with associated direct sum decomposition $V = S_1 \oplus S_2$, where S_1 is the subspace generated by $\{|00\rangle, |11\rangle\}$, the subspace in which the two bits are equal, and S_2 is the subspace generated by $\{|10\rangle, |01\rangle\}$, the subspace in which the two bits are not equal. Let P_1 and P_2 be the projection operators onto S_1 and S_2 respectively. When a system in state $|\psi\rangle = a_{00}|00\rangle + a_{01}|01\rangle + a_{10}|10\rangle + a_{11}|11\rangle$ is measured in this way, with probability $|P_i|\psi\rangle|^2 = \langle\psi|P_i|\psi\rangle$, the state after measurement becomes $\frac{P_i|\psi\rangle}{|P_i|\psi\rangle|}$.

Let $c_1 = \langle\psi|P_1|\psi\rangle = \sqrt{|a_{00}|^2 + |a_{11}|^2}$ and $c_2 = \langle\psi|P_w|\psi\rangle = \sqrt{|a_{01}|^2 + |a_{10}|^2}$. After measurement the state will be $|u\rangle = \frac{1}{c_1}(a_{00}|00\rangle + a_{11}|11\rangle)$ with probability $|c_1|^2 = |a_{00}|^2 + |a_{11}|^2$ and $|v\rangle = \frac{1}{c_2}(a_{01}|01\rangle + a_{10}|10\rangle)$ with probability $|c_2|^2 = |a_{01}|^2 + |a_{10}|^2$. If the first outcome happens, then we know that the two bit values are equal, but we do not know whether they are 0 or 1. If the second case happens, we know that the two bit values are not equal, but we do not know which one is 0 and which one is 1. Thus, the measurement does not determine the value of the two bits, only whether the two bits are equal.

As in the case of single-qubit states, most states are a superposition with respect to a measurement's subspace decomposition. In the previous example, a state that is a superposition containing components with both equal and unequal bit values is transformed by measurement either to a state (generally still a superposition of standard basis elements), in which in all components the bit values are equal, or to a state in which the bit values are not equal in all of the components.

Before further developing the formalism used to describe quantum measurement, we give an additional example, one in which the associated subspaces are not generated by subsets of the standard basis elements.

Example 4.2.6 *Measuring a two-qubit state with respect to the Bell basis decomposition.* Recall from section 3.2 the four Bell states $|\Phi^+\rangle = \frac{1}{\sqrt{2}}(|00\rangle + |11\rangle)$, $|\Phi^-\rangle = \frac{1}{\sqrt{2}}(|00\rangle - |11\rangle)$, $|\Psi^+\rangle = \frac{1}{\sqrt{2}}(|01\rangle + |10\rangle)$, and $|\Psi^-\rangle = \frac{1}{\sqrt{2}}(|01\rangle - |10\rangle)$. Let $V = S_{\Phi^+} \oplus S_{\Phi^-} \oplus S_{\Psi^+} \oplus S_{\Psi^-}$ be the direct sum decomposition into the subspaces generated by the Bell states. Measurement of the state $|00\rangle$ with respect to this decomposition yields $|\Phi^+\rangle$ with probability $1/2$ and $|\Phi^-\rangle$ with probability $1/2$, because $|00\rangle = \frac{1}{\sqrt{2}}(|\Phi^+\rangle + |\Phi^-\rangle)$. The reader can determine the outcomes and their probabilities for the three other standard basis elements, and a general two-qubit state.

The next section continues developing the standard formalism used throughout the quantum mechanics literature to describe quantum measurement.

4.3 Hermitian Operator Formalism for Measurement

Instead of explicitly writing out the subspace decomposition associated with a measurement, including the definition of each subspace of the decomposition in terms of a generating set, a mathematical shorthand is used. Certain operators, called Hermitian operators, define a unique orthogonal subspace decomposition, their eigenspace decomposition. Moreover, for every such decomposition, there exists a Hermitian operator whose eigenspace decomposition is this decomposition. Given this correspondence, Hermitian operators can be used to describe measurements. We begin by reminding our readers of definitions and facts about eigenspaces and Hermitian operators.

Let $O : V \rightarrow V$ be a linear operator. Recall from linear algebra that if $O\vec{v} = \lambda\vec{v}$ for some non-zero vector $\vec{v} \in V$, then λ is an *eigenvalue* and $\vec{v}$ is a λ-*eigenvector* of O. If both $\vec{v}$ and $\vec{w}$ are λ-eigenvectors of O, then $\vec{v} + \vec{w}$ is also a λ-eigenvector, so the set of all λ-eigenvectors forms a subspace of V called the λ-*eigenspace* of O. For an operator with a diagonal matrix representation, the eigenvalues are simply the values along the diagonal.

An operator $O : V \rightarrow V$ is *Hermitian* if it is equal to its adjoint, $O^\dagger = O$. The eigenspaces of Hermitian operators have special properties. Suppose λ is an eigenvalue of an Hermitian operator O with eigenvector $|x\rangle$. Since

$$\lambda\langle x|x\rangle = \langle x|\lambda|x\rangle = \langle x_\lambda|(O|x_\lambda\rangle) = ((\langle x|O^\dagger)|x\rangle = \bar{\lambda}\langle x|x\rangle,$$

$\lambda = \bar{\lambda}$, which means that all eigenvalues of a Hermitian operator are real.

To give the connection between Hermitian operators and orthogonal subspace decompositions, we need to show that the eigenspaces $S_{\lambda_1}, S_{\lambda_2}, \ldots, S_{\lambda_k}$ of a Hermitian operator are orthogonal and satisfy $S_{\lambda_1} \oplus S_{\lambda_2} \oplus \cdots \oplus S_{\lambda_k} = V$. For any operator, two distinct eigenvalues have disjoint eigenspaces since, for any unit vector $|x\rangle$, $O|x\rangle = \lambda_0|x\rangle$ and $O|x\rangle = \lambda_1|x\rangle$ imply $(\lambda_0 - \lambda_1)|x\rangle = 0$, which implies that $\lambda_0 = \lambda_1$. For any Hermitian operator, the eigenvectors for distinct eigenvalues must be orthogonal. Suppose $|v\rangle$ is a λ-eigenvector and $|w\rangle$ is a μ-eigenvector with $\lambda \neq \mu$. Then

$$\lambda\langle v|w\rangle = (\langle v|O^\dagger)|w\rangle = \langle v|(O|w\rangle) = \mu\langle v|w\rangle.$$

Since λ and μ are distinct eigenvalues, $\langle v|w\rangle = 0$. Thus, S_{λ_i} and S_{λ_j} are orthogonal for $\lambda_i \neq \lambda_j$. Exercise 4.16 shows that the direct sum of all of the eigenspaces for a Hermitian operator $O : V \to V$ is the whole space V.

Let V be an N-dimensional vector space, and let $\lambda_1, \lambda_2, \ldots, \lambda_k$ be the $k \leq N$ distinct eigenvalues of an Hermitian operator $O : V \to V$. We have just shown that $V = S_{\lambda_1} \oplus \cdots \oplus S_{\lambda_k}$, where S_{λ_i} is the eigenspace of O with eigenvalue λ_i. This direct sum decomposition of V is called the *eigenspace decomposition* of V for the Hermitian operator O. Thus, any Hermitian operator $O : V \to V$ uniquely determines a subspace decomposition for V. Furthermore, any decomposition of a vector space V into the direct sum of subspaces $S_1, \ldots, S_k$ can be realized as the eigenspace decomposition of a Hermitian operator $O : V \to V$: let P_i be the projectors onto the subspaces S_i, and let $\lambda_1, \lambda_2, \ldots, \lambda_k$ be any set of distinct real values; then $O = \sum_{i=1}^{k} \lambda_i P_i$ is a Hermitian operator with the desired direct sum decomposition. Thus, when describing a measurement, instead of directly specifying the associated subspace decomposition, we can specify a Hermitian operator whose eigenspace decomposition is that decomposition.

Any Hermitian operator with the appropriate direct sum decomposition can be used to specify a given measurement; in particular, the values of the λ_i are irrelevant as long as they are distinct. The λ_i should be thought of simply as labels for the corresponding subspaces, or equivalently as labels for the measurement outcomes. In quantum physics, these labels are often chosen to represent a shared property, such as the energy, of the eigenstates in the corresponding eigenspace. For our purposes, we do not need to assign labels with meaning; any distinct set of eigenvalues will do.

Specifying a measurement in terms of a Hermitian operator is standard practice throughout the quantum-mechanics and quantum-information-processing literature. It is important to recognize, however, that quantum measurement is not modeled by the *action* of a Hermitian operator on a state. The projectors P_j associated with a Hermitian operator O, not O itself, act on a state. Which projector acts on the state depends on the probabilities $p_j = \langle\psi|P_j|\psi\rangle$. For example, the result of measuring $|\psi\rangle = a|0\rangle + b|1\rangle$ according to the Hermitian operator $Z = |0\rangle\langle 0| - |1\rangle\langle 1|$ does *not* result in the state $a|0\rangle - b|1\rangle$, even though

$$\begin{pmatrix} 1 & 0 \\ 0 & -1 \end{pmatrix} \begin{pmatrix} a \\ b \end{pmatrix} = \begin{pmatrix} a \\ -b \end{pmatrix}.$$

Direct multiplication by a Hermitian operator generally does not even result in a well-defined state; for example,

$$\begin{pmatrix} 0 & 0 \\ 0 & 1 \end{pmatrix} |0\rangle = \begin{pmatrix} 0 & 0 \\ 0 & 1 \end{pmatrix} \begin{pmatrix} 1 \\ 0 \end{pmatrix} = \begin{pmatrix} 0 \\ 0 \end{pmatrix}.$$

The Hermitian operator is only a convenient bookkeeping trick, a concise way of specifying the subspace decomposition associated with the measurement.

4.3.1 The Measurement Postulate

Many aspects of our model of quantum mechanics are not directly observable by experiment. For example, as we saw in section 2.3, given a single instance of an unknown single-qubit state $a|0\rangle + b|1\rangle$, there is no way to determine experimentally what state it is in; we cannot directly observe the quantum state. It is only the results of measurements that we can directly observe. For this reason, the Hermitian operators we use to specify measurements are called *observables*.

The measurement postulate of quantum mechanics states that:

- Any quantum measurement can be specified by a Hermitian operator O called an observable.

- The possible outcomes of measuring a state $|\psi\rangle$ with an observable O are labeled by the eigenvalues of O. Measurement of state $|\psi\rangle$ results in the outcome labeled by the eigenvalue λ_i of O with probability $|P_i|\psi\rangle|^2$ where P_i is the projector onto the λ_i-eigenspace.

- (*Projection*) The state after measurement is the normalized projection $P_i|\psi\rangle / |P_i|\psi\rangle|$ of $|\psi\rangle$ onto the λ_i-eigenspace S_i. Thus the state after measurement is a unit length eigenvector of O with eigenvalue λ_i.

We should make clear that what we have described here is a mathematical formalism for measurement. It does not tell us what measurements can be done in practice, or with what efficiency. Some measurements that may be mathematically simple to state may not be easy to implement. Furthermore, the eigenvalues of physically realizable measurements may have meaning—for example, as the position or energy of a particle—but for us the eigenvalues are just arbitrary labels.

While a Hermitian operator uniquely specifies a subspace decomposition, for a given subspace decomposition there are many Hermitian operators whose eigenspace decomposition is that decomposition. In particular, since the eigenvalues are simply labels for the subspaces or possible outcomes, the specific values of the eigenvalues are irrelevant; it matters only which ones are distinct. For example, measuring with the Hermitian operator $|0\rangle\langle 0| - |1\rangle\langle 1|$ results in the same states with the same probabilities as measuring with $100|0\rangle\langle 0| - 100|1\rangle\langle 1|$, but these outcomes do not agree with the outcomes of the trivial measurement corresponding to the Hermitian operator $|0\rangle\langle 0| + |1\rangle\langle 1|$ or $42|0\rangle\langle 0| + 42|1\rangle\langle 1|$.

Example 4.3.1 *Hermitian operator formalism for measurement of a single qubit in the standard basis.* Using the description in example 4.2.3 of measurement of a single-qubit system in the standard basis, let us build up a Hermitian operator that specifies this measurement. The subspace decomposition corresponding to this measurement is $V = S \oplus S'$, where S is the subspace generated by $|0\rangle$ and S' is generated by $|1\rangle$. The projectors associated with S and S' are $P = |0\rangle\langle 0|$ and $P' = |1\rangle\langle 1|$ respectively. Let λ and λ' be any two distinct real values, say $\lambda = 2$ and $\lambda' = -3$. Then the operator

$$O = 2|0\rangle\langle 0| - 3|1\rangle\langle 1| = \begin{pmatrix} 2 & 0 \\ 0 & -3 \end{pmatrix}$$

is a Hermitian operator specifying the measurement of a single-qubit state in the standard basis.

Any other distinct values for λ and λ' could have been used. We will generally use either

$$|1\rangle\langle 1| = \begin{pmatrix} 0 & 0 \\ 0 & 1 \end{pmatrix} \text{ or}$$

$$Z = |0\rangle\langle 0| - |1\rangle\langle 1| = \begin{pmatrix} 1 & 0 \\ 0 & -1 \end{pmatrix}$$

to specify single-qubit measurements in the standard basis.

Example 4.3.2 *Hermitian operator formalism for measurement of a single qubit in the Hadamard basis.* We wish to construct a Hermitian operator corresponding to measurement of a single qubit in the Hadamard basis $\{|+\rangle, |-\rangle\}$. The subspaces under consideration are S_+, generated by $|+\rangle$, and S_-, generated by $|-\rangle$, with associated projectors $P_+ = |+\rangle\langle +| = \frac{1}{2}(|0\rangle\langle 0| + |0\rangle\langle 1| + |1\rangle\langle 0| + |1\rangle\langle 1|)$ and $P_- = |-\rangle\langle -| = \frac{1}{2}(|0\rangle\langle 0| - |0\rangle\langle 1| - |1\rangle\langle 0| + |1\rangle\langle 1|)$. We are free to choose λ_+ and λ_- any way we like as long as they are distinct. If we take $\lambda_+ = 1$ and $\lambda_- = -1$, then

$$X = |0\rangle\langle 1| + |1\rangle\langle 0| = \begin{pmatrix} 0 & 1 \\ 1 & 0 \end{pmatrix}$$

is a Hermitian operator for single-qubit measurement in the Hadamard basis.

Example 4.3.3 The Hermitian operator $A = |01\rangle\langle 01| + 2|10\rangle\langle 10| + 3|11\rangle\langle 11|$ has matrix representation

$$\begin{pmatrix} 0 & 0 & 0 & 0 \\ 0 & 1 & 0 & 0 \\ 0 & 0 & 2 & 0 \\ 0 & 0 & 0 & 3 \end{pmatrix}$$

with respect to the standard basis in the standard order $\{|00\rangle, |01\rangle, |10\rangle, |11\rangle\}$. The eigenspace decomposition for A consists of four subspaces, each generated by one of the standard basis vectors $|00\rangle$, $|01\rangle$, $|10\rangle$, $|11\rangle$. The operator A is one of many Hermitian operators that specify measurement with respect to the full standard basis decomposition described in example 4.2.4. The Hermitian operator $A' = 73|00\rangle\langle00| + 50|01\rangle\langle01| - 3|10\rangle\langle10| + 23|11\rangle\langle11|$ is another.

Example 4.3.4 The Hermitian operator

$$B = |00\rangle\langle00| + |01\rangle\langle01| + \pi(|10\rangle\langle10| + |11\rangle\langle11|) = \begin{pmatrix} 1 & 0 & 0 & 0 \\ 0 & 1 & 0 & 0 \\ 0 & 0 & \pi & 0 \\ 0 & 0 & 0 & \pi \end{pmatrix}$$

specifies measurement of a two-qubit system with respect to the subspace decomposition $V = S_0 \oplus S_1$, where S_0 is generated by $\{|00\rangle, |01\rangle\}$ and S_1 is generated by $\{|10\rangle, |11\rangle\}$, so B specifies measurement of the first qubit in the standard basis as described in example 3.3.3.

Example 4.3.5 The Hermitian operator

$$C = 2(|00\rangle\langle00| + |11\rangle\langle11|) + 3(|01\rangle\langle01| + |10\rangle\langle10|) = \begin{pmatrix} 2 & 0 & 0 & 0 \\ 0 & 3 & 0 & 0 \\ 0 & 0 & 3 & 0 \\ 0 & 0 & 0 & 2 \end{pmatrix}$$

specifies measurement with respect to the subspace decomposition $V = S_2 \oplus S_3$, where S_2 is generated by $\{|00\rangle, |11\rangle\}$ and S_3 is generated by $\{|01\rangle, |10\rangle\}$, so C specifies the measurement for bit equality described in example 4.2.5.

Given the subspace decomposition for a Hermitian operator O, it is possible to find an orthonormal eigenbasis of V for O. If O has n distinct eigenvalues, as in the general case, the eigenbasis is unique up to length one complex factors. If O has fewer than n eigenvalues, some of the eigenvalues are associated with an eigenspace of more than one dimension. In this case, a random orthonormal basis can be chosen for each eigenspace S_i. The matrix for the Hermitian operator O with respect to any of these eigenbases is diagonal.

Any Hermitian operator O with eigenvalues λ_j can be written as $O = \sum_j \lambda_j P_j$, where P_j are the projectors for the λ_j-eigenspaces of O. Every projector is Hermitian with eigenvalues 1 and 0 where the 1-eigenspace is the image of the operator. For an m-dimensional subspace S of V spanned by the basis $\{|i_1\rangle, \ldots, |i_m\rangle\}$, the associated projector

$$P_S = \sum_{j=1}^{m} |i_j\rangle\langle i_j|$$

maps vectors in V into S. If P_S and P_T are projectors for orthogonal subspaces S and T, the projector for the direct sum $S \oplus T$ is $P_S + P_T$. If P is a projector onto subspace S then $\mathbf{tr}(P)$, the sum of the diagonal elements of any matrix representing P, is the dimension of S. This argument applies to any basis since the trace is basis independent. Box 10.1 describes this and other properties of the trace.

Given linear operators A and B on vector spaces V and W respectively, the *tensor product* $A \otimes B$ acts on elements $v \otimes w$ of the tensor product space $V \otimes W$ as follows:

$$(A \otimes B)(v \otimes w) = Av \otimes Bw.$$

It follows from this definition that

$$(A \otimes B)(C \otimes D) = AC \otimes BD.$$

Let O_0 and O_1 be Hermitian operators on spaces V_0 and V_1 respectively. Then $O_0 \otimes O_1$ is a Hermitian operator on the space $V_0 \otimes V_1$. Furthermore, if O_i has eigenvalues λ_{ij} with associated eigenspaces S_{ij}, then $O_0 \otimes O_1$ has eigenvalues $\lambda'_{jk} = \lambda_{0j}\lambda_{1k}$. If an eigenvalue $\lambda'_{jk} = \lambda_{0j}\lambda_{1k}$ is unique, then its associated eigenspace S'_{jk} is the tensor product of S_{0j} and S_{1k}. In general, the eigenvalues λ'_{jk} need not be distinct. An eigenvalue λ' of $O_0 \otimes O_1$ that is the product of eigenvalues of O_0 and O_1 in multiple ways, $\lambda'_i = \lambda'_{j_1 k_1} = \cdots = \lambda'_{j_m k_m}$, has eigenspace $S = (S_{0j_1} \otimes S_{1k_1}) \oplus \cdots \oplus (S_{0j_m} \otimes S_{1k_m})$.

Most Hermitian operators O on $V_0 \otimes V_1$ cannot be written as a tensor product of two Hermitian operators O_1 and O_2 acting on V_0 and V_1 respectively. Such a decomposition is possible only if each subspace in the subspace decomposition described by O can be written as $S = S_0 \otimes S_1$ for S_0 and S_1 in the subspace decompositions associated to O_1 and O_2 respectively. While for most Hermitian operators this condition does not hold, it does hold for all of the observables we have described so far. For example,

$$\begin{pmatrix} 1 & 0 \\ 0 & -1 \end{pmatrix} \otimes \begin{pmatrix} 2 & 0 \\ 0 & 3 \end{pmatrix} = (|0\rangle\langle 0| - |1\rangle\langle 1|) \otimes (2|0\rangle\langle 0| + 3|1\rangle\langle 1|)$$

$$= 2|00\rangle\langle 00| + 3|01\rangle\langle 01| - 2|10\rangle\langle 10| - 3|11\rangle\langle 11|$$

specifies the full measurement in the standard basis, but with a different Hermitian operator from the one used in example 4.3.3. The operator

$$\begin{pmatrix} 1 & 0 \\ 0 & \pi \end{pmatrix} \otimes I = |00\rangle\langle 00| + |01\rangle\langle 01| + \pi(|10\rangle\langle 10| + |11\rangle\langle 11|)$$

specifies measurement of the first qubit in the standard basis as described in example 4.3.4, as does $Z \otimes I$, where $Z = |0\rangle\langle 0| - |1\rangle\langle 1|$. The Hermitian operator

$$Z \otimes Z = |00\rangle\langle 00| - |01\rangle\langle 01| - |10\rangle\langle 10| + |11\rangle\langle 11|$$

specifies the measurement for bit equality described in example 4.3.5. We now give an example of a two-qubit measurement that cannot be expressed as the tensor product of two single-qubit measurements.

Example 4.3.6 *Not all measurements are tensor products of single-qubit measurements.* Consider a two-qubit state. The observable M with matrix representation

$$M = \begin{pmatrix} 0 & 0 & 0 & 0 \\ 0 & 0 & 0 & 0 \\ 0 & 0 & 0 & 0 \\ 0 & 0 & 0 & 1 \end{pmatrix}$$

determines whether both bits are set to one. Measurement with the operator M results in a state contained in one of the two subspaces S_0 and S_1, where S_1 is the subspace spanned by $\{|11\rangle\}$ and S_0 is spanned by $\{|00\rangle, |01\rangle, |10\rangle\}$.

Measuring with M is quite different from measuring both qubits in the standard basis and then performing the classical AND operation. For instance, the state $|\psi\rangle = 1/\sqrt{2}(|01\rangle + |10\rangle)$ remains unchanged when measured with M, but measuring both qubits of $|\psi\rangle$ would result in either the state $|01\rangle$ or $|10\rangle$.

Any Hermitian operator $Q_1 \otimes Q_2$ on a two-qubit system is said to be composed of single-qubit measurements if Q_1 and Q_2 are Hermitian operators on the single-qubit systems. Furthermore, any Hermitian operator of the form $Q \otimes I$ or $I \otimes Q'$ on a two-qubit system is said to be a measurement on a single qubit of the system. More generally, a Hermitian operator of the form

$$I \otimes \cdots \otimes I \otimes Q \otimes I \otimes \cdots \otimes I$$

on an n-qubit system is said to be a single-qubit measurement of the system. Any Hermitian operator of the form $A \otimes I$ on a system $V \otimes W$, where A is a Hermitian operator acting on V is said to be a measurement of subsystem V.

Section 5.1 shows that measurement operators in the standard basis, when combined with quantum state transformations, are sufficient to perform arbitrary quantum measurements. In particular, there are quantum operations taking any basis to any other, so we can get all possible subspace decompositions of the state space by starting with a subspace decomposition in which all of the subspaces are generated by standard basis vectors and transforming. Understanding the effects of quantum measurement in different bases is crucial for a thorough understanding of entangled states and quantum information processing generally. Sections 2.4 and 3.4 illustrate the power of measuring in different bases as a key aspect of these quantum key distribution schemes. The next section turns to Bell's theorem, which further illustrates this point while at the same time giving deeper insight into nonclassical properties of entangled states.

When talking about measurement of an n-qubit system, there are two totally distinct types of decompositions of the vector space V under consideration: the tensor product decomposition into the n separate qubits and the direct sum decomposition into $k \leq 2^n$ subspaces associated with the measuring device. These decompositions could not be more different. In particular, a tensor component V_i of $V = V_1 \otimes \cdots \otimes V_n$ is not a subspace of V. Similarly, the subspaces associated with measurements do not correspond to the subsystems, such as individual qubits, of the whole system.

Section 2.3 mentioned that only one classical bit of information can be extracted from a single qubit. We can now both generalize this statement and make it more precise. Since any observable on an n-qubit system has at most 2^n distinct eigenvalues, there are at most 2^n possible results of a given measurement. Thus, a single measurement of an n-qubit system will reveal at most n bits of classical information. Since, in general, the measurement changes the state, any further measurements give information about the new state, not the original one. In particular, if the observable has 2^n distinct eigenvalues, measurement sends the state to an eigenvector, and further measurement cannot extract any additional information about the original state.

4.4 EPR Paradox and Bell's Theorem

In 1935, Albert Einstein, Boris Podolsky, and Nathan Rosen wrote a paper entitled "Can quantum-mechanical description of physical reality be considered complete?". The paper contained a thought experiment that inspired the simpler thought experiment, due to David Bohm, that we describe here. The experiment involves a pair of photons in the state $\frac{1}{\sqrt{2}}(|00\rangle + |11\rangle)$. Pairs of particles in such a state are called EPR pairs in honor of Einstein, Podolsky, and Rosen, even though such states did not appear in their paper.

Imagine a source that generates EPR pairs $\frac{1}{\sqrt{2}}(|00\rangle + |11\rangle)$ and sends the first particle to Alice and the second to Bob. Alice and Bob can be arbitrarily far apart. Each person can perform measurements only on the particle he or she receives. More precisely, Alice can use only observables of the form $O \otimes I$ to measure the system, and Bob can use only observables of the form $I \otimes O'$, where O and O' are single-qubit observables.

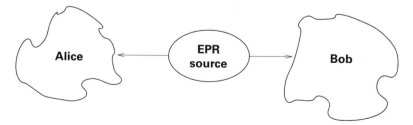

As we saw when we analyzed the Ekert91 quantum key distribution protocol in section 3.4, if Alice measures her particle in the standard single-qubit basis, and observes the state $|0\rangle$, the effect of this measurement is to project the state of the quantum system onto that part of the state compatible with the results of Alice's measurement, so the combined state will now be $|00\rangle$. If Bob now measures his particle, he will always observe $|0\rangle$. Thus it appears that Alice's measurement has affected the state of Bob's particle. Similarly, if Alice measures $|1\rangle$, so will Bob. By symmetry, if Bob were to measure his qubit first, Alice would observe the same result as Bob. When measuring in the standard basis, Alice and Bob will always observe the same results, regardless of the relative timing. The probability that either qubit is measured to be $|0\rangle$ is $1/2$, but the two results are always correlated.

If these particles are far enough apart and the measurements happen close in time (more specifically, if the measurements are *relativistically spacelike separated*), it may sound as if an interaction between these particles is happening faster than the speed of light. We said earlier that a measurement performed by Alice appears to affect the state of Bob's particle, but this wording is misleading. Following special relativity, it is incorrect to think of one measurement happening first and causing the results of the other; it is possible to set up the EPR scenario so that one observer sees Alice measure first, then Bob, while another observer sees Bob measure first, then Alice. According to relativity, physics must explain equally well the observations of both observers. While the causal terminology we used cannot be compatible with both observers' observations, the actual experimental values are invariant under change of observer; the experimental results can be explained equally well by Bob measuring first and then Alice as the other way around. This symmetry shows while there is correlation between the two particles, Alice and Bob cannot use their EPR pair to communicate faster than the speed of light. All that can be said is that Alice and Bob will observe correlated random behavior.

Even though the results themselves are perfectly compatible with relativity theory, the behavior remains mysterious. If Alice and Bob had a large number of EPR pairs that they measure in sequence, they would see an odd mixture of correlated and probabilistic results: each of their sequences of measurements appear completely random, but if Alice and Bob compare their results, they see that they witnessed the same random sequence from their two separate particles. Their sequence of entangled pairs behaves like a pair of magic coins that always land the same way up when tossed together, but whether they both land heads or both land tails is completely random. So far, quantum mechanics is not the only theory that can explain these results; they could also be explained by a *classical* theory that postulates that particles have an internal *hidden* state that determines the result of the measurement, and that this *hidden* state is identical in two particles generated at the same time by the EPR source, but varies randomly over time as the pairs are generated. According to such a classical theory, the reason we see random instead of deterministic results is simply because we, as of yet, have no way of accessing these hidden states. The hope of proponents of such theories was that eventually physics would advance to a stage in which this hidden state would be known to us. Such theories are known as *local hidden-variable theories*. The *local* part comes from the assumption that the hidden variables are internal to each of the particles and do not depend on external influences; in particular, the hidden variables do not depend on the state of faraway particles or measuring devices.

Is it possible to construct a local hidden-variable theory that agrees with all of the experimental results we use quantum mechanics to model? The answer is "no," but it was not until Bell's work of 1964 that anyone realized that it was possible to construct experiments that could distinguish quantum mechanics from all local hidden-variable theories. Since then such experiments have been done, and all of the results have agreed with those predicted by quantum mechanics. Thus, no local hidden-variable theory whatsoever can explain how nature truly works.

Bell showed that any local hidden variable theory predicts results that satisfy an inequality, known as *Bell's inequality*. Section 4.4.1 presents the setup. Section 4.4.2 describes the results predicted by quantum theory. Section 4.4.3 establish Bell's inequality for any local hidden variable theory in a special case. Section 4.4.4 gives Bell's inequality in full generality.

4.4.1 Setup for Bell's Theorem

Imagine an EPR source that emits pairs of photons whose polarizations are in an entangled state $|\psi\rangle = \frac{1}{\sqrt{2}}(|\uparrow\uparrow\rangle + |\rightarrow\rightarrow\rangle)$, where we are using the notation $|\uparrow\rangle$ and $|\rightarrow\rangle$ for photon polarization of section 2.1.2. We suppose that the two photons travel in opposite directions, each toward a polaroid (polarization filter). These polaroids can be set at three different angles. In the special case we consider first, the polaroids can be set to vertical, $+60°$ off vertical, and $-60°$ off vertical.

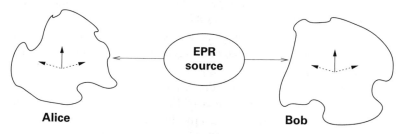

Alice   Bob

4.4.2 What Quantum Mechanics Predicts

Let O_θ be a single-qubit observable with 1-eigenspace generated by $|v\rangle = \cos\theta|0\rangle + \sin\theta|1\rangle$ and -1-eigenspace generated by $|v^\perp\rangle = -\sin\theta|0\rangle + \cos\theta|1\rangle$. Quantum mechanics predicts that measurement of $|\psi\rangle$ with $O_{\theta_1} \otimes O_{\theta_2}$ results in a state with eigenvalue 1 with probability $\cos^2(\theta_1 - \theta_2)$. In other words, the probability that the state ends up in the subspace generated by $\{|v_1\rangle|v_2\rangle, |v_1^\perp\rangle|v_2^\perp\rangle\}$, and not the -1-eigenspace generated by $\{|v_1\rangle|v_2^\perp\rangle, |v_1^\perp\rangle|v_2\rangle\}$, is $\cos^2(\theta_1 - \theta_2)$. Proving this fact is the subject of exercise 4.20. Here we describe its surprising nonclassical implications.

The three different settings for each polaroid, $-60°$, vertical, and $+60°$, correspond to three observables, $M_\nearrow$, $M_\uparrow$, and $M_\nwarrow$, each with two possible outcomes: either the photon passes through the polaroid, an outcome we will denote with P, or it is absorbed, an outcome we will denote with A. Using the fact that measurement with observable $O_{\theta_1} \otimes O_{\theta_2}$ results in a state with eigenvalue 1 with probability $\cos^2(\theta_1 - \theta_2)$, we can compute the probability that measurement of two photons, by polaroids set at angles θ_1 and θ_2, give the same result, PP or AA. If both polaroids are set at the same angle, then both photon measurements give the same results with probability $\cos^2 0 = 1$: both photons will pass through the polaroids, or both will be absorbed. When the polaroid on the right is set to vertical, and the one on the left is set to $+60°$, both measurements agree with probability $\cos^2 60 = 1/4$. Unless the two polaroids are set at the same angle, the difference between the angles is either 60 or 120 degrees, so in all of these cases the two measurements agree $1/4$ of the time and disagree $3/4$ of the time.

If the polaroids are set randomly for a series of EPR pairs emanating from the source, then

• with probability $1/3$ the polaroid orientation will be the same and the measurements will agree, and

• with probability $2/3$ the polaroid orientation will differ and the measurements will agree with probability $1/4$.

Thus, overall, the measurements will agree half the time and disagree half the time. When such an experiment is performed, these are indeed the probabilities that are seen.

4.4.3 Special Case of Bell's Theorem: What Any Local Hidden Variable Theory Predicts

This section shows that no local hidden-variable theory can give these probabilities. Suppose there is some hidden state associated with each photon that determines the result of measuring the photon with a polaroid in each of the three possible settings. We do not know the nature of such a state, but there are only 2^3 binary combinations in which these states can respond to measurement by polaroids in the 3 orientations. We label these 8 possibilities $h_0, \ldots, h_7$.

	↗	↑	↖
h_0	P	P	P
h_1	P	P	A
h_2	P	A	P
h_3	P	A	A
h_4	A	P	P
h_5	A	P	A
h_6	A	A	P
h_7	A	A	A

We can think of h_i as the equivalence class of all hidden states, however these might look, that give the indicated measurement results. Experimentally, it has been established that both polaroids, when set at the same angle, always give the same result when measuring the photons of an EPR pair $|\psi\rangle$. For a local hidden-variable theory to have any chance of modeling experimental results, it must predict that both photons of the entangled pair be in the same equivalence class of hidden states h_i. For example, if the photon on the right responds to the three polaroid positions ↗, ↑, ↖ with PAP, then so must the photon on the left.

Now consider the 9 possible combinations of orientations of the two polaroids

$$\{(\nearrow, \nearrow), (\nearrow, \uparrow), \ldots, (\nwarrow, \nwarrow)\}$$

and the expected agreement of the measurements for photon pairs in each hidden state h_i. Measurements on hidden states h_0 and h_7 ({PPP, PPP} and {AAA, AAA}) agree for all possible pairs of orientations, giving 100 percent agreement. Measurements of the hidden state h_1, {PPA, PPA}, agree in five of the nine possible orientations and disagree in the others. The other six cases are similar to h_1, giving 5/9 agreement and 4/9 disagreement. No matter with what probability distribution the EPR source emits photons with hidden states, the expected agreement between the two measurements will be at least 5/9. Thus, no local hidden-variable theory can give the 50–50 agreement predicted by quantum theory and seen in experiments.

4.4.4 Bell's Inequality

Bell's inequality is an elegant generalization of the preceding argument. The more general setup also has a sequence of EPR pairs emanating from a photon source toward two polaroids, with three possible settings. We now consider polaroids that can be set at any triple of three distinct angles a, b, and c.

If we record the results of repeated measurements at random settings of the polaroids, chosen from the settings above, we can count the number of times that the measurements match for any pair of settings. Let P_{xy} denote the sum of the observed probability that either

- the two photons interact in the same way with both polaroids (either both pass through, or both are absorbed) when the first polaroid is set at angle x and the second at angle y, or
- the two photons interact in the same way with both polaroids when the first polaroid is set at angle y and the second at angle x.

Since whenever the two polaroids are on the same setting, the measurement of the photons will always give the same result $P_{xx} = 1$ for any setting x. We now show that the inequality,

$$P_{ab} + P_{ac} + P_{bc} \geq 1,$$

known as *Bell's inequality*, holds for any local hidden-variable theory and any sequence of settings for each of the polaroids.

We establish this inequality by showing that the inequality holds for the probabilities associated with any one equivalence class of hidden states, from which we deduce that it holds for any distribution of these equivalence classes. According to any local hidden-variable theory, the result of measuring a photon by a polaroid in each of the three possible settings is determined by a local hidden state h of the photon. Again, we think of h as an equivalence class of all hidden states that give the indicated measurement results. The fact that both polaroids, when set at the same angle, always give the same result when measuring the photons in an EPR state $|\psi\rangle$ means that both photons of the entangled pair must be in the same equivalence class of hidden states h. For example, if the photon on the right responds to the three polaroid positions a, b, c with PAP, then so must the photon on the left. Let P_{xy}^h be 1 if the result of the two measurements agree on states with hidden variable h, and 0 otherwise. Since any measurement has only two possible results, P and A, simple logic tells us that the result of measuring a photon, with a given hidden state h, in each of the three polaroid settings, a, b, and c, will be the same for at least one of the settings. Thus, since the two photons of state $|\psi\rangle$ are in the same hidden state, for any h,

$$P_{ab}^h + P_{ac}^h + P_{bc}^h \geq 1.$$

Let w_h be the probability with which the source emits photons of kind h. Then the sum of the observed probabilities $P_{ab} + P_{ac} + P_{bc}$ is a weighted sum, with weights w_h, of the results for photons of each hidden kind h:

$$P_{ab} + P_{ac} + P_{bc} = \sum_h w_h (P_{ab}^h + P_{ac}^h + P_{bc}^h).$$

The weighted average of numbers all greater than 1 is greater than 1, so since $P_{ab}^h + P_{ac}^h + P_{bc}^h \geq 1$ for any h, we may conclude that

$$P_{ab} + P_{ac} + P_{bc} \geq 1.$$

This inequality holds for any local hidden-variable theory and gives us a testable requirement.

By exercise 4.20, quantum theory predicts that the probability that the two results will be the same is the square of the cosine of the angle between the two polaroid settings. If we take the angle between settings a and b to be θ and the angle between settings b and c to be ϕ, then the inequality becomes

$$\cos^2 \theta + \cos^2 \phi + \cos^2(\theta + \phi) \geq 1.$$

For the special case of section 4.4.3, quantum theory tells us that for $\theta = \phi = 60°$ each term is $1/4$. Since $3/4 < 1$, these probabilities violate Bell's inequality, and therefore we can conclude that no local, deterministic theory can give the same predictions as quantum mechanics. Furthermore, experiments similar to but somewhat more sophisticated than the setup described here have been done, and their results confirm the prediction of quantum theory and nature's violation of Bell-like inequalities.

Bell's theorem shows that it is not possible to model entangled states and their measurement with a local hidden-variable theory. Strictly speaking, entangled states should not be talked about in terms of local hidden states or cause and effect. But since there are some situations in which entanglement can be safely talked about in one or the other of these ways, and since both are more familiar than the sort of quantum correlation that actually exists, terminology suggesting either of these modes of thinking persists in the literature.

4.5 References

The original Einstein, Podolsky, Rosen paper [109] is worth reading for an account of their thinking. The first formulation of the paradox as we presented it here is due to Bohm [54].

Our account of Bell's inequalities is loosely based on Penrose's excellent account [225] of a special case of Bell's theorem for spin-1/2 particles. Greenstein and Zajonc [140] give a detailed description, accessible to nonphysicists, of Bell's theorem and the EPR paradox, experimental techniques for generating entangled photon pairs, and Aspect's experiments testing for quantum violation of Bell's inequalities. Detailed results of the experiments by Aspect et al. are published in [25, 26, 24].

Stronger statements than the ones we presented can be made about the sorts of theories that Bell's inequality rules out. The issues here can be relatively subtle. Mermin's article [208] gives a readable account of some of these issues. Peres's book [226] delves into these issues in detail. For a discussion of the various interpretations of quantum mechanics and their perceived strengths and weaknesses, see Sudbery's book [267] and Bub's book [71].

4.6 Exercises

Exercise 4.1. Give the matrix, in the standard basis, for the following operators

a. $|0\rangle\langle 0|$.

b. $|+\rangle\langle 0| - \mathbf{i}|-\rangle\langle 1|$.

c. $|00\rangle\langle 00| + |01\rangle\langle 01|$.

d. $|00\rangle\langle 00| + |01\rangle\langle 01| + |11\rangle\langle 01| + |10\rangle\langle 11|$.

e. $|\Psi^+\rangle\langle\Psi^+|$ where $|\Psi^+\rangle = \frac{1}{\sqrt{2}}(|00\rangle + |11\rangle)$.

Exercise 4.2. Write the following operators in bra/ket notation

a. The Hadamard operator $H = \begin{pmatrix} \frac{1}{\sqrt{2}} & \frac{1}{\sqrt{2}} \\ \frac{1}{\sqrt{2}} & -\frac{1}{\sqrt{2}} \end{pmatrix}$.

b. $X = \begin{pmatrix} 0 & 1 \\ 1 & 0 \end{pmatrix}$.

c. $Y = \begin{pmatrix} 0 & 1 \\ -1 & 0 \end{pmatrix}$.

d. $Z = \begin{pmatrix} 1 & 0 \\ 0 & -1 \end{pmatrix}$.

e. $\begin{pmatrix} 23 & 0 & 0 & 0 \\ 0 & -5 & 0 & 0 \\ 0 & 0 & 0 & 0 \\ 0 & 0 & 0 & 9 \end{pmatrix}$.

f. $X \otimes X$.

g. $X \otimes Z$.

h. $H \otimes H$.

i. The projection operators $P_1 : V \to S_1$ and $P_2 : V \to S_2$, where S_1 is spanned by $\{|+\rangle|+\rangle, |-\rangle|-\rangle\}$ and S_2 is spanned by $\{|+\rangle|-\rangle, |-\rangle|+\rangle\}$.

Exercise 4.3. Show that any projection operator is its own adjoint.

Exercise 4.4. Rewrite example 3.3.2 on page 42 in terms of projection operators.

Exercise 4.5. Rewrite example 3.3.3 on page 42 in terms of projection operators.

Exercise 4.6. Rewrite example 3.3.4 on page 43 in terms of projection operators.

Exercise 4.7. Using the projection operator formalism

a. compute the probability of each of the possible outcomes of measuring the first qubit of an arbitrary two-qubit state in the Hadamard basis $\{|+\rangle, |-\rangle\}$.

b. compute the probability of each outcome for such a measurement on the state $|\Psi^+\rangle = \frac{1}{\sqrt{2}}(|00\rangle + |11\rangle)$.

c. for each possible outcome in (b), describe the possible outcomes if we now measure the second qubit in the standard basis.

d. for each possible outcome in (b), describe the possible outcomes if we now measure the second qubit in the Hadamard basis.

Exercise 4.8. Show that $(A|x\rangle)^\dagger = \langle x|A^\dagger$.

Exercise 4.9. Design a measurement on a three-qubit system that distinguishes between states in which all bit values are equal and those in which they are not, and gives no other information. Write all operators in bra/ket notation.

Exercise 4.10. Design a measurement on a three-qubit system that distinguishes between states in which the number of 1 bits is even, and those in which the number of 1 bits is odd, and gives no other information. Write all operators in bra/ket notation.

Exercise 4.11. Design a measurement on a three-qubit system that distinguishes between states with different numbers of 1 bits and gives no other information. Write all operators in bra/ket notation.

Exercise 4.12. Suppose O is a measurement operator corresponding to a subspace decomposition $V = S_1 \oplus S_2 \oplus S_3 \oplus S_4$ with projection operators P_1, P_2, P_3, and P_4. Design a measurement operator for the subspace decomposition $V = S_5 \oplus S_6$, where $S_5 = S_1 \oplus S_2$ and $S_6 = S_3 \oplus S_4$.

Exercise 4.13.

a. Let O be any observable specifying a measurement of an n-qubit system. Suppose that after measuring $|\psi\rangle$ according to O, we obtain $|\phi\rangle$. Show that if we now measure $|\phi\rangle$ according to O, we simply obtain $|\phi\rangle$ again, with certainty.

b. Reconcile the result of (a) with the fact that for most observables O it is not true that $O^2 = O$.

Exercise 4.14.

a. Give the outcomes and their probabilities for measurement of each of the standard basis elements with respect to the Bell decomposition of example 4.2.6.

b. Give the outcomes and their probabilities for measurement of a general two-qubit state $|\psi\rangle = a_{00}|00\rangle + a_{01}|01\rangle + a_{10}|10\rangle + a_{11}|11\rangle$ with respect to the Bell decomposition.

Exercise 4.15.

a. Show that the operator B of example 4.3.4 is of the form $Q \otimes I$, where Q is a (2×2)-Hermitian operator.

b. Show that any operator of the form $Q \otimes I$, where Q is a (2×2)-Hermitian operator and I is the (2×2)-identity operator, specifies a measurement of a two-qubit system. Describe the subspace decomposition associated with such an operator.

c. Describe the subspace decomposition associated with an operator of the form $I \otimes Q$ where Q is a (2×2)-Hermitian operator and I is the (2×2)-identity operator, and give a high-level description of such measurements.

Exercise 4.16.   This exercise shows that for any Hermitian operator $O : V \to V$, the direct sum of all eigenspaces of O is V.

A *unitary* operator U satisfies $U^\dagger U = I$.

a. Show that the columns of a unitary matrix U form an orthonormal set.

b. Show that if O is Hermitian, then so is UOU^{-1} for any unitary operator U.

c. Show that any operator has at least one eigenvalue λ and λ-eigenvector v_λ.

d. Use the result of (c) to show that for any matrix $A : V \to V$, there is a unitary operator U such that the matrix for UAU^{-1} is upper triangular (meaning all entries below the diagonal are zero).

e. Show that for any Hermitian operator $O : V \to V$ with eigenvalues $\lambda_1, \ldots, \lambda_k$, the direct sum of the λ_i-eigenspaces S_{λ_i} gives the whole space:

$$V = S_{\lambda_1} \oplus S_{\lambda_2} \oplus \cdots \oplus S_{\lambda_k}.$$

Exercise 4.17.

a. Show that any state resulting from measuring an unentangled state with a single-qubit measurement is still unentangled.

b. Can other types of measurement produce an entangled state from an unentangled one? If so, give an example. If not, give a proof.

c. Can an unentangled state be obtained by measuring a single qubit of an entangled state?

Exercise 4.18.   Show that if there is no measurement of one of the qubits that gives a single result with certainty, then the two qubits are entangled.

Exercise 4.19. Give an explicit description of the observable O_θ of section 4.4.2 in both bra/ket and matrix notation.

Exercise 4.20. Let O_{θ_1} be the single-qubit observable with $+1$-eigenvector

$$|v_1\rangle = \cos\theta_1|0\rangle + \sin\theta_1|1\rangle$$

and -1-eigenvector

$$|v_1^\perp\rangle = -\sin_1\theta|0\rangle + \cos\theta_1|1\rangle.$$

Similarly, let O_{θ_2} be the single-qubit observable with $+1$-eigenvector

$$|v_2\rangle = \cos\theta_2|0\rangle + \sin\theta_2|1\rangle$$

and -1-eigenvector

$$|v_2^\perp\rangle = -\sin\theta_2|0\rangle + \cos\theta_2|1\rangle.$$

Let O be the two-qubit observable $O_{\theta_1} \otimes O_{\theta_2}$. We consider various measurements on the EPR state $|\psi\rangle = \frac{1}{\sqrt{2}}(|00\rangle + |11\rangle)$. We are interested in the probability that the measurements $O_{\theta_1} \otimes I$ and $I \otimes O_{\theta_2}$, if they were performed on the state $|\psi\rangle$, would *agree* on the two qubits in that either both qubits are measured in the 1-eigenspace or both are measured in -1-eigenspace of their respective single-qubit observables. As in example 4.2.5, we are not interested in the specific outcome of the two measurements, just whether or not they would agree. The observable $O = O_{\theta_1} \otimes O_{\theta_2}$ gives exactly this information.

a. Find the probability that the measurements $O_{\theta_1} \otimes I$ and $I \otimes O_{\theta_2}$, when performed on $|\psi\rangle$, would agree in the sense of both resulting in a $+1$ eigenvector or both resulting in a -1 eigenvector. (Hint: Use the trigonometric identities $\cos(\theta_1 - \theta_2) = \cos(\theta_1)\cos(\theta_2) + \sin(\theta_1)\sin(\theta_2)$ and $\sin(\theta_1 - \theta_2) = \sin(\theta_1)\cos(\theta_2) - \cos(\theta_1)\sin(\theta_2)$ to obtain a simple form for your answer.)

b. For what values of θ_1 and θ_2 do the results always agree?

c. For what values of θ_1 and θ_2 do the results never agree?

d. For what values of θ_1 and θ_2 do the results agree half the time?

e. Show that whenever $\theta_1 \neq \theta_2$ and θ_1 and θ_2 are chosen from $\{-60°, 0°, 60°\}$, then the results agree $1/4$ of the time and disagree $3/4$ of the time.

Exercise 4.21.

a. Most of the time the effect of performing two measurements, one right after the other, cannot be achieved by a single measurement. Find a sequence of two measurements whose effect cannot be achieved by a single measurement, and explain why this property is generally true for most pairs of measurements.

b. Describe a sequence of two distinct nontrivial measurements that can be achieved by a single measurement.

c. For each of the measurements specified by the operators A, B, C, and M from examples 4.3.3, 4.3.4, 4.3.5, and 4.3.6, say whether the measurement can be achieved as a sequence of single-qubit measurements.

d. How does performing the sequence of measurements $Z \otimes I$ followed by $I \otimes Z$ compare with performing the single measurement $Z \otimes Z$?

Exercise 4.22.   Show that no matter in which basis the first qubit of an EPR pair $\frac{1}{\sqrt{2}}(|00\rangle + |11\rangle)$ is measured, the two possible outcomes have equal probability.

5 Quantum State Transformations

The last two chapters discussed encoding information in quantum states and some of the uniquely quantum properties of such *quantum information*, such as entangled states, the exponential state space, and quantum measurement. This chapter develops the basic mechanisms for computing on quantum information. Computation on quantum information takes place through dynamic transformation of quantum systems. In order to understand quantum computation, we must understand which sorts of transformations nature allows and which it does not. This chapter focuses on transformations of a closed quantum system, transformations that map the state space of the quantum system to itself. Measurement is not a transformation in this sense. Chapter 10 discusses more general transformations, transformations of a subsystem that is part of a larger quantum system.

This chapter begins with a brief discussion of transformations on general quantum systems, and it then focuses on multiple-qubit systems. Section 5.1 discusses the unitarity requirement on quantum state transformations and the *no-cloning* principle. The no-cloning restriction is central to both the limitations and the advantages of encoding information in quantum states; for example, it underlies the security of quantum cryptographic protocols such as the ones described in sections 2.4 and 3.4, and it is also vital to the argument of section 4.3.1 that no more than n classical bits worth of information can be extracted from an n-qubit system.

After discussing considerations for transformations of general quantum systems, the chapter restricts discussion to n-qubit systems and develops building blocks for the *standard circuit model* of quantum computation. Part II uses this model to describe quantum algorithms. All quantum transformations on n-qubit quantum systems can be expressed as a sequence of transformations on single-qubit and two-qubit subsystems. Some quantum state transformations can be implemented in terms of these basic gates more easily than others. The efficiency of a quantum transform is quantified in terms of the number of one- and two-qubit gates used. Section 5.2 looks at single-qubit and two-qubit transformations, ways of combining them, and a graphical notation for describing sequences of transformations. Section 5.3 describes applications of these simple gates to two communication problems: dense coding and quantum state teleportation. Section 5.4 is devoted to showing that any quantum transformation can be realized as a sequence of one- and two-qubit transformations. Section 5.5 discusses finite sets of gates that can be used to approximate all quantum transformations universally. The chapter concludes with a definition of the standard circuit model for quantum computation.

5.1 Unitary Transformations

In this book, *quantum transformation* will mean a mapping from the state space of a quantum system to itself. Measurements are not quantum transformations in this sense; there are only finitely many outcomes, and the result of applying a measurement to a specific state is only probabilistic. Chapter 10 considers open quantum systems, systems that are subsystems of a larger quantum system, and studies the transformations of subsystems induced by transformations of the larger system. In this chapter, we concern ourselves only with transformations of closed quantum systems.

Nature does not allow arbitrary transformations of a quantum system. Nature forces these transformations to respect properties connected to quantum measurement and quantum superposition. The transformations must be linear transformations of the vector space associated with the state space so that a state that is a superposition of other states goes to the superposition of their images; more precisely, linearity means that for any quantum transformation U,

$$U(a_1|\psi_1\rangle + \cdots + a_k|\psi_k\rangle) = a_1 U|\psi_1\rangle + \cdots + a_k U|\psi_k\rangle$$

on any superposition $|\psi\rangle = a_1|\psi_1\rangle + \cdots + a_k|\psi_k\rangle$. Unit length vectors must go to unit length vectors, which implies that orthogonal subspaces go to orthogonal subspaces. These properties ensure that measuring and then applying a transform to the outcome gives the same result as first applying the transform and then measuring in the transformed basis. Specifically, the probability of obtaining outcome $U|\phi\rangle$ by first applying U to $|\psi\rangle$ and then measuring with respect to the decomposition $\oplus U S_i$ is the same as the probability of obtaining $U|\phi\rangle$ by measuring $|\psi\rangle$ with respect to the decomposition $\oplus S_i$ and then applying U. These properties hold if U preserves the inner product; for any $|\psi\rangle$ and $|\phi\rangle$, the inner product of their images, $U|\psi\rangle$ and $U|\phi\rangle$, must be the same as the inner product between $|\psi\rangle$ and $|\phi\rangle$:

$$\langle\phi|U^\dagger U|\psi\rangle = \langle\phi|\psi\rangle.$$

A straightforward mathematical argument shows that this condition holds for all $|\psi\rangle$ and $|\phi\rangle$ only if $U^\dagger U = I$. In other words, for any quantum transformation U, its adjoint $U^\dagger$ must be equal to its inverse, precisely the condition, $U^\dagger = U^{-1}$, for a linear transformation to be *unitary*. Furthermore, this condition is sufficient; the set of allowed transformations of a quantum system corresponds exactly to the set of unitary operators on the complex vector space associated with the state space of the quantum system. Since unitary operators preserve the inner product, they map orthonormal bases to orthonormal bases. In fact, the converse is true: any linear transformation that maps an orthonormal basis to an orthonormal basis is unitary.

Geometrically, all quantum state transformations are rotations of the complex vector space associated with the quantum state space. The ith column of the matrix is the image $U|i\rangle$ of the ith basis vector, so for a unitary transformation given in matrix form, U is unitary if and only if the set of columns of its matrix representation are orthonormal. Since $U^\dagger$ is unitary if and only if U is, it follows that U is unitary if and only if its rows are orthonormal. The product $U_1 U_2$ of two unitary transformations is again unitary. The tensor product $U_1 \otimes U_2$ is a unitary transformation of the space $X_1 \otimes X_2$ if U_1 and U_2 are unitary transformations of X_1 and X_2 respectively. Linear combinations of unitary operators, however, are not in general unitary.

The unitarity condition simply ensures that the operator does not violate any general principles of quantum theory. It does not imply that a transformation can be implemented efficiently; most unitary operators cannot be efficiently implemented, even approximately. In later chapters, particularly when we examine quantum algorithms, we will concern ourselves with questions about the efficiency of certain quantum transformations.

An obvious consequence of the unitary condition is that every quantum state transformation is reversible. Chapter 6 describes work of Charles Bennett, Edward Fredkin, and Tommaso Toffoli, done prior to the development of quantum information processing, that shows that all classical computations can be made reversible with only a negligible loss of efficiency. Thus, the reversibility requirement does not impose an unworkably strict restriction on quantum algorithms.

In the standard circuit model of quantum computation, all computation is carried out by quantum transformations, with measurement used only at the end to read out the results. Since measurement can effect changes in quantum states, the dynamics of measurement, rather than quantum state transformations, provide an alternative means to achieve computation. Section 13.4 describes an alternate, but equally powerful, model of quantum computation in which all computation takes place by measurement.

The phrases *quantum transformation* or *quantum operator* refer to unitary operators acting on the state space, not measurement operators. While measurements are modeled by operators, the behavior of measurement is not modeled by the direct action of the measurement's Hermitian operator on the state space, but rather by the indirect, probabilistic procedure described by the measurement postulate of section 4.3.1. One of the least satisfactory aspects of quantum theory is that there are two distinct classes of manipulations of quantum states: quantum transformations and measurement. Section 10.3 describes a tighter, but still unsatisfactory, relation between the two.

5.1.1 Impossible Transformations: The No-Cloning Principle

This section describes a simple, but important, consequence of the unitary condition: unknown quantum states cannot be copied or cloned. In fact, the linearity of unitary transformations alone implies the result. Suppose U is a unitary transformation that *clones*, in that $U(|a\rangle|0\rangle) = |a\rangle|a\rangle$ for all quantum states $|a\rangle$. Let $|a\rangle$ and $|b\rangle$ be two orthogonal quantum states. That U clones means $U(|a\rangle|0\rangle) = |a\rangle|a\rangle$ and $U(|b\rangle|0\rangle) = |b\rangle|b\rangle$. Consider $|c\rangle = \frac{1}{\sqrt{2}}(|a\rangle + |b\rangle)$. By linearity,

$$U(|c\rangle|0\rangle) = \frac{1}{\sqrt{2}}(U(|a\rangle|0\rangle) + U(|b\rangle|0\rangle))$$

$$= \frac{1}{\sqrt{2}}(|a\rangle|a\rangle + |b\rangle|b\rangle).$$

But if U is a cloning transformation then

$$U(|c\rangle|0\rangle) = |c\rangle|c\rangle = 1/2(|a\rangle|a\rangle + |a\rangle|b\rangle + |b\rangle|a\rangle + |b\rangle|b\rangle),$$

which is not equal to $(1/\sqrt{2})(|a\rangle|a\rangle + |b\rangle|b\rangle)$. Thus, there is no unitary operation that can reliably clone all quantum states.

The no-cloning theorem tells us that it is impossible to clone a specific unknown quantum state reliably. It does not preclude the construction of a known quantum state from a known quantum state. It is possible to perform an operation that appears to be copying the state in one basis but does not do so in others. For example, it is possible obtain n particles in an entangled state $a|00\ldots0\rangle + b|11\ldots1\rangle$ from an unknown state $a|0\rangle + b|1\rangle$. But it is not possible to create the n particle state $(a|0\rangle + b|1\rangle) \otimes \cdots \otimes (a|0\rangle + b|1\rangle)$ from an unknown state $a|0\rangle + b|1\rangle$.

5.2 Some Simple Quantum Gates

Just as for classical computation, it is a boon to quantum computation, both for implementation and analysis, that arbitrarily complex computations can be achieved by composing simple elements. Section 5.4 shows that any quantum state transformation on an n-qubit system can be realized using a sequence of one- and two-qubit quantum state transformations. We will call any quantum state transformation that acts on only a small number of qubits a *quantum gate*. Sequences of quantum gates are called *quantum gate arrays* or *quantum circuits*.

In the quantum-information-processing literature, gates are mathematical abstractions useful for describing quantum algorithms; quantum gates do not necessarily correspond to physical objects, as they do in the classical case. So the gate terminology and its accompanying graphical notation must not be taken too literally. For solid state or optical implementations, there may be actual physical gates, but in NMR and ion trap implementations, the qubits are stationary particles, and the *gates* are operations on these particles using magnetic fields or laser pulses. For these implementations, gates operate on a physical register of qubits.

From a practical point of view, the standard description of computation in terms of one- and two-qubit gates leaves something to be desired. Ideally, we would write all our computations in terms of gates that are easy to implement physically and are robust, but we do not yet know which ones these are. Furthermore, in order to realize physically a quantum computer capable of performing arbitrary quantum transformations, it would be convenient to have only finitely many gates that could generate all unitary transformations. Unfortunately, such a set is impossible; there are uncountably many quantum transformations, and a finite set of generators can only generate countably many elements. Section 5.5 shows that it is possible, however, for finite sets of gates to generate arbitrarily close approximations to all unitary transformations. A number of such sets are known, but it is unclear which of these will be most practical from a physical implementation point of view. For analyzing quantum algorithms, it is useful to have a standard set of gates with which to analyze the efficiency of quantum algorithms. The set we use includes all one-qubit gates together with the two-qubit gate described in section 5.2.4.

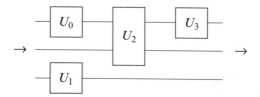

Figure 5.1
A sample graphical representation for a three-qubit quantum gate array. Data flow left to right through the circuit.

Graphical notation, representing series of quantum state transformations acting on various combinations of qubits, is commonly used to describe sequences of transformations and to analyze the resulting algorithms. Simple transformations are graphically represented by appropriately labeled boxes which are connected to form more complex circuits. A sample graphical representation is shown in figure 5.1. Each horizontal line corresponds to a qubit. The transformations on the left are performed first, and the processing proceeds from left to right. The boxes labeled with U_0, U_1, and U_3 correspond to single-qubit transformations, while the one labeled U_2 corresponds to a two-qubit transformation. When we talk about applying an operator U to qubit i of an n-qubit quantum system, we mean that we apply the operator $I \otimes \cdots \otimes I \otimes U \otimes I \otimes \cdots \otimes I$ to the entire system, where I is the single-qubit identity operator, applied to each of the other qubits of the system.

The remainder of this section describes a variety of frequently used quantum gates.

5.2.1 The Pauli Transformations
The Pauli transformations are the most commonly used single-qubit transformations:

$$I : |0\rangle\langle0| + |1\rangle\langle1| \quad \begin{pmatrix} 1 & 0 \\ 0 & 1 \end{pmatrix}$$

$$X : |1\rangle\langle0| + |0\rangle\langle1| \quad \begin{pmatrix} 0 & 1 \\ 1 & 0 \end{pmatrix}$$

$$Y : -|1\rangle\langle0| + |0\rangle\langle1| \quad \begin{pmatrix} 0 & 1 \\ -1 & 0 \end{pmatrix}$$

$$Z : |0\rangle\langle0| - |1\rangle\langle1| \quad \begin{pmatrix} 1 & 0 \\ 0 & -1 \end{pmatrix},$$

where I is the identity transformation, X is negation (the classical NOT operation on $|0\rangle$ and $|1\rangle$ viewed as classical bits), Z changes the relative phase of a superposition in the standard basis, and $Y = ZX$ is a combination of negation and phase change. In graphical notation, these gates are represented by boxes

labeled appropriately.

There is variation in the literature as to which transformations are the Pauli transformations, and in the notation used. The main discrepancy is whether $-i(|0\rangle\langle 1| - |1\rangle\langle 01|)$ is considered the Pauli transformation instead of $Y = |0\rangle\langle 1| - |1\rangle\langle 0|$, as we do here. The operator iY is Hermitian, which is a useful property in some settings, for example, if we wanted to use it to describe measurement. Also, sometimes the notation σ_x, σ_y, and σ_z is used instead. Throughout this book, we use I, X, Y, and Z for the Pauli operators representing single-qubit transformations. In chapter 10, we use the notation $\sigma_x = X$, $\sigma_y = -iY$, and $\sigma_z = Z$ when the Pauli operators are used to describe quantum states.

5.2.2 The Hadamard Transformation

Another important single-qubit transformation is the Hadamard transformation

$$H = \frac{1}{\sqrt{2}}(|0\rangle\langle 0| + |1\rangle\langle 0| + |0\rangle\langle 1| - |1\rangle\langle 1|),$$

or

$$H : |0\rangle \rightarrow |+\rangle = \tfrac{1}{\sqrt{2}}(|0\rangle + |1\rangle))$$
$$|1\rangle \rightarrow |-\rangle = \tfrac{1}{\sqrt{2}}(|0\rangle - |1\rangle),$$

which produces an even superposition of $|0\rangle$ and $|1\rangle$ from either of the standard basis elements. Note $HH = I$. In the standard basis, the matrix for the Hadamard transformation is

$$H = \frac{1}{\sqrt{2}}\begin{pmatrix} 1 & 1 \\ 1 & -1 \end{pmatrix}.$$

5.2.3 Multiple-Qubit Transformations from Single-Qubit Transformations

Multiple-qubit transformations can be constructed as tensor products of single-qubit transformations. These transformations are uninteresting as multiple-qubit transformations in the sense that they are equivalent to performing the single-qubit transformations on each of the qubits separately in some order. For example, $U \otimes V$ can be obtained by first applying $U \otimes I$ and then $I \otimes V$.

More interesting are those multiple-qubit transformations that can change the entanglement between qubits of the system. Entanglement is not a local property in the sense that transformations that act separately on two or more subsystems cannot affect the entanglement between those subsystems. More precisely, let $|\psi\rangle$ be a two-qubit state and U and V be single-qubit unitary transformations. Then $(U \otimes V)|\psi\rangle$ is entangled if and only if $|\psi\rangle$ is. The widely used class of two-qubit controlled gates discussed in the next section illustrates the effects transformations can have on entanglement.

5.2.4 The Controlled-NOT and Other Singly Controlled Gates

The controlled-NOT gate, C_{not}, acts on the standard basis for a two-qubit system, with $|0\rangle$ and $|1\rangle$ viewed as classical bits, as follows: it flips the second bit if the first bit is 1 and leaves it unchanged otherwise. The C_{not} transformation has representation

$$C_{not} = |0\rangle\langle 0| \otimes I + |1\rangle\langle 1| \otimes X$$

$$= |0\rangle\langle 0| \otimes (|0\rangle\langle 0| + |1\rangle\langle 1|) + |1\rangle\langle 1| \otimes (|1\rangle\langle 0| + |0\rangle\langle 1|)$$

$$= |00\rangle\langle 00| + |01\rangle\langle 01| + |11\rangle\langle 10| + |10\rangle\langle 11|,$$

from which it is easy to read off its effect on the standard basis elements:

$$C_{not} : |00\rangle \rightarrow |00\rangle$$
$$|01\rangle \rightarrow |01\rangle$$
$$|10\rangle \rightarrow |11\rangle$$
$$|11\rangle \rightarrow |10\rangle.$$

The matrix representation (in the standard basis) for C_{not} is

$$\begin{pmatrix} 1 & 0 & 0 & 0 \\ 0 & 1 & 0 & 0 \\ 0 & 0 & 0 & 1 \\ 0 & 0 & 1 & 0 \end{pmatrix}.$$

Observe that C_{not} is unitary and is its own inverse. Furthermore, the C_{not} gate cannot be decomposed into a tensor product of two single-qubit transformations.

The importance of the C_{not} gate for quantum computation stems from its ability to change the entanglement between two qubits. For example, it takes the unentangled two-qubit state $\frac{1}{\sqrt{2}}(|0\rangle + |1\rangle)|0\rangle$ to the entangled state $\frac{1}{\sqrt{2}}(|00\rangle + |11\rangle)$:

$$C_{not}\left(\frac{1}{\sqrt{2}}(|0\rangle + |1\rangle) \otimes |0\rangle\right) = C_{not}\left(\frac{1}{\sqrt{2}}(|00\rangle + |10\rangle)\right)$$

$$= \frac{1}{\sqrt{2}}(|00\rangle + |11\rangle).$$

Similarly, since it is its own inverse, it can take an entangled state to an unentangled one.

The controlled-NOT gate is so common that it has its own graphical notation.

The open circle indicates the control bit, the $\times$ indicates negation of the target bit, and the line between them indicates that the negation is conditional, depending on the value of the control bit. Some authors use a solid circle to indicate negative control, in which the target bit is toggled when the control bit is 0 instead of 1.

A useful class of two-qubit controlled gates, which generalizes the C_{not} gate, consists of gates that perform a single-qubit transformation Q on the second qubit when the first qubit is $|1\rangle$ and do nothing when it is $|0\rangle$. These controlled gates have graphical representation

We use the following shorthand for these transformations:

$$\bigwedge Q = |0\rangle\langle 0| \otimes I + |1\rangle\langle 1| \otimes Q.$$

The transformation C_{not}, for example, becomes $\bigwedge X$ in this notation. In the standard computational basis, the two-qubit operator $\bigwedge Q$ is represented by the 4×4 matrix

$$\begin{pmatrix} I & 0 \\ 0 & Q \end{pmatrix}.$$

Let us look in more depth at one of these controlled gates, the controlled phase shift $\bigwedge e^{i\theta}$, where $e^{i\theta}$ is shorthand for $e^{i\theta} I$. In the standard basis, the controlled phase shift changes the phase of the second bit if and only if the control bit is one:

$$\bigwedge e^{i\theta} = |00\rangle\langle 00| + |01\rangle\langle 01| + e^{i\theta}|10\rangle\langle 10| + e^{i\theta}|11\rangle\langle 11|.$$

Its effect on the standard basis elements is as follows:

$$
\begin{aligned}
\bigwedge e^{i\theta} : |00\rangle &\rightarrow |00\rangle \\
|01\rangle &\rightarrow |01\rangle \\
|10\rangle &\rightarrow e^{i\theta}|10\rangle \\
|11\rangle &\rightarrow e^{i\theta}|11\rangle
\end{aligned}
$$

and it has matrix representation

$$\begin{pmatrix} 1 & 0 & 0 & 0 \\ 0 & 1 & 0 & 0 \\ 0 & 0 & e^{i\theta} & 0 \\ 0 & 0 & 0 & e^{i\theta} \end{pmatrix}.$$

The controlled phase shift makes use of a single-qubit transformation that was a physically meaningless global phase shift when applied to a single-qubit system, but when used as part of a conditional transformation, this phase shift becomes nontrivial, changing the relative phase between elements of a superposition. For example, it takes

$$\frac{1}{\sqrt{2}}(|00\rangle + |11\rangle) \rightarrow \frac{1}{\sqrt{2}}(|00\rangle + e^{i\theta}|11\rangle).$$

Graphical icons can be combined into quantum circuits. The following circuit, for instance, swaps the value of the two bits.

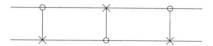

In other words, this *swap circuit* takes

$|00\rangle \mapsto |00\rangle$

$|01\rangle \mapsto |10\rangle$

$|10\rangle \mapsto |01\rangle$

$|11\rangle \mapsto |11\rangle$,

and $|\psi\rangle|\phi\rangle \mapsto |\phi\rangle|\psi\rangle$ for all single-qubit states $|\psi\rangle$ and $|\phi\rangle$.

Three cautions are in order. The first concerns the use of a basis to specify the transformation. The second concerns the basis dependence of the notion of *control*. The third suggests care in interpreting the graphical notation for quantum circuits.

Caution 1: *Phases in Specifications of Transformations*   Section 3.1.3 discussed the important distinction between the quantum state space (projective space) and the associated complex vector space. We need to keep this distinction in mind when interpreting the standard ways quantum state transformations are specified. A unitary transformation on the complex vector space is completely determined by its action on a basis. The unitary transformation is not completely determined by specifying what states the states corresponding to basis states are sent to, a subtle distinction. For example, the controlled phase shift takes the four quantum states represented by $|00\rangle$, $|01\rangle$, $|10\rangle$, and $|11\rangle$ to themselves; $|10\rangle$ and $e^{i\theta}|10\rangle$ represent exactly the same quantum state, and so do $|11\rangle$ and $e^{i\theta}|11\rangle$. As we saw above, however, this transformation is not the identity transformation since it takes $\frac{1}{\sqrt{2}}(|00\rangle + |11\rangle)$ to $\frac{1}{\sqrt{2}}(|00\rangle + e^{i\theta}|10\rangle)$. To avoid mistakes, remember that notation such as

$|00\rangle \rightarrow |00\rangle$

$|01\rangle \rightarrow |01\rangle$

$|10\rangle \rightarrow e^{i\theta}|10\rangle$

$|11\rangle \rightarrow e^{i\theta}|11\rangle$

is used to specify a unitary transformation on the complex vector space in terms of vectors in that vectors space, not in terms of the states corresponding to these vectors. Specifying that the vector $|0\rangle$ goes to the vector $-|1\rangle$ is different from specifying that $|0\rangle$ goes to $|1\rangle$ because the two vectors $-|1\rangle$ and $|1\rangle$ are different vectors even if they correspond to the same state. The quantum transformation on the state space is easily derived from the unitary transformation on the associated complex vector space.

Caution 2: *Basis Dependence of the Notion of Control*   The notion of the *control bit* and the *target bit* is a carryover from the classical gate and should not be taken too literally. In the standard basis, the C_{not} operator behaves exactly as the classical gate does on classical bits. However, one should not conclude that the *control bit* is never changed. When the input qubits are not one of the standard basis elements, the effect of the controlled gate can be somewhat counterintuitive. For example, consider the C_{not} gate in the Hadamard basis $\{|+\rangle, |-\rangle\}$:

$$C_{not} : |{+}{+}\rangle \rightarrow |{+}{+}\rangle$$
$$|{+}{-}\rangle \rightarrow |{-}{-}\rangle$$
$$|{-}{+}\rangle \rightarrow |{-}{+}\rangle$$
$$|{-}{-}\rangle \rightarrow |{+}{-}\rangle.$$

In the Hadamard basis, it is the state of the second qubit that remains unchanged, and the state of the first qubit that is flipped depending on the state of the second bit. Thus, in this basis the sense of which bit is the *control bit* and which the *target bit* has been reversed. But we have not changed the transformation at all, only the way we are thinking about it. Furthermore, in most bases, we do not see a *control bit* or a *target bit* at all. For example, as we have seen, the controlled-NOT transforms $\frac{1}{\sqrt{2}}(|0\rangle + |1\rangle)|0\rangle$ to $\frac{1}{\sqrt{2}}(|00\rangle + |11\rangle)$. In this case the controlled-NOT entangles the qubits so that it is not possible to talk about their states separately.

A related fact, which we will use in constructing algorithms and in quantum error correction, is that the following two circuits are equivalent:

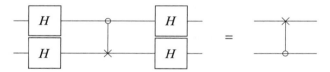

Caution 3: *Reading circuit diagrams*   The graphical representation of quantum circuits can be misleading if one is not careful to interpret it properly. In particular, one cannot determine the effect the transformation has on the input qubits, even if they are all in standard basis states, by simply looking at the line in the diagram corresponding to that qubit. Let us look at the circuit

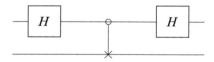

acting on the input state $|0\rangle|0\rangle$. Since the Hadamard transformation is its own inverse, it might at first appear that the first qubit's state would remain unchanged by the transformation. But it does not. Recall from caution 2 that the controlled-NOT gate does not leave the first qubit unaffected in general. In fact, this circuit takes the input state $|00\rangle$ to $1/2(|00\rangle + |10\rangle + |01\rangle - |11\rangle)$, an effect that cannot be seen immediately from the circuit and so must be explicitly calculated.

5.3 Applications of Simple Gates

For many years, EPR pairs, and entanglement more generally, were viewed as quantum mechanical oddities of merely theoretical interest. Quantum information processing changes that perception by providing practical applications of entanglement. Two communications applications, dense coding and teleportation, illustrate the usefulness of EPR pairs when used together with a few simple quantum gates.

Dense coding uses one quantum bit together with a shared EPR pair to encode and transmit two classical bits. Since EPR pairs can be distributed ahead of time, only one qubit needs to be physically transmitted to communicate two bits of information. This result is surprising, since, as section 2.3 explained, only one classical bit's worth of information can be extracted from a qubit. Teleportation is the opposite of dense coding in that it uses two classical bits to transmit the state of a single qubit. Teleportation is surprising in two respects. In spite of the no-cloning principle of quantum mechanics, there exists a mechanism for the transmission of an unknown quantum state. Also, teleportation shows that two classical bits suffice to communicate a qubit state that can be in any one of an infinite number of possible states.

The key to both dense coding and teleportation is the use of entangled particles. The initial setup is the same for both processes. Alice and Bob wish to communicate. Each is sent one of the entangled particles making up an EPR pair

$$|\psi_0\rangle = \frac{1}{\sqrt{2}}(|00\rangle + |11\rangle).$$

Suppose Alice is sent the first particle, and Bob the second:

$$|\psi_0\rangle = \frac{1}{\sqrt{2}}(|0_A\rangle|0_B\rangle + |1_A\rangle|1_B\rangle).$$

Alice can perform transformations only on her particle, and Bob can perform transformations only on his, until Alice sends Bob her particle or vice versa. In other words, until a particle is transmitted between them, Alice can perform transformations only of the form $Q \otimes I$ on the EPR pair, where Q is a single-qubit transformation, and Bob transformations only of the form $I \otimes Q$. More generally, for $K = 2^k$, let $I^{(K)}$ be the $2^k \times 2^k$ identity matrix. If Alice has n qubits and Bob has m qubits, then Alice can perform transformations only of the form $U \otimes I^{(M)}$, where U is an n-qubit transformation, and Bob can perform transformations only of the form $I^{(N)} \otimes U$.

5.3.1 Dense Coding

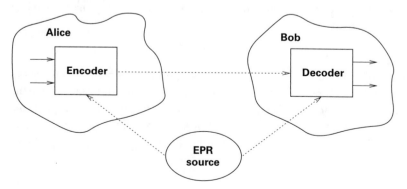

Alice Alice wishes to transmit the state of two classical bits encoding one of the numbers 0 through 3. Depending on this number, Alice performs one of the Pauli transformations $\{I, X, Y, Z\}$ on her qubit of the entangled pair $|\psi_0\rangle$. The resulting state is shown in the following table.

Value	Transformation	New state
0	$|\psi_0\rangle = (I \otimes I)|\psi_0\rangle$	$\frac{1}{\sqrt{2}}(|00\rangle + |11\rangle)$
1	$|\psi_1\rangle = (X \otimes I)|\psi_0\rangle$	$\frac{1}{\sqrt{2}}(|10\rangle + |01\rangle)$
2	$|\psi_2\rangle = (Z \otimes I)|\psi_0\rangle$	$\frac{1}{\sqrt{2}}(|00\rangle - |11\rangle)$
3	$|\psi_3\rangle = (Y \otimes I)|\psi_0\rangle$	$\frac{1}{\sqrt{2}}(-|10\rangle + |01\rangle)$

Alice then sends her qubit to Bob.

Bob To decode the information, Bob applies a controlled-NOT to the two qubits of the entangled pair and then applies the Hadamard transformation H to the first qubit:

$$
\left.
\begin{array}{l}
\frac{1}{\sqrt{2}}(|00\rangle + |11\rangle) \\[4pt]
\frac{1}{\sqrt{2}}(|10\rangle + |01\rangle) \\[4pt]
\frac{1}{\sqrt{2}}(|00\rangle - |11\rangle) \\[4pt]
\frac{1}{\sqrt{2}}(-|10\rangle + |01\rangle)
\end{array}
\right\}
\quad
\begin{array}{c}
C_{not} \\ \longrightarrow
\end{array}
\quad
\left\{
\begin{array}{l}
\frac{1}{\sqrt{2}}(|00\rangle + |10\rangle) \\[4pt]
\frac{1}{\sqrt{2}}(|11\rangle + |01\rangle) \\[4pt]
\frac{1}{\sqrt{2}}(|00\rangle - |10\rangle) \\[4pt]
\frac{1}{\sqrt{2}}(-|11\rangle + |01\rangle)
\end{array}
\right\}
$$

$$
=
\left\{
\begin{array}{l}
\frac{1}{\sqrt{2}}(|0\rangle + |1\rangle) \otimes |0\rangle \\[4pt]
\frac{1}{\sqrt{2}}(|1\rangle + |0\rangle) \otimes |1\rangle \\[4pt]
\frac{1}{\sqrt{2}}(|0\rangle - |1\rangle) \otimes |0\rangle \\[4pt]
\frac{1}{\sqrt{2}}(-|1\rangle + |0\rangle) \otimes |1\rangle
\end{array}
\right.
$$

$$
\begin{array}{c}
H \otimes I \\ \longrightarrow
\end{array}
\quad
\left\{
\begin{array}{l}
|00\rangle \\
|01\rangle \\
|10\rangle \\
|11\rangle.
\end{array}
\right.
$$

Bob then measures the two qubits in the standard basis to obtain the two-bit binary encoding of the number Alice wished to send.

5.3.2 Quantum Teleportation

The objective of teleportation is to transmit enough information, using only classical bits, about the quantum state of a particle that a receiver can reconstruct the exact quantum state. Since the no-cloning principle of quantum mechanics means that a quantum state cannot be copied, the quantum state of the original particle cannot be preserved. It is this property—that the original state at the source must be destroyed in the course of creating the state at the target—that gives quantum teleportation its name.

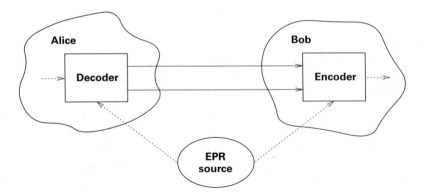

Alice   Alice has a qubit whose state $|\phi\rangle = a|0\rangle + b|1\rangle$ she does not know. She wants to send this state to Bob through classical channels. As in the setup for the dense coding application, Alice and Bob each possess one qubit of an entangled pair

$$|\psi_0\rangle = \frac{1}{\sqrt{2}}(|00\rangle + |11\rangle).$$

The starting state is the three-qubit quantum state

$$|\phi\rangle \otimes |\psi_0\rangle = \frac{1}{\sqrt{2}}\big(a|0\rangle \otimes (|00\rangle + |11\rangle) + b|1\rangle \otimes (|00\rangle + |11\rangle)\big)$$

$$= \frac{1}{\sqrt{2}}\big(a|000\rangle + a|011\rangle + b|100\rangle + b|111\rangle\big).$$

Alice controls the first two qubits and Bob controls the last one.

Alice applies the decoding step used by Bob in the dense coding scenario to the combined state of the qubit $|\phi\rangle$ to be transmitted and her half of the entangled pair. In other words, Alice now applies $C_{not} \otimes I$ followed by $H \otimes I \otimes I$ to this state to obtain

$$(H \otimes I \otimes I)(C_{not} \otimes I)(|\phi\rangle \otimes |\psi_0\rangle)$$

$$= (H \otimes I \otimes I)\frac{1}{\sqrt{2}}\big(a|000\rangle + a|011\rangle + b|110\rangle + b|101\rangle\big)$$

$$= \frac{1}{2}\big(a(|000\rangle + |011\rangle + |100\rangle + |111\rangle) + b(|010\rangle + |001\rangle - |110\rangle - |101\rangle)\big)$$

$$= \frac{1}{2}\big(|00\rangle(a|0\rangle + b|1\rangle) + |01\rangle(a|1\rangle + b|0\rangle) + |10\rangle(a|0\rangle - b|1\rangle) + |11\rangle(a|1\rangle - b|0\rangle)\big).$$

Alice measures the first two qubits and obtains one of the four standard basis states $|00\rangle$, $|01\rangle$, $|10\rangle$, and $|11\rangle$ with equal probability. Depending on the result of her measurement, the quantum state of Bob's qubit is projected to $a|0\rangle + b|1\rangle$, $a|1\rangle + b|0\rangle$, $a|0\rangle - b|1\rangle$, or $a|1\rangle - b|0\rangle$. Alice sends the result of her measurement as two classical bits to Bob.

After these transformations, crucial information about the original state $|\phi\rangle$ is contained in Bob's qubit. There is now nothing Alice can do on her own to reconstruct the original state of her qubit. In fact, the no-cloning principle implies that at any given time, only one of Alice or Bob can reconstruct the original quantum state.

Bob When Bob receives the two classical bits from Alice, he knows how the state of his half of the entangled pair compares to the original state of Alice's qubit. Bob can reconstruct the original state of Alice's qubit, $|\phi\rangle$, by applying the appropriate decoding transformation to his qubit, originally part of the entangled pair. The following table shows the state of Bob's qubit before the decoding has taken place and the decoding operator Bob should use depending on the value of the bits he received from Alice.

State	Bits received	Decoding
$a|0\rangle + b|1\rangle$	00	I
$a|1\rangle + b|0\rangle$	01	X
$a|0\rangle - b|1\rangle$	10	Z
$a|1\rangle - b|0\rangle$	11	Y

After decoding, Bob's qubit will be in the quantum state, $a|0\rangle + b|1\rangle$, in which Alice's qubit started. This decoding step is the encoding step of dense coding, and the encoding step was the decoding step of dense coding, so teleportation and dense coding are in some sense inverses of each other.

5.4 Realizing Unitary Transformations as Quantum Circuits

This section shows how arbitrary unitary transformations can be implemented from a set of primitive transformations. The primitive set we consider includes the two-qubit C_{not} gate, in addition to three kinds of single-qubit gates. Using just these four types of operations, any arbitrary n-qubit unitary transformation can be implemented. Section 5.4.1 shows that general single-qubit transformations can be decomposed into products of the three kinds of primitive single-qubit operators. Sections 5.4.2 and 5.4.3 show how to construct multiple-qubit controlled versions of single-qubit transformations. Section 5.4.4 uses these transformations to construct arbitrary unitary transformations.

This chapter merely shows that all quantum transformations can be implemented in terms of simple gates; we are not yet concerned with the efficiency of such implementations. Most quantum transformations do not have an efficient implementation in terms of simple gates. Much of the rest of the book will be devoted to understanding which quantum transformations have efficient implementations and how these can be used to solve computational problems.

5.4.1 Decomposition of Single-Qubit Transformations
This section shows that all single-qubit transformations can be written as a combination of three types of transformations, phase shifts $K(\delta)$, rotations $R(\beta)$, and phase rotations $T(\alpha)$.

$$K(\delta) = e^{i\delta} I \qquad \text{A phase shift by } \delta$$

$$R(\beta) = \begin{pmatrix} \cos\beta & \sin\beta \\ -\sin\beta & \cos\beta \end{pmatrix} \qquad \text{A rotation by } \beta$$

$$T(\alpha) = \begin{pmatrix} e^{i\alpha} & 0 \\ 0 & e^{-i\alpha} \end{pmatrix} \qquad \text{A phase rotation by } \alpha.$$

Note that

$$K(\delta_1 + \delta_2) = K(\delta_1)K(\delta_2),$$

$$R(\beta_1 + \beta_2) = R(\beta_1)R(\beta_2),$$

and

$$T(\alpha_1 + \alpha_2) = T(\alpha_1)T(\alpha_2),$$

and that the operator K commutes with K, T, and R.

Rather than write $K(\delta)$, we frequently just write the scalar factor $e^{i\delta}$. Even though, as a transformation on a single-qubit system, $K(\delta)$ performs a global phase change, and thus is equivalent to the identity on the single-qubit system, we include it here because we will use it later as part of multiple-qubit conditional transformations in which this factor becomes a relative phase shift that is physically relevant. The transformation $R(\alpha)$ and $T(\alpha)$ are rotations by 2α about the y- and z-axis of the Bloch sphere respectively.

This paragraph shows that any single-qubit unitary transformation Q can be decomposed into a sequence of transformations of the form $Q = K(\delta)T(\alpha)R(\beta)T(\gamma)$. Since the $K(\delta)$ is a global phase shift with no physical effect, the space of all single-qubit transformations has only three real dimensions. Given the transformation

$$Q = \begin{pmatrix} u_{00} & u_{01} \\ u_{10} & u_{11} \end{pmatrix},$$

it follows immediately from the unitarity condition $QQ^\dagger = I$ that $|u_{00}|^2 + |u_{01}|^2 = 1$, $u_{00}\overline{u_{10}} + u_{01}\overline{u_{11}} = 0$, and $|u_{11}|^2 + |u_{10}|^2 = 1$. A short calculation gives $|u_{00}| = |u_{11}|$ and $|u_{01}| = |u_{10}|$. So the magnitudes of the coefficients u_{ij} can be written as the sine and cosine of some angle β; we can write Q as

$$Q = \begin{pmatrix} e^{i\theta_{00}}\cos(\beta) & e^{i\theta_{01}}\sin(\beta) \\ -e^{i\theta_{10}}\sin(\beta) & e^{i\theta_{11}}\cos(\beta) \end{pmatrix}.$$

Furthermore, the phases are not independent: $u_{10}\overline{u_{00}} + u_{11}\overline{u_{01}} = 0$ implies that $\theta_{10} - \theta_{00} = \theta_{11} - \theta_{01}$. Since

$$K(\delta)T(\alpha)R(\beta)T(\gamma) = \begin{pmatrix} e^{i(\delta+\alpha+\gamma)}\cos\beta & e^{i(\delta+\alpha-\gamma)}\sin\beta \\ -e^{i(\delta-\alpha+\gamma)}\sin\beta & e^{i(\delta-\alpha-\gamma)}\cos\beta \end{pmatrix},$$

we can find δ, α, γ for a given Q by solving the equations

$$\delta + \alpha + \gamma = \theta_{00},$$

$$\delta + \alpha - \gamma = \theta_{01},$$

$$\delta - \alpha + \gamma = \theta_{10}.$$

Using $\theta_{11} = \theta_{10} - \theta_{00} + \theta_{01}$, it is easy to see that this solution also satisfies $\delta - \alpha - \gamma = \theta_{11}$.

5.4.2 Singly Controlled Single-Qubit Transformations

Let $Q = K(\delta)T(\alpha)R(\beta)T(\gamma)$ be an arbitrary single-qubit unitary transformation. The controlled gate $\bigwedge Q$ can be implemented by first constructing $\bigwedge K(\delta)$ and implementing $\bigwedge Q'$ for $Q' = T(\alpha)R(\beta)T(\gamma)$. Then $\bigwedge Q = (\bigwedge K(\delta))(\bigwedge Q')$. We now show how to implement these two transformations in terms of basic gates.

The conditional phase shift can be implemented by primitive single-qubit operations:

$$\bigwedge K(\delta) = |0\rangle\langle 0| \otimes I + |1\rangle\langle 1| \otimes K(\delta)$$

$$= |0\rangle\langle 0| \otimes I + e^{i\delta}|1\rangle\langle 1| \otimes I$$

$$= (K(\delta/2)T(-\delta/2)) \otimes I.$$

Graphically, the implementation looks like

It may appear surprising that the conditional phase shift $K(\delta)$ can be realized by a circuit acting on the first qubit only, with no transformations acting directly on the second qubit. The reason that transformations on the first qubit suffice is that a phase shift affects the whole quantum state, not just a single qubit. In particular, $|x\rangle \otimes a|y\rangle = a|x\rangle \otimes |y\rangle$.

Implementing $\bigwedge Q'$ is slightly more involved. For $Q' = T(\alpha)R(\beta)T(\gamma)$, define the following transformations:

$$Q_0 = T(\alpha)R(\beta/2),$$

$$Q_1 = R(-\beta/2)T\left(\frac{-\gamma - \alpha}{2}\right),$$

$$Q_2 = T\left(\frac{\gamma - \alpha}{2}\right).$$

The claim is that $\bigwedge Q'$ can be defined as

$$\bigwedge Q' = (I \otimes Q_0)C_{not}(I \otimes Q_1)C_{not}(I \otimes Q_2)$$

or graphically

It is easy to see that this circuit performs the following transformation:

$$|0\rangle \otimes |x\rangle \rightarrow |0\rangle \otimes Q_0 Q_1 Q_2 |x\rangle.$$

$$|1\rangle \otimes |x\rangle \rightarrow |1\rangle \otimes Q_0 X Q_1 X Q_2 |x\rangle.$$

Using $R(\beta)R(-\beta) = I$ and $T(\alpha)T(\gamma) = T(\alpha + \gamma)$, the property $Q_0 Q_1 Q_2 = I$ follows immediately from the definition of the Q_i. To show that $Q_0 X Q_1 X Q_2 = Q'$, use $X R(\beta) X = R(-\beta)$ and $X T(\alpha) X = T(-\alpha)$. Then

$$Q_0 X Q_1 X Q_2 = T(\alpha) R(\beta/2)(X R(-\beta/2) X)(X T(-\frac{\gamma + \alpha}{2}) X) T(\frac{\gamma - \alpha}{2})$$

$$= Q'.$$

In this way, we can realize a version of an arbitrary single-qubit transformation controlled by a single qubit.

5.4.3 Multiply Controlled Single-Qubit Transformations

The graphical notation of sections 5.2.4 and 5.4.2 for controlled operations generalizes to more than one control bits. Let $\bigwedge_k Q$ be the $(k+1)$-qubit transformation that applies Q to qubit 0 when qubits 1 through k are all 1. For example, the *controlled-controlled*-NOT gate or *Toffoli gate* $\bigwedge_2 X$, which negates the last bit of three if and only if the first two are both 1, has the following graphical representation.

The subscript 2 in the notation $\bigwedge_2 X$ indicates that there are two control bits. We write the C_{not} gate as both $\bigwedge X$ and $\bigwedge_1 X$.

The construction of 5.4.2 can be iterated to obtain arbitrary single-qubit transformations controlled by k qubits. To implement $\bigwedge_2 Q$, a three-qubit gate that applies Q controlled by two qubits, start by replacing each of Q_0, Q_1, and Q_2 in the previous construction with a single-qubit controlled version,

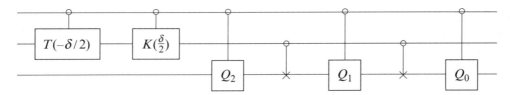

This circuit can be expanded, as in the previous section, into single-qubit and controlled-NOT gates, for a total of twenty five single-qubit gates and 12 controlled-NOT gates. Repeating this process leads to circuits for controlled versions of single-qubit transformations with k control bits, $\bigwedge_k Q$, with 5^k single-qubit transformations and $\frac{1}{2}(5^k - 1)$ controlled-NOT gates. As section 6.4.2 shows, significantly more efficient implementations of $\bigwedge_k Q$ are known.

All of the controlled gates seen so far are executed when the control bits are 1. To implement a singly controlled gate that is executed when the control bit is 0, the control bit can be negated, as in

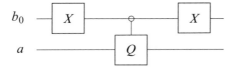

For any length k bit-string s, temporarily negating the appropriate control qubits in this way, enables the realization of a controlled gate that applies Q to qubit 0 exactly when the other k qubits are in the pattern s. More precisely, let $|s\rangle$ be the k-qubit standard basis vector labeled with bit-string s. This construction implements the $(k + 1)$–qubit controlled gate that applies the single-qubit transformation Q to qubit 0 when qubits 1 though k are in the basis state $|s\rangle$ and does nothing to qubit 0 when qubits 1 though k are in a different basis state. Such constructions can be further generalized to $(k + 1)$-qubit controlled gates that apply the single-qubit transformation Q to qubit i when the other qubits are in a specific basis state and do nothing when they are in a different basis state. In other words, this transformation applies Q to the two-dimensional subspace spanned by the two basis vectors $|x_k \ldots x_i \ldots x_0\rangle$ and $|x_k \ldots \hat{x}_i \ldots x_0\rangle$, where $\hat{x}_i = x_i \oplus 1$, that differ only in bit i, and it leaves the orthogonal subspace invariant.

Section 5.4.4 uses such gates to exhibit an explicit implementation of an arbitrary unitary transformation. The construction of section 5.4.4 uses two different transformations related to a pair consisting of a k-bit bit-string s and a single-qubit transformation Q: the first applies Q to the ith qubit with the standard ordering of the basis $\{|0\rangle, |1\rangle\}$ when the other k qubits are in state $|s\rangle$, and the second applies Q to the ith qubit with the basis in the other order. In other words, this second transformation applies XQX to qubit i when the other qubits are in state $|s\rangle$. We use the notation $\bigwedge_x^i Q$, or where x is a $(k + 1)$-bit bit-string such that $x_k \ldots x_{i+1}x_{i-1} \ldots x_0 = s_{k-1} \ldots s_0$, to represent both of these transformations depending on the value of x_i. When x_i is 0, the single-qubit transformation Q is applied. When x_i is 1, the transformation XQX is applied. When i is specified, the notation $\hat{x}$ means that the ith bit of a bit-string x has been flipped: $\hat{x} = x \oplus 2^i$. For any single-qubit transformation Q, the transformation $\bigwedge_{\hat{x}}^i Q = \bigwedge_x^i \hat{Q}$, where $\hat{Q} = XQX$. Geometrically $\bigwedge_x^i Q$ is a rotation in the two-dimensional complex subspace spanned by standard basis vectors $|x\rangle$ and $|\hat{x}\rangle$.

Example 5.4.1 On a two-qubit system $|b_1 b_0\rangle$, $\bigwedge_{10}^0 X$ is the standard C_{not}, with b_1 being the control bit and b_0 being the target. The notation $\bigwedge_{11}^0 X$ also represents the C_{not} transformation because X is invariant under reversing the order of the basis for qubit b_0: $X = XXX$. The notation $\bigwedge_{00}^0 X$ is a controlled-NOT transformation except that now X is performed only when b_1 has value 0. The notation $\bigwedge_{01}^1 X$ describes the standard C_{not} but with b_0 as the control bit and b_1 as the target.

This section showed how to implement multiply controlled single-qubit gates using a number of basic gates that is exponential in the number of qubits. Section 6.4.2 shows how to implement efficiently any multiply controlled single-qubit operation. That construction uses linearly many basic gates and a single additional qubit.

5.4.4 General Unitary Transformations

This section presents a systematic way to implement an arbitrary unitary transformation on the 2^n-dimensional vector space associated with the state space of an n-qubit system. The intuitive idea behind the construction is that any unitary transformation is simply a rotation of the 2^n-dimensional complex vector space underlying the n-qubit quantum state space, and that any rotation can be obtained by a sequence of rotations in two-dimensional subspaces.

Let $N = 2^n$. This section writes all matrices in the standard basis, but with a *nonstandard ordering* $\{|x_0\rangle, \ldots, |x_{N-1}\rangle\}$ such that successive basis elements differ by only one bit. Such a sequence of binary numbers is called a *Gray code*. Any Gray code will do. For $0 \le i \le N - 2$, let j_i be the bit on which $|x_i\rangle$ and $|x_{i+1}\rangle$ differ, and B_i be the shared pattern of all the other bits in $|x_i\rangle$ and $|x_{i+1}\rangle$. The next few paragraphs show how to realize an arbitrary unitary operator U as a sequence of multiply controlled single-qubit operators $\bigwedge_{x_i}^{j_i} Q$ that perform a series of rotations, each in a two-dimensional subspace spanned by successive basis elements.

Consider transformations U_m of the form

$$U_m = \begin{pmatrix} I^{(m)} & 0 \\ 0 & V_{N-m} \end{pmatrix},$$

where $I^{(m)}$ is the $m \times m$ identity matrix and V_{N-m} is an $(N - m) \times (N - m)$-unitary matrix with $0 \le m \le N - 2$. We wish to show that given any $(N \times N)$-matrix $U_{m-1}, 0 < m \le N - 2$, of this form there exist operators C_m, the product of multiply controlled single-qubit operators, and a U_m, now with a larger identity component $I^{(m)}$, such that $U_{m-1} = C_m U_m$. Then, taking $V_N = U$, the unitary operator U can be written as

$$U = U_0 = C_1 \cdots C_{N-2} U_{N-2}.$$

The transformation U_{N-2} has the form

$$U_{N-2} = \begin{pmatrix} I^{(N-2)} & 0 \\ 0 & V_2 \end{pmatrix},$$

which is simply the operation $\bigwedge_x^j V_2$ where $x = x_{N-2}$ and, using the Gray code condition, $j = j_{N-2}$ is the bit in which the last two basis vectors $|x_{N-2}\rangle$ and $|x_{N-1}\rangle$ differ. So once we show how to implement the C_m using multiply controlled single-qubit operators, we will have succeeded in showing that any unitary operator can be expressed in terms of such operators, and thus can be implemented using only C_{not}, $K(\delta)$, $R(\beta)$, and $T(\alpha)$.

The basis vector $|x_m\rangle$ is the first basis vector on which U_{m-1} acts nontrivially. Write

$$|v_m\rangle = U_{m-1}|x_m\rangle = a_m |x_m\rangle + \cdots + a_N |x_N\rangle.$$

We may assume that a_N is real, since we can multiply U_{m-1} by a global phase. If we can find a unitary transformation W_m, composed only of multiply controlled single-qubit transformations, that takes $|v_m\rangle$ to $|x_m\rangle$ and does not affect any of the first m elements of the basis, $W_m U_{m-1}$ would have the desired form, so we would take $U_m = W_m U_{m-1}$ and $C_m = W_m^{-1}$. To define W_m, begin by rewriting the coefficients of the last two components of $|v_m\rangle$:

$$|v_m\rangle = a_m |x_m\rangle + \cdots + c_{N-1} \cos(\theta_{N-1}) e^{i\phi_{N-1}} |x_{N-1}\rangle + c_{N-1} \sin(\theta_{N-1}) |x_N\rangle,$$

where

$$a_{N-1} = |a_{N-1}| e^{i\phi_{N-1}},$$

$$c_{N-1} = \sqrt{|a_{N-1}|^2 + |a_N|^2},$$

$$\cos(\theta_{N-1}) = |a_{N-1}|/c_{N-1},$$

$$\sin(\theta_{N-1}) = |a_N|/c_{N-1}.$$

Then

$$\bigwedge_{x_{N-1}}^{j_{N-1}} R(\theta_{N-1}) \bigwedge_{x_{N-1}}^{j_{N-1}} K(-\phi_{N-1})$$

takes $|v_m\rangle$ to

$$a_m |x_m\rangle + \cdots + a'_{N-1} |x_{N-1}\rangle,$$

where $a'_{N-1} = c_{N-1}$, since $\bigwedge_{x_{N-1}}^{j_{N-1}} K(-\phi_{N-1})$ cancels the $e^{i\phi_{N-1}}$ factor, and $\bigwedge_{x_{N-1}}^{j_{N-1}} R(\theta_{N-1})$ rotates so that all of the amplitude that was in $|x_N\rangle$ is now in $|x_{N-1}\rangle$. None of the other basis vectors are affected because the controlled part of the operators ensure that only basis vectors with bits in pattern B_{N-1} are affected. To obtain the rest of W_m, we iterate this procedure over all pairs of coordinates $\{a_{N-2}, a'_{N-1}\}$ through $\{a_m, a'_{m+1}\}$ to obtain the operator

$$W_m = \bigwedge_{x_m}^{j_m} R(\theta_m) \bigwedge_{x_m}^{j_m} K(-\phi_m) \cdots \bigwedge_{x_{N-1}}^{j_{N-1}} R(\theta_{N-1}) \bigwedge_{x_{N-1}}^{j_{N-1}} K(-\phi_{N-1}),$$

which takes $|v_m\rangle$ to $a'_m |x_m\rangle$, where

$$a_i = |a_i| e^{i\phi_i},$$

$$a'_i = c_i,$$

$$c_i = \sqrt{|a_i|^2 + |a_{i+1}|^2},$$

$$\cos(\theta_i) = |a_i|/c_i,$$

$$\sin(\theta_i) = |a'_{i+1}|/c_i.$$

The coefficient $a'_m = 1$, since the image of $|v_m\rangle$ must be a unit vector, and the final $\bigwedge_{x_m}^{j_m} K(-\phi_m)$ ensures that it is a positive real.

While this procedure provides an implementation for any unitary operator U in terms of simple transformations, the number of gates needed is exponential in the number of qubits. For this reason, it has limited practical value in that more efficient implementations are needed for realistic computations. Most unitary operators do not have efficient realizations in terms of simple gates; the art of quantum algorithm design is in finding useful unitary operators that have efficient implementations.

5.5 A Universally Approximating Set of Gates

Section 5.4 showed that all unitary transformations can be realized as a sequence of single-qubit transformations and controlled-NOT gates. From a practical point of view, we would prefer to deal with a finite set of gates. It is easy to show that for any finite set of gates there are unitary transformations that cannot be realized as a combination of these gates, but there are finite sets of gates that can approximate any unitary transformation to arbitrary accuracy. Furthermore, for any desired level of accuracy 2^{-d}, this approximation can be done efficiently; there is a polynomial $p(d)$ such that any single-qubit unitary transformation can be approximated to within 2^{-d} by a sequence of no more than $p(d)$ gates from the finite set. We will not prove this efficiency result, known as the Solovay-Kitaev theorem, but we will exhibit a finite set of gates that can be used to approximate all unitary transformations.

Since any unitary transformation can be realized using single-qubit and C_{not} gates, it suffices to find a finite set of gates that can approximate all single-qubit transformations. Consider the set consisting of the Hadamard gate H, the phase gate $P_{\frac{\pi}{2}}$, the $\pi/8$-gate $P_{\frac{\pi}{4}}$, and the C_{not} gate where

$$P_{\frac{\pi}{2}} = \begin{pmatrix} 1 & 0 \\ 0 & e^{i\pi/2} \end{pmatrix} = |0\rangle\langle0| + i|1\rangle\langle1|$$

and

$$P_{\frac{\pi}{4}} = \begin{pmatrix} 1 & 0 \\ 0 & e^{\frac{i\pi}{4}} \end{pmatrix} = |0\rangle\langle0| + e^{\frac{i\pi}{4}}|1\rangle\langle1|.$$

Recall from section 5.4.1 the single-qubit operator $T(\theta) = e^{i\theta}|0\rangle\langle0| + e^{-i\theta}|1\rangle\langle1|$. The $\pi/8$-gate $P_{\frac{\pi}{4}}$ got its name because, up to a global phase, it acts in the same way as the gate $T(-\frac{\pi}{8})$,

$$P_{\frac{\pi}{4}} = e^{\frac{i\pi}{8}} T\left(-\frac{\pi}{8}\right),$$

and unfortunately the name stuck in spite of the confusion it causes. (When used on their own, it does not matter whether $P_{\frac{\pi}{4}}$ or $T(-\frac{\pi}{8})$ is used, since they differ only in a global phase, but when used as part of a controlled gate construction, this phase becomes a physically relevant relative phase.)

A rotation R is a rational rotation if, for some integer m, $R^m = I$. If no such m exists, then R is an irrational rotation. It may seem surprising that a set of gates consisting only of rational rotations on the Bloch sphere can approximate all single-qubit transformations. Don't we need an irrational rotation? In fact, the proof proceeds by using these gates to construct an irrational rotation. Such a construction is possible because the group of rotations of a sphere differs from the group of rotations of a Euclidean plane. In the Euclidean plane, the product of two rational rotations is always rational, but the analogous statement is not true for rotations of the sphere. Exercise 5.21 guides the reader through proofs of the relevant properties of groups of rotations of the sphere and the Euclidean plane.

Exercises 5.19–5.22 develop the steps in the following spherical geometry argument in more detail. The gate $P_{\frac{\pi}{4}}$ is a rotation by $\pi/4$ about the z-axis of the Bloch sphere. The transformation $S = HP_{\frac{\pi}{4}}H$ is a rotation by $\pi/4$ about the x-axis. It is a good exercise in spherical geometry to show that $V = P_{\frac{\pi}{4}}S$ is an irrational rotation. Since V is irrational, any rotation W about the same axis can be approximated to within arbitrary precision 2^{-d} by some power of V. Recall from section 5.4.1 that any single-qubit transformation may be achieved (up to global phase) by combining rotations about the y- and z-axes: for every single-qubit operation W there exist angles α, β, γ, and δ such that

$$W = K(\delta)T(\alpha)R(\beta)T(\gamma),$$

where $T(\alpha)$ rotates by angle α about the z-axis and $R(\alpha)$ rotates by angle α about the y-axis. The set of rotations about any two distinct axes can achieve arbitrary single-qubit transformations.

Since HVH has a different axis from V, the two transformations H and V generate all single-qubit operators. Other universally approximating finite sets, with varying advantages and disadvantages, exit.

5.6 The Standard Circuit Model

A *circuit model* for quantum computation describes all computations in terms of a circuit composed of simple gates followed by a sequence of measurements. The simple gates are drawn either from a universal set of simple gates or a universally approximating set of quantum gates. The *standard circuit model* for quantum computation takes as its gate set the C_{not} gate together with all single-qubit transformations, and it takes as its set of measurements single-qubit measurements in the standard basis. So all computations in the standard model consist of a sequence of single-qubit and C_{not} gates followed by a sequence of single-qubit measurements in the standard basis. While a finite set of gates would be more realistic than the infinite set of all single-qubit transformations, the infinite set is easier to work with and, by the results of Solovay and Kitaev, the infinite set does not yield significantly greater computational power. For conceptual clarity, the n qubits of the computation are often organized into registers, subsets of the n qubits.

Other models of quantum computation exist. Each model provides its own insights into the workings of quantum computation, and each has contributed to the growth of the field through new algorithms, new approaches to robust quantum computation, or new approaches to building quantum computers. The most significant of these models will be discussed in section 13.4.

One of the strengths of the standard circuit model is that it makes finding quantum analogs of classical computation straightforward. That is the subject of the next chapter. Finding quantum analogs of reversible classical circuits is easy; all of the technical difficulties involve the entirely classical problem of converting an arbitrary classical circuit into a reversible classical circuit. The results of section 5.4 show that any quantum transformation can be realized in terms of the basic gates of the standard circuit model. But it says nothing about efficiency. Chapter 6 finds not only a quantum analog for any classical computation, but also a quantum analog with comparable efficiency. Part II explores the design of quantum algorithms, which involves finding quantum transformations that can be efficiently implemented in terms of the basic gates of the standard circuit model and figuring out how to use them to solve certain problems more efficiently than is possible classically.

5.7 References

The no-cloning theorem is due to Wootters and Zurek [286]. Both dense coding and quantum teleportation were discovered in the early 1990s, dense coding by Bennett and Wiesner [46] and quantum teleportation by Bennett et al. [44]. Single-qubit teleportation has been realized in several experiments, see for example, [57], [221], and [56].

An outline for a proof of the Solovay-Kitaev theorem was given in [173]. Dawson and Nielsen provide a pedagogical review of this result in [95]. A related issue, namely how much precision is needed to carry out a quantum computation of k steps is answered by Bernstein and Vazirani [49]: a precision of $O(\log k)$ bits suffices. (See box 6.1 for the $O(t)$ notation.)

The implementation of complex unitary transformations from basic ones is described in a paper by Barenco et al. [31].

A proof that most quantum transformations cannot be implemented efficiently and exactly in terms of two-qubit gates can be found in Knill's *Approximation by Quantum Circuits* [177]. Deutsch found a single three-qubit gate that by itself can produce arbitrarily good approximations to any unitary transformation [100]. Later, Deutsch, Barenco, and Ekert showed that almost any two-qubit gate could accomplish the same thing [101]. Others have found other small sets of generators.

5.8 Exercises

Exercise 5.1. Show that any linear transformation U that takes unit vectors to unit vectors preserves orthogonality: if subspaces S_1 and S_2 are orthogonal, then so are $U S_1$ and $U S_2$.

Exercise 5.2. For which sets of states is there a cloning operator? If the set has a cloning operator, give the operator. If not, explain your reasoning.

a. $\{|0\rangle, |1\rangle\}$,

b. $\{|+\rangle, |-\rangle\}$,

c. $\{|0\rangle, |1\rangle, |+\rangle, |-\rangle\}$,

d. $\{|0\rangle|+\rangle, |0\rangle|-\rangle, |1\rangle|+\rangle, |1\rangle|-\rangle\}$,

e. $\{a|0\rangle + b|1\rangle\}$, where $|a|^2 + |b|^2 = 1$.

Exercise 5.3. Suppose Eve attacks the BB84 quantum key distribution of section 2.4 as follows. For each qubit she intercepts, she prepares a second qubit in state $|0\rangle$, applies a C_{not} from the transmitted qubit to her prepared qubit, sends the first qubit on to Bob, and measures her qubit. How much information can she gain, on average, in this way? What is the probability that she is detected by Alice and Bob when they compare s bits? How do these quantities compare to those of the direct measure-and-transmit strategy discussed in section 2.4?

Exercise 5.4. Prove that the following are decompositions for some of the standard gates.

$$I = K(0)T(0)R(0)T(0)$$

$$X = -iT(\pi/2)R(\pi/2)T(0)$$

$$H = -iT(\pi/2)R(\pi/4)T(0)$$

Exercise 5.5. A vector $|\psi\rangle$ is *stabilized* by an operator U if $U|\psi\rangle = |\psi\rangle$. Find the set of vectors stabilized by

a. the Pauli operator X,

b. the Pauli operator Y,

c. the Pauli operator Z,

d. $X \otimes X$,

e. $Z \otimes X$,

f. C_{not}.

Exercise 5.6.

a. Show that $R(\alpha)$ is a rotation of 2α about the y-axis of the Bloch sphere.

b. Show that $T(\beta)$ is a rotation of 2β about the z-axis of the Bloch sphere.

c. Find a family of single-qubit transformations that correspond to rotations of 2γ about the x axis.

Exercise 5.7. Show that the Pauli operators form a basis for all linear operators on a two-dimensional space.

Exercise 5.8. What measurement does the operator $\mathbf{i}Y$ describe?

Exercise 5.9. How can the circuit of figure 5.2 be used to measure the qubits b_0 and b_1 for equality without learning anything else about the state of b_0 and b_1? (Hint: you are free to chose any initial state on the register consisting of qubits a_0 and a_1.)

Exercise 5.10. An n-qubit cat state is the state $\frac{1}{\sqrt{2}}(|00\ldots0\rangle + |11\ldots1\rangle)$. Design a circuit that, upon input of $|00\ldots0\rangle$, constructs a cat state.

Exercise 5.11. Let

$$|W_n\rangle = \frac{1}{\sqrt{n}}(|0\ldots001\rangle + |0\ldots010\rangle + |0\ldots100\rangle + \cdots + |1\ldots000\rangle).$$

Design a circuit that, upon input of $|00\ldots0\rangle$, constructs $|W_n\rangle$.

Exercise 5.12. Design a circuit that constructs the Hardy state

$$\frac{1}{\sqrt{12}}(3|00\rangle + |01\rangle + |10\rangle + |11\rangle).$$

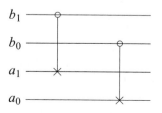

Figure 5.2
Circuit for exercise 5.9.

Exercise 5.13. Show that the swap circuit of section 5.2.4 does indeed swap two single-qubit values in that it sends $|\psi\rangle|\phi\rangle$ to $|\phi\rangle|\psi\rangle$ for all single-qubit states $|\psi\rangle$ and $|\phi\rangle$.

Exercise 5.14. Show how to implement the Toffoli gate $\bigwedge_2 X$ in terms of single-qubit and C_{not} gates.

Exercise 5.15. Design a circuit that determines if two single qubits are in the same quantum state. The circuit may include an ancilla qubit to be measured. The measurement should give a positive answer if the two-qubit states are identical, a negative answer if the two-qubit states are orthogonal, and be more likely to give a positive answer the closer the states are to being identical.

Exercise 5.16. Design a circuit that permutes the values of three qubits in that it sends $|\psi\rangle|\phi\rangle|\eta\rangle$ to $|\phi\rangle|\eta\rangle|\psi\rangle$ for all single-qubit states $|\psi\rangle$, $|\phi\rangle$, and $|\eta\rangle$.

Exercise 5.17. Compare the effect of the following two circuits

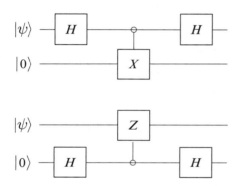

Exercise 5.18. Show that for any finite set of gates there must exist unitary transformations that cannot be realized as a sequence of transformations chosen from this set.

Exercise 5.19. Let R be an irrational rotation about some axis of a sphere. Show that for any other rotation R' about the same axis and for any desired level of approximation 2^{-d} there is some power of R that approximates R' to the desired level of accuracy.

Exercise 5.20. Show that the set of rotations about any two distinct axes of the Bloch sphere generate all single-qubit transformations (up to global phase).

Exercise 5.21.

a. In the Euclidean plane, show that a rotation of angle θ may be achieved by composing two reflections.

b. Use part (a) to show that a clockwise rotation of angle θ about a point P followed by a clockwise rotation of angle ϕ about a point Q results in a clockwise rotation of angle $\theta + \phi$ around the point R, where R is the intersection point of the two rays, one through P at angle $\theta/2$ from the line between P and Q, and the other through point Q at an angle of $\phi/2$ from the line between P and Q.

c. Show that the product of any two rational rotations of the Euclidean plane is also rational.

d. On a sphere of radius 1, a triangle with angles θ, ϕ, and η has area $\theta + \phi + \eta$ (where θ, ϕ, and η are in radians). Use this fact to describe the result of rotating clockwise by angle θ around a point P followed by rotating clockwise by angle ϕ around a point Q in terms of the area of a triangle.

e. Prove that on the sphere the product of two rational rotations may be an irrational rotation.

Exercise 5.22.

a. Show that the gates H, $P_{\frac{\pi}{2}}$ and $P_{\frac{\pi}{4}}$ are all (up to global phase) rational rotations of the Bloch sphere. Give the axis of rotation and the angle of rotation for each of these gates, and also the gate $S = H P_{\frac{\pi}{4}} H$.

b. Show that the transformation $V = P_{\frac{\pi}{4}} S$ is an irrational rotation of the Bloch sphere.

6 Quantum Versions of Classical Computations

This chapter constructs, for any classical computation, a quantum circuit that can perform the same computation with comparable efficiency. This result proves that quantum computation is at least as powerful as classical computation. In addition, many quantum algorithms begin by using this construction to compute a classical function on a superposition of values prior to using nonclassical means for efficiently extracting information from this superposition.

The construction of quantum analogs to all classical computations relies on a classical result that constructs a reversible analog to any classical computation. Section 6.1 describes relations between classical reversible computation and both general classical computation and quantum computation. Section 6.1.1 exhibits reversible versions of Boolean logic gates and quantum analogs of these reversible versions. Given a classical reversible circuit composed of reversible Boolean logic gates, simple substitution of the analogous quantum gates for the reversible gates gives the desired quantum circuit. The hard step in proving that every classical computation has a comparably efficient quantum analog is proving that every classical computation has a reversible version of comparable efficiency. Although this construction is purely classical, it is of such fundamental importance to quantum computation that we present it here. Section 6.2 provides this construction. Section 6.3 describes the language that section 6.4 uses to specify explicit quantum circuits for several classical functions such as arithmetic operations.

6.1 From Reversible Classical Computations to Quantum Computations

Any sequence of quantum transforms effects a unitary transformation U on the quantum system. As long as no measurements are made, the initial quantum state of the system prior to a computation can be recovered from the final quantum state $|\psi\rangle$ by running $U^{-1} = U^\dagger$ on $|\psi\rangle$. Thus, any quantum computation is *reversible* prior to measurement in the sense that the input can always be computed from the output.

In contrast, classical computations are not in general reversible: it is not usually possible to compute the input from the output. For example, while the classical NOT operation is reversible, the AND, OR, and NAND are not. Every classical computation does, however, have a classical reversible analog that takes only slightly more computational resources. Section 6.1.1 shows how to make basic Boolean gates reversible. Section 6.2.2 shows how to make entire Boolean circuits reversible in a resources efficient way, considering space, the number of bits required, and the number of primitive gates. This construction of efficient classical reversible versions of arbitrary Boolean circuits easily generalizes to a construction of quantum circuits that efficiently implement general classical circuits.

Any classical reversible computation with n input and n output bits simply permutes the $N = 2^n$ bit strings. Thus, for any such classical reversible computation there is a permutation $\pi : \mathbf{Z}_N \to \mathbf{Z}_N$ sending an input bit string to its output bit string. This permutation can be used to define a quantum transformation

$$U_\pi : \sum_{x=0}^{N-1} a_x |x\rangle \mapsto \sum_{x=0}^{N-1} a_x |\pi(x)\rangle,$$

that behaves on the standard basis vectors, viewed as classical bit strings, exactly as π did. The transformation U_π is unitary, since it simply reorders the standard basis elements.

Any classical computation on n input and m output bits defines function

$$f : \mathbf{Z}_N \to \mathbf{Z}_M$$

$$x \mapsto f(x)$$

mapping the $N = 2^n$ input bit strings to the $M = 2^m$ output bit strings. Such a function can be extended in a canonical way to a reversible function π_f acting on $n + m$ bits partitioned into two registers, the n-bit input register and the m-bit output register:

$$\pi_f : \mathbf{Z}_L \to \mathbf{Z}_L$$

$$(x, y) \mapsto (x, y \oplus f(x)),$$

where $\oplus$ denotes the bitwise exclusive-OR. The function π_f acts on the $L = 2^{n+m}$ bit strings, each made up of an n-bit bit string x and an m-bit bit string y. For $y = 0$, the function π acts like f, except that the output appears in the output register and the input register retains the input. There are many other ways of making a classical computation reversible, and for a particular classical computation, there may be a reversible version that requires fewer bits, but this construction always works.

Since π_f is reversible, there is a corresponding unitary transformation $U_f : |x, y\rangle \to |x, y \oplus f(x)\rangle$. Graphically the transformation U_f is depicted as

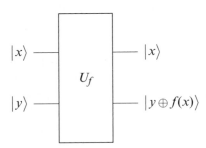

Section 5.4 showed how to implement any unitary operation in terms of simple gates. For most unitary transformations, that implementation is highly inefficient. While most unitary operators do not have an efficient implementation, U_f has an efficient implementation as long as there is a classical circuit that computes f efficiently. The method for constructing an efficient implementation of U_f from an efficient classical circuit for f has two parts. The first part constructs an efficient reversible classical circuit that computes f. The second part substitutes quantum gates for each of the reversible gates that make up the reversible classical circuit. Section 6.1.1 defines reversible Boolean logic gates and covers the easy second part of the construction. Section 6.2 explains the involved construction of an efficient reversible classical circuit for any efficient classical circuit.

6.1.1 Reversible and Quantum Versions of Simple Classical Gates

This section describes reversible versions of the Boolean logic gates NOT, XOR, AND, and NAND. Quantum versions of these gates act like the reversible gates on elements of the standard basis. Their action on other input states is prescribed by the linearity of quantum operations; the action of a gate on a superposition is the linear combination of the action of the gate on the standard basis elements making up the superposition. In this way, the behavior of a reversible gate fully defines the behavior of its quantum analog, and vice versa. The tight connection between the two allows us to use the same notation for both gates with the understanding that the quantum gates can be applied to arbitrary superpositions, whereas the classical reversible gates are applied to bit strings that correspond to the standard basis elements.

Let b_1 and b_0 be two binary variables, variables taking on only values 0 or 1. We define the following quantum gates:

NOT   The NOT gate is already reversible. We will use X to refer to both the classical reversible gate and the single-qubit operator $X = |0\rangle\langle 1| + |1\rangle\langle 0|$ of section 5.2, which performs a classical NOT operation on classical bits encoded as the standard basis elements.

XOR   The controlled negation performed by the $C_{not} = \bigwedge_1 X$ gate amounts to an XOR operation on its input values. It retains the value of the first bit b_1, and replaces the value of the bit b_0 with the XOR of the two values.

$$|b_1\rangle \quad\longrightarrow\quad |b_1\rangle$$
$$|b_0\rangle \quad\longrightarrow\quad |b_1 \oplus b_0\rangle$$

The quantum version behaves like the reversible version on the standard basis vectors, and its behavior on all other states can be deduced from the linearity of the operator.

AND   It is impossible to perform a reversible AND operation with only two bits. The three-bit controlled-controlled-NOT gate, or Toffoli gate, $T = \bigwedge_2 X$ can be used to perform a reversible AND operation.

$T|b_1, b_0, 0\rangle = |b_1, b_0, b_1 \wedge b_0\rangle,$

where $\wedge$ is notation for the classical AND of the two bit values.

The Toffoli gate is defined for all input: when the value of the third bit is 1,

$T|b_1, b_0, 1\rangle = |b_1, b_0, 1 \oplus b_1 \wedge b_0\rangle.$

By varying the values of input bits, the Toffoli gate T can be used to construct a complete set of Boolean connectives, not just the classical AND. Thus, any combinatorial circuit can be constructed from Toffoli gates alone. The Toffoli gate computes NOT, AND, XOR, and NAND in the following way:

$T|1, 1, x\rangle = |1, 1, \neg x\rangle$

$T|x, y, 0\rangle = |x, y, x \wedge y\rangle$

$T|1, x, y\rangle = |1, x, x \oplus y\rangle$

$T|x, y, 1\rangle = |x, y, \neg(x \wedge y)\rangle,$

where $\neg$ indicates the classical NOT acting on the bit value.

An alternative to the Toffoli gate, the Fredkin gate F, acts as a *controlled swap*:

$F = \bigwedge_1 S,$

where S is the two-bit swap operation

$S : |xy\rangle \rightarrow |yx\rangle.$

The Fredkin gate F, like the Toffoli gate T, can implement a complete set of classical Boolean operators:

$F|x, 0, 1\rangle = |x, x, \neg x\rangle$

$F|x, y, 1\rangle = |x, y \vee x, y \vee \neg x\rangle$

$F|x, 0, y\rangle = |x, y \wedge x, y \wedge \neg x\rangle,$

where $\vee$ is notation for the classical OR of the two bit values.

Because a complete set of classical Boolean connectives can be implemented using just the Toffoli gate T, or the Fredkin gate F, these gates can be combined to realize arbitrary Boolean circuits. Section 6.2 describes explicit implementations of certain classical functions. As the equations for the Toffoli gate illustrate, the operations C_{not} and X can be implemented by Toffoli gates with the addition of one or two bits permanently set to 1. For clarity, we use C_{not} and X gates in our construction, but all constructions can be done using only Toffoli gates, since we can replace all uses of C_{not} and X with Toffoli gates that have additional input bits with their

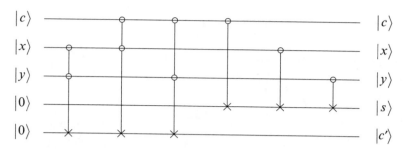

Figure 6.1
One-bit full adder.

input values set appropriately. For example, the circuit shown in figure 6.1 implements a one-bit full adder using Toffoli and controlled-NOT gates, where x and y are the data bits, s is their sum (modulo 2), c is the incoming carry bit, and c' is the new carry bit. Several one-bit adders can be strung together to achieve full n-bit addition.

6.2 Reversible Implementations of Classical Circuits

This section develops systematic ways to turn arbitrary classical Boolean circuits into reversible classical circuits of comparable computational efficiency in terms of the number of bits and the number of gates. The resulting reversible circuits are composed entirely of Toffoli and negation gates. A quantum circuit with the same efficiency as the classical reversible circuit is obtained by the trivial substitution of quantum Toffoli and X gates for classical Toffoli and negation gates. Thus, as soon as we have an efficient version of a computation in terms of Toffoli gates, we immediately know how to obtain a quantum implementation of the same efficiency.

6.2.1 A Naive Reversible Implementation
Rather than start with arbitrary Boolean circuits, we consider a classical machine that consists of a register of bits and a processing unit. The processing unit performs simple Boolean operations or gates on one or two of the bits in the register at a time and stores the result in one of the register's bits. We assume that, for a given size input, the sequence of operations and their order of execution are fixed and do not depend on the input data or on other external control. In analogy with quantum circuits, we draw bits of the register as horizontal lines. A simple program (for four-bit conjunction) for this kind of machine is depicted in figure 6.2.

An arbitrary Boolean circuit can be transformed into a sequence of operations on a large enough register to hold input, output, and intermediate bits. The space complexity of a circuit is the size of the register.

Computations performed by this machine are not reversible in general; by reusing bits in the register, the machine erases information that cannot be reconstructed later. A trivial, but highly

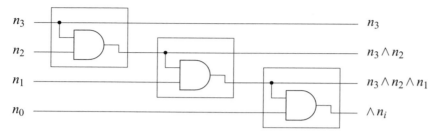

Figure 6.2
Irreversible classical circuit for four-bit conjunction.

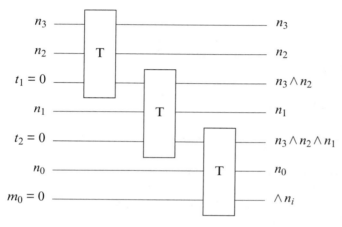

Figure 6.3
Reversible classical circuit for four-bit conjunction.

space inefficient, solution to this problem is not to reuse bits during the entire computation. Figure 6.3 illustrates how the circuit can be made reversible by assigning the results of each operation to a new bit. The operation that reversibly computes the conjunction and leaves the result in a bit initially set to 0 is, of course, the Toffoli gate. Since the NOT gate is reversible, and NOT together with AND form a complete set of Boolean operations, this construction can be generalized to turn any computation using Boolean logic operations into one using only reversible gates. This implementation, however, needs an additional bit for every AND performed, so if the original computation takes t steps, then a reversible one constructed in this naive way requires up to t additional bits of space.

Furthermore, this additional space is no longer in the 0 state and cannot be directly reused, for example, to compose two reversible circuits. Reusing temporary bits will be crucial to keeping the space requirements close to that of the original nonreversible classical computation. Resetting a bit to zero is not as trivial as it might seem. A transformation that resets a bit to 0, regardless of whether it was 0 or 1 before, is not reversible (it loses information), so it cannot be used as part of a reversible computation. Reversible computations cannot reclaim space through a simple *reset* operation. They can, however, *uncompute* any bit set during the course of a reversible computation by reversing the part of the computation that computed the bit.

Example 6.2.1 Consider the computation of figure 6.3. Bits t_1 and t_0 are temporarily used to obtain the output in bit m_0. Figure 6.4 shows how to uncompute these bits, resetting them to their original 0 value by reversing all but the last step of the circuit in figure 6.3, so that they may be reused as part of a continuing computation. Here the temporary bits are reclaimed at the cost of roughly doubling the number of steps.

We can reduce the number of qubits needed by uncomputing them and reusing them in the course of the algorithm. The method of uncomputing bits by performing all of the steps in reverse order, except those giving the output, works for any classical Boolean subcircuit. Consider a classical Boolean subcircuit of t gates operating on a s-bit register. The naive construction requires up to t additional bits in the register.

Example 6.2.2 Suppose we want to construct the conjunction of eight bits. Simply reversing the steps, generalizing the approach shown in figure 6.3, would require six additional temporary bits and one bit for the final output. We can save space by using the four-bit AND circuit of figure 6.4 four times and then combining the results as shown in figure 6.5. This construction

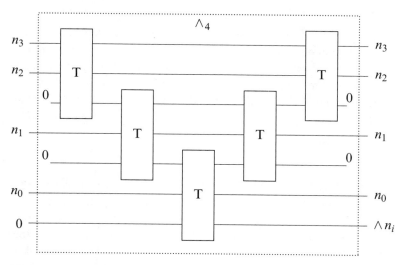

Figure 6.4
Reversible circuit that reclaims temporary bits.

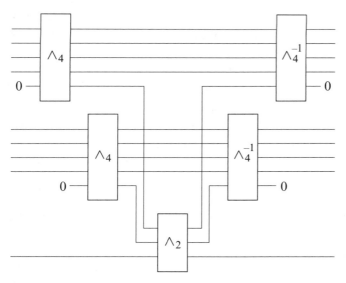

Figure 6.5
Combining reversible four-bit AND-circuits of figure 6.4 to construct an eight-way conjunction.

uses two temporary bits in addition to the two temporary bits used in each of the four-bit ANDs. Since each of the four-bit ANDs uncomputes its temporary bits, these bits can be reused by the subsequent four-bit ANDs. This circuit uses only a total of four additional temporary bits, though it does require more gates.

There is an art to deciding when to uncompute which bits to maintain efficiency and to retain subresults used subsequently in the computation. The key ideas of this section, adding bits to obtain reversibility and uncomputing their values so that they may be reused, are the main ingredients of the general construction described in section 6.2.2. By choosing carefully when and what to uncompute, it is possible to make a positive tradeoff, sacrificing some additional gates to obtain a much more efficient use of space. Examples, such as an explicit efficient implementation of an m-way AND, are given in section 6.4.

6.2.2 A General Construction

This section shows how, by carefully choosing which bits to uncompute when, a reversible version of any classical computation can be achieved with only minor increases in the number of gates and bits. We show that any classical circuit using t gates and s bits, has a reversible counterpart using only $O(t^{1+\epsilon})$ gates and $O(s \log t)$ bits. (See box 6.1 for the $O(t)$ notation.) For $t \gg s$, this construction uses significantly less space than the $(s + t)$ space of the naive approach described in section 6.2.1 at only a small increase in the number of gates.

Box 6.1
Notation for Efficiency Bounds

$O(f(n))$ is the set of functions bounded by f. Formally,

$g \in O(f(n))$ if and only if there exist constants k and n_0 such that $|g(n)| \leq k|f(n)|$
for all $n > n_0$.

Similarly, $\Omega(f(n))$ is the set of functions such that

$g \in \Omega(f(n))$ if and only if there exist constants k and n_0 such that $|g(n)| \geq k|f(n)|$
for all $n > n_0$.

Finally, the class of functions bounded by f from above and below is

$\Theta(f(n)) = O(f(n)) \cap \Omega(f(n)).$

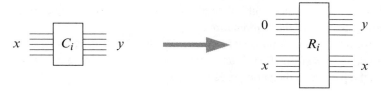

Figure 6.6
Converting circuits C_i into reversible ones R_i.

In order to understand how to obtain these bounds, we must consider carefully how many bits are being used and in what way. Let C be a classical circuit, composed of AND and NOT gates, that uses no more than t gates and s bits. The circuit C can be partitioned in time into $r = \lceil t/s \rceil$ subcircuits each containing s or fewer consecutive gates $C = C_1 C_2 \ldots C_r$. Each subcircuit C_i has s input and s output bits, some of which may be unchanged.

Using techniques from section 6.2.1, each circuit C_i can be replaced by a reversible circuit R_i that uses at most s additional bits as shown in figure 6.6. The circuit R_i returns its input as well as the s output values used in the subsequent computation. The input values will be used to uncompute and recompute R_i in order to save space.

More than s gates may be required to construct R_i. In general, R_i can be constructed using at most $3s$ gates. While other more efficient constructions are possible, the following three steps always work.

• **Step 1** Compute all of the output values in a reversible way. For every AND or NOT gate in the original circuit C_i, the circuit R_i has a Toffoli or NOT gate. This step uses the same number of gates, s, as C_i, and uses no more than s additional bits.

- **Step 2** Copy all of the output values, the values used in subsequent parts of the computation, to the output register, a set of no more than s additional bits.

- **Step 3** Perform the sequence of gates used to carry out step 1, but this time in reverse order. In this way all bits, except those in the output register, are reset to their original values. Specifically, all temporary bits are returned to 0, and we have recovered all of the input values.

The circuits $R_1 \ldots R_r$, when combined as in figure 6.7, perform the computation C in a reversible but space-inefficient way. The subcircuits R_i can be combined in a special way that uses space more efficiently by uncomputing and reusing some of the bits. Uncomputing requires additional gates, so we must choose carefully when to uncompute in order to reduce the usage of space without needing too many more gates. First, we show how to obtain a reversible version using $O(t^{\log_2 3})$ gates and $O(s \log t)$ bits, and then we improve on this method to obtain $O(t^{1+\epsilon})$ gates and $O(s \log t)$ bit bounds.

The basic principle for combining the $r = \lceil t/s \rceil$ circuits R_i is indicated in figure 6.8. The idea is to uncompute and recompute parts of the state selectively to reuse the space. We systematically modify the computation $R_1 R_2 \ldots R_r$ to reduce both the total amount of space used and to reset all the temporary bits to zero by the end of the computation.

To simplify the analysis, we take r to be a power of two, $r = 2^k$. For $1 \leq i \leq k$, let $r_i = 2^i$. We perform the following recursive transformation $\mathcal{B}$ that breaks a sequence into two equal-sized parts, recursively transforms the parts, and then composes them in the way shown:

$$\mathcal{B}(R_1, \ldots, R_{r_{i+1}}) = \mathcal{B}(R_1, \ldots, R_{r_i})\mathcal{B}(R_{1+r_i}, \ldots, R_{r_{i+1}}) \left(\mathcal{B}(R_1, \ldots, R_{r_i}) \right)^{-1}$$

$$\mathcal{B}(R) = R,$$

where $(\mathcal{B}(R_1, \ldots, R_{t_i}))^{-1}$ acts on exactly the same bits as $\mathcal{B}(R_1, \ldots, R_{t_i})$ and so requires no additional space.

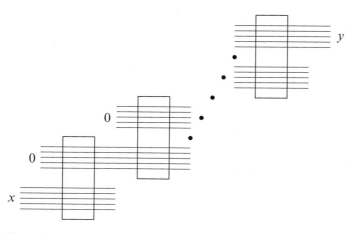

Figure 6.7
Composing the circuits R_i to obtain a reversible, but inefficient, version of the circuit C.

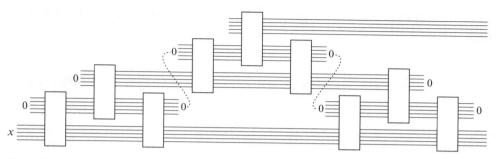

Figure 6.8
Composition of reversible computation circuits R_i in a way that reclaims storage.

The transformed computation uncomputes all space except the output of the last step, so the additional space usage is bounded by s. Thus, $B(R_1, \ldots, R_{r_i})$ requires at most s more space than $B(R_1, \ldots, R_{r_{i-1}})$. We can write the space $S(i)$ required for each of the $k = \log_2 r$ steps i in the recursion in terms of the space requirements of the previous step: $S(i) \leq s + S(i-1)$ with $S(1) \leq 2s$. The recursion ends after $k = \log_2 r$ steps, so the final computation $B(R_1, \ldots, R_r)$ requires at most $S(r) \leq (k+1)s = s(\log_2 r + 1)$ space. From the definition of B, it follows immediately that $T(i)$, the number of circuits R_j executed by the computation $B(R_1, \ldots, R_{r_i})$, is $T(i) = 3T(i-1)$ with $T(1) = 1$. By assumption $r = 2^k$, so the reversible version of C we constructed uses

$$T(2^k) = 3T(2^{k-1}) = 3^k = 3^{\log_2 r} = r^{\log_2 3}$$

reversible circuit R_i, each of which requires fewer than $3s$ gates. Thus, any classical computation of t steps and s bits can be done reversibly in $O(t^{\log_2 3})$ steps and $O(s \log_2 t)$ bits.

To obtain the $O(t^{1+\epsilon})$ bound, instead of using a binary decomposition, consider the following m-ary decomposition. To simplify the analysis, suppose that r is a power of m, $r = m^k$. For $1 \leq i \leq k$, let $r_i = m^i$. Abbreviating $R_{1+(x-1)r_i}, \ldots, R_{xr_i}$ as $\vec{R}_{x,i}$, then

$$B(\vec{R}_{1,i+1}) = B(\vec{R}_{1,i}, \vec{R}_{2,i}, \ldots \vec{R}_{m,i})$$

$$= B(\vec{R}_{1,i}), B(\vec{R}_{2,i}), \ldots B(\vec{R}_{m-1,i}),$$

$$B(\vec{R}_{m,i}),$$

$$B(\vec{R}_{m-1,i})^{-1}, \ldots B(\vec{R}_{2,i})^{-1}, B(\vec{R}_{1,i})^{-1}$$

$$B(R) = R.$$

In each step of the recursion, each block is split into m pieces and replaced with $2m - 1$ blocks. We may assume without loss of generality that $r = m^k$ for some k, in which case we stop recursing after k steps. At this point $r = m^k$ subcircuits C_1 have been replaced by $(2m - 1)^k$ reversible circuits R_i, so the total number of circuits R_i for the final computation is $(2m - 1)^k$, which we rewrite in terms of r:

$$(2m - 1)^{\log_m r} = r^{\log_m (2m-1)} \approx r^{\log_m 2m} = r^{1+\frac{1}{\log_2 m}}.$$

The number of primitive gates in R_i is bounded by $3s$ and $r = \lceil t/s \rceil$. The total number of gates for a reversible circuit of t gates is

$$T(t) \approx 3s \left(\frac{t}{s}\right)^{1+\frac{1}{\log_2 m}} < 3t^{1+\frac{1}{\log_2 m}}.$$

Thus, for any $\epsilon > 0$, it is possible to choose m sufficiently large that the number of gates required for the reversible computation is $O(t^{1+\epsilon})$. The space bound remains the same as before, $O(s \log_2 t)$.

Reversible versions of classical Boolean circuits constructed in this manner can be turned directly into quantum circuits consisting entirely of Toffoli and X gates. While our argument was given in terms of Boolean circuits, Bennett used the same argument to show that any classical Turing machine can be turned into a reversible one. Based on these arguments, from any classical circuit for f, an implementation of U_f can be constructed of comparable number of gates and bits.

The care needed in uncomputing and reusing bits generalizes to qubits where the need for uncomputing values is even greater: uncomputing ensures that temporary qubits are no longer entangled with output qubits. This need to unentangle temporary values at the end of a computation is one of the differences between classical and quantum implementations. Quantum transformations, being reversible, cannot simply reset qubits. Naively, one might think that temporary qubits could be reset by measuring the qubit and then, depending on the measurement outcome, performing a transformation to set them to $|0\rangle$. But if the temporary qubits were entangled with qubits containing the desired result, or results used later in the computation, measuring the temporary qubits may alter those results. Uncomputing temporary qubits disentangles them from the rest of the system without affecting the state of the rest of the system. The circuits of section 6.4 contain a number of examples of uncomputing temporary qubits.

The next section sets up the language for quantum implementations used in section 6.4 to describe explicit implementations of certain arithmetic functions. These implementations are often more efficient than the general construction just given, but they all have analogous classical implementations of comparable efficiency. Part II is devoted to truly quantum algorithms, algorithms with no classical analog.

6.3 A Language for Quantum Implementations

The quantum circuits we have discussed provide one way of describing a sequence of quantum gates acting on registers of qubits. We now give an alternate way of describing quantum circuits that is more compact and easier to reason about. We use this notation to describe quantum implementations for some specific arithmetic functions. When we talk about the efficiency of these implementations, we simply count the number of simple gates in the quantum circuits they describe. This quantity is called the *circuit complexity*. We explain the relation of circuit complexity with other notions of complexity in section 7.2.

The notation up to this point is standard and frequently used in the literature. Here we describe a language that we developed for describing quantum circuits or sequences of simple quantum gates. We use a program-like notation to give concise descriptions of quantum circuits that are cumbersome in graphical notation. Moreover, a single program in this notation can describe precisely a whole class of circuits acting on variable numbers of qubits as input (and classes that depend on other varying classical parameters). For example, just as in the classical case, a quantum circuit for adding 24-bit numbers differs from a circuit for adding 12-bit numbers, though they may be related. This programlike notation enables us to describe the relation precisely, while the graphical notation, though it may be suggestive, remains imprecise.

The notation uses both classical and quantum variables. Classical control structures such as iteration, recursion, and conditionals are used to define the order in which quantum state transformations are to be applied. Classical information can be used in the construction of a quantum state or as parameters of quantum state transformations, but quantum information cannot be used in classical control structures. The programs we write are simply classical prescriptions for sequences of quantum gates that operate on a single global quantum register.

6.3.1 The Basics

Quantum variables are names for registers, subsets of qubits of a single global quantum register. If x is the variable name for an n-qubit register, we may write $x[n]$ if we wish to make the number of qubits in x explicit. We use x_i to refer to the ith qubit of x, and $x_i \cdots x_k$ for qubits i through k of the register denoted by x. We will generally order the qubits of a register from highest index to lowest index so that if register x contains a standard basis vector $|b\rangle$, then $b = \sum_i x_i 2^i$. If U is a unitary transformation on n qubits, and x, y, and z are names for registers with a combined total of n qubits, then the program step $U|x, y, z\rangle = U|x\rangle|y\rangle|z\rangle$ means "apply U to the qubits denoted by the register names in the order given." It is illegal to use any qubit twice in this notation, so the registers x, y, and z must be disjoint; this restriction is necessary because "wiring" different input values to the same quantum bit is not possible or even meaningful. We are abusing the ket notation slightly here in that it is sometimes being used to stand for a placeholder, a qubit, that can contain a qubit value, a quantum state, and sometimes for the qubit value itself, but context should keep the two uses clear.

Example 6.3.1 The Toffoli gate with control bits b_5 and b_3 and target bit b_2 has the following graphical representation

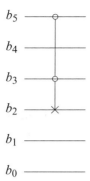

which is awkward to represent in the standard tensor product notation because the qubits it acts on are not adjacent. In our notation, this transformation can be written as $T|b_5, b_3, b_2\rangle$. The notation $T|b_2, b_3, b_2\rangle$ is not allowed, since it repeats qubits. The notation $(T \otimes C_{not} \otimes H)|x_5 \cdots x_3\rangle|x_1, x_0\rangle|x_7\rangle$, for a transformation acting on six qubits of a ten-qubit register $x = x_9 x_8 \cdots x_0$, is just another way of representing the transformation $I \otimes I \otimes H \otimes I \otimes T \otimes I \otimes C_{not}$, where the separate kets indicate which qubits the transformation making up the tensor product is acting upon; the Toffoli gate T acts on qubits x_5, x_4, and x_3, the C_{NOT} on qubits x_1 and x_0, and the Hadamard gate H on qubit x_7. The notation $(T \otimes C_{not} \otimes H)|x_5 \cdots x_3\rangle|x_4, x_0\rangle|x_7\rangle$ is illegal because the first and second registers are not disjoint: they share qubit $|x_4\rangle$.

Controlled operations are so frequently used that we give them their own notation; the notation $|b\rangle$ **control** $U|x\rangle$, where b and x are disjoint registers, means that on any standard basis vector the operator U is applied to the contents of register x only if all of the bits in b are 1. Writing $\neg|b\rangle$ **control** $U|x\rangle$ is a convenient shorthand for the sequence

$X \otimes \cdots \otimes X|b\rangle$
$|b\rangle$ **control** $U|x, y\rangle$
$X \otimes \cdots \otimes X|b\rangle$.

If we write a sequence of state transformations, they are intended to be applied in order.

We allow programs to declare local temporary registers using **qubit** $t[n]$, provided that the program restores the qubits in these registers to their initial $|0\rangle$ state. This condition ensures that temporary qubits can be reused for different executions of the program and that the overall storage requirement is bounded. Furthermore, it ensures that the temporary qubits do not remain entangled with the other registers.

6.3.2 Functions
We allow the introduction of new names for sequences of program steps. Unlike commands such as **control**, the command **define** does not do anything to the qubits; it simply defines

Box 6.2
Language Summary

Terms

U	Name for a unitary transform
U^{-1}	Name for the inverse of U
x	Name for a register of qubits
$x[k]$	Indicates number of qubits in register x
qubit $x[k]$	Indicates x is a name for a register of temporary qubits initially set to $|0\rangle$
qubit t	Indicates t is a name for a temporary qubit initially set to $|0\rangle$
x_i	Name for the ith qubit of register x
$x_i \ldots x_j$	A sequence of qubits of register x
$|r\rangle$	Indicates use of qubits named r

Statements   (Γ stands for an abstract statement)

$U|r\rangle$	Apply U to qubits named r
$|b\rangle$ **control** Γ	Controlled form of statement Γ with control qubits b
$\neg|b\rangle$ **control** Γ	Statement Γ controlled by negation of qubits b
$|b_1\rangle|b_0\rangle$ **control** Γ	Statement Γ controlled by two qubits named b_1 and b_0
for $i \in [a..b]$ $\qquad$ $\Gamma(i)$	Perform the sequence of statements $\Gamma(a), \Gamma(a+1), \ldots, \Gamma(b)$, which depend on the classical parameter i
define $Name|x[k]\rangle =$ $\qquad$ $\Gamma_0, \Gamma_1, \ldots \Gamma_n$	Introduce $Name$ as a name for a statement that performs statements Γ_0 through Γ_n to a k-qubit register x
$Name|r\rangle$	Applies the steps described in the definition of $Name$ to register r.
$Name^{-1}|r\rangle$	Applies the inverse of all the steps described in the definition of $Name$ in reverse order to register r. Since all quantum transformations are reversible, this transformation is always well defined.

a new function by telling the machine what sequence of commands a new function variable name represents. For example, addition modulo 2 with an incoming carry bit can be defined as

$$Sum : |c, a, b\rangle \rightarrow |c, a, (a+b+c) \bmod 2\rangle$$

define $Sum\ |c\rangle|a\rangle|b\rangle =$
   $|a\rangle$ **control** $X|b\rangle$
   $|c\rangle$ **control** $X|b\rangle$.

It operates on three single qubits by adding the value of a and the value of the carry c to the value of b. The program would be drawn as the circuit

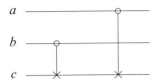

A corresponding carry operator is of the form

$$Carry : |c, a, b, c'\rangle \rightarrow |c, a, b, c' \oplus C(a, b, c)\rangle,$$

where the carry $C(a, b, c)$ is 1 if two or more of the bits a, b, c, are 1, that is, $C(a, b, c) = (a \wedge b) \oplus (c \wedge (a \oplus b))$. A program for Carry might look like

define $Carry\ |c, a, b, c'\rangle =$

$|a\rangle|b\rangle$ **control** $X|c'\rangle$	Compute $a \wedge b$ in register c'	(1)
$|a\rangle$ **control** $X|b\rangle$	Compute $a \oplus b$ in register b	(2)
$|c\rangle|b\rangle$ **control** $X|c'\rangle$	Toggle result c' if c and current value of b	(3)
$|a\rangle$ **control** $X|b\rangle$	Reset b to original value	(4)

In this program, the register b temporarily holds, starting in step (2), the XOR of the original values of a and b. In step (3), this value of register b means that c' is toggled if c and exactly one of the original values of a and b is 1. Register b is reset to its original value in step (4).

Repetition and conditional execution of sequences of quantum state transformations can be controlled using classical programming constructs. Only classical, not quantum, information can be used in the control structure. However, in quantum algorithms there is a choice as to which classical input values are placed in quantum registers and which are used simply as part of the classical control structure. For instance, one program to add x to itself n times might take classical input n and use it only as part of the classical control, while another might place n in an additional quantum register. The two programs would be of the form

$$A_n : |x, 0\rangle \mapsto |x, nx\rangle$$

and

$$A : |x, n, 0\rangle \mapsto |x, n, nx\rangle$$

respectively. This distinction will be more important when we consider quantum algorithms that act on superpositions of input values; only input values placed in quantum registers, not input values that are part of the classical control structure, can be in superposition.

The definition of a new program may use the same program recursively provided that the recursion can be unwound classically: recursive application of functions is allowed only as a shorthand for a classically specified sequence of quantum transformations. We can use the **qubit** $t[n]$ construction recursively as long as the recursion depth is bounded by a static classical constant.

6.4 Some Example Programs for Arithmetic Operations

The programs in this section implement quantum circuits for modular arithmetic and supporting operations. The operations shown are more general (though less efficient) than the modular arithmetic implementations used as part of Shor's algorithm; here the modulus, M, is placed in a quantum register so these algorithms can act on superpositions of different moduli as well as on superpositions of other input.

6.4.1 Efficient Implementation of AND

We give a linear implementation of an m-way AND computed into an output qubit using just one additional temporary qubit. First we define a supporting transformation $Flip$ that generalizes the Toffoli gate T. The transformation $Flip$ acts on an m-qubit register $a = |a_{m-1} \ldots a_0\rangle$ and an $(m-1)$-qubit register $b = |b_{m-2} \ldots b_0\rangle$ and negates qubit b_i exactly when the $(i+2)$-conjunction $\bigwedge_{j=0}^{i+1} a_j$ is true. We define $Flip$ in terms of Toffoli gates T that perform bit flips on some of the qubits of register b depending on the contents of register a.

define $Flip\ |a[2]\rangle|b[1]\rangle =$ (base case $m = 2$)
   $T\,|a_1\rangle|a_0\rangle|b\rangle$

define $Flip\ |a[m]\rangle|b[m-1]\rangle =$ (general case $m \geq 3$)
   $T\,|a_{m-1}\rangle|b_{m-3}\rangle|b_{m-2}\rangle$
   $Flip\ |a_{m-2} \ldots a_0\rangle|b_{m-3} \ldots b_0\rangle$
   $T\,|a_{m-1}\rangle|b_{m-3}\rangle|b_{m-2}\rangle$

An inductive argument shows that $Flip$, when defined in this way, behaves as described. The transformation $Flip$, when applied to an m-qubit register a and an $(m-1)$-qubit register b, uses $2(m-2)+1$ Toffoli gates T.

Next we define an $AndTemp$ operation on a $(2m-1)$-qubit computational state that uses $m-2$ additional qubits to compute an AND on m bits. We will shortly use $AndTemp$ to construct an AND operation that makes more efficient use of qubits. The operation $AndTemp$ places the conjunction of the bits in register a in the single-qubit register b, making temporary use of the qubits in register c.

define $AndTemp\ |a[2]\rangle|b[1]\rangle =$ (base case $m = 2$)
   $T\,|a_1\rangle|a_0\rangle|b\rangle$

define $AndTemp\ |a[m]\rangle|b[1]\rangle|c[m-2]\rangle =$ (general case $m \geq 3$)
   $Flip\ |a\rangle\,(|b\rangle|c\rangle)$ Compute conjunction in b (1)
   $Flip\ |a_{m-2} \ldots a_0\rangle|c\rangle$ Reset c (2)

The parentheses in $Flip\ |a\rangle\,(|b\rangle|c\rangle)$ indicate that $Flip$ is applied to the m-qubit register a and the $m-1$ qubit register that is the concatenation of registers b and c. By the definition of $Flip$, step (1) leaves the conjunction of the a_j in b but changes the contents of c in the process. Step (2) undoes these changes to c. Since the first *Flip* uses $2(m-2)+1$ Toffoli gates and the second *Flip* uses $2(m-3)+1$ Toffoli gates, *AndTemp* requires $4m-8$ gates. An attractive feature of this construction of *AndTemp* is that the $m-2$ additional qubits in register c can be in any state at the start of the computation, and they will be returned to their original states by the end of the computation so that we can use these to compute the m-way AND if there are sufficiently many computational qubits already ($n \geq 2m-2$). Clever use of this property of *AndTemp* will allow us to define an *And* on up to n qubits that uses only 1 additional temporary qubit.

To construct the conjunction using less space, we recursively use *AndTemp* on one half of the qubits, using the other half temporarily and vice versa. Thus, a general *And* operator that requires a single temporary qubit can be defined as follows: Let $k = \lfloor m/2 \rfloor$, and $j = k-2$ for even m, $j = k-1$ for odd m. The operator *And* has the effect of flipping b if and only if all bits of a are 1.

define *And* $|a[1]\rangle|b[1]\rangle =$ Trivial unary case, $m = 1$
  $C_{not}|a_0\rangle|b\rangle$

define *And* $|a[2]\rangle|b[1]\rangle =$ Binary case, $m = 2$
  $T|a_1\rangle|a_0\rangle|b\rangle$

define *And* $|a[m]\rangle|b\rangle =$ General case, $3 \leq m$
  qubit $t[1]$ use a temporary qubit
  $AndTemp\ |a_{m-1} \ldots a_k\rangle\ |t\rangle\ |a_j \ldots a_0\rangle$ (1)
  $AndTemp\ (|t\rangle|a_j \ldots a_0\rangle)\ |b\rangle\ |a_{k+j-2} \ldots a_k\rangle$ (2)
  $AndTemp\ |a_{m-1} \ldots a_k\rangle\ |t\rangle\ |a_j \ldots a_0\rangle$ (3)

Step (1) computes the conjunction of the high-order bits using the low-order bits temporarily. In step (2) we compute the conjunction of the low-order bits using the high-order bits temporarily. Since *AndTemp* uses a linear number of gates, so does *And*.

6.4.2 Efficient Implementation of Multiply Controlled Single-Qubit Transformations

The linear implementation of *And* given in the last section enables a linear implementation of the multiply controlled single-qubit transformations $\bigwedge_x^i Q$ of section 5.4.3. Given an m-bit bit string z, let $X^{(z)}$ be the transformation

$$X^{(z)} = X \otimes I \cdots \otimes X \otimes X,$$

which contains an X at any position where z has a 0 bit, and an I at any position where z has a 1 bit. We implement the transformation *Conditional*(z, Q), which acts on qubit b with single-qubit transformation Q if and only if the bits of register a match bit string z.

define *Conditional*$(z, Q)\ |a[m]\rangle|b[1]\rangle =$
  qubit t use a temporary qubit   (1)

$X^{(z)}|a\rangle$	if a and z match, a becomes all 1's	(2)
$And\ |a\rangle|t\rangle$	AND bits of a	(3)
$|t\rangle$ **control** $Q|b\rangle$	if a matched z, apply Q to b	(4)
$And\ |a\rangle|t\rangle$	uncompute AND	(5)
$X^{(z)}|a\rangle$	uncompute match	(6)

This construction uses 2 additional qubits and only $O(m)$ simple gates. When z is $11\ldots 1$ and $Q = X$, then $Conditional(z, Q)$ is simply the And operator of the previous section.

6.4.3 In-Place Addition

We define an Add transformation that adds two n-bit binary numbers. The transformation

$$Add : |c\rangle|a\rangle|b\rangle \rightarrow |c\rangle|a\rangle|(a + b + c)\ \mathrm{mod}\ 2^{n+1}\rangle,$$

where a and c are n-qubit registers and b is an $(n + 1)$-qubit register, adds two n-bit numbers, placed in registers a and b, and puts the result in register b when register c and the highest order bit, b_n, of register b are initially 0.

The implementation of Add uses n recursion steps, where n is the number of bits in the numbers to be added. The ith step in the recursion adds the $n - i$ highest bits, with the carry in the lowest of these $n - i$ highest bits having first been computed. The construction uses Sum and $Carry$ defined in section 6.3.2. We consider the two cases $n = 1$ and $n > 1$:

define $Add\ |c\rangle|a\rangle|b[2]\rangle =$	base case $n = 1$	
$\quad Carry\ |c\rangle|a\rangle|b_0\rangle|b_1\rangle$	carry in high bit of b	(1)
$\quad Sum\ |c\rangle|a\rangle|b_0\rangle$	sum in low bit of b	(2)

define $Add\ |c[n]\rangle|a[n]\rangle|b[n + 1]\rangle =$	general case $n > 1$	
$\quad Carry\ |c_0\rangle|a_0\rangle|b_0\rangle|c_1\rangle$	compute the carry for low bits	(3)
$\quad Add\ |c_{n-1}\cdots c_1\rangle|a_{n-1}\cdots a_1\rangle|b_n\cdots b_1\rangle$	add $n - 1$ highest bits	(4)
$\quad Carry^{-1}|c_0\rangle|a_0\rangle|b_0\rangle|c_1\rangle$	uncompute the carry	(5)
$\quad Sum\ |c_0\rangle|a_0\rangle|b_0\rangle$	compute the low order bit	(6)

Step (5) is needed to ensure that the carry register is reset to its initial value. The $Carry^{-1}$ operator is implemented by running, in reverse order, the inverse of each transformation in the definition of the $Carry$ operator.

6.4.4 Modular Addition

The following program defines modular addition for n-bit binary numbers a and b.

$$AddMod\ |a\rangle|b\rangle|M\rangle \rightarrow |a\rangle|(b + a)\ \mathrm{mod}\ M\rangle|M\rangle,$$

where the registers a and M have n qubits and b is an $n + 1$-qubit register. When the highest order bit, b_n, of register b is initially 0, the transformation $AddMod$ replaces the contents of register

b with $b + a \bmod M$, where M is the contents of register M. The contents of registers a and M (and the temporaries c and t) are unchanged by $AddMod$. The construction makes use of the Add transformation we defined in the previous section.

define $AddMod \; |a[n]\rangle|b[n+1]\rangle|M[n]\rangle =$

qubit t	use a temporary bit	(1)
qubit $c[n]$	storage for the n-bit carry	(2)
$Add \; |c\rangle|a\rangle|b\rangle$	add a to b	(3)
$Add^{-1}|c\rangle|M\rangle|b\rangle$	subtract M from b	(4)
$|b_n\rangle$ **control** $X|t\rangle$	toggle t when underflow	(5)
$|t\rangle$ **control** $Add \; |c\rangle|M\rangle|b\rangle$	when underflow, add M back to b	(6)
$Add^{-1}|c\rangle|a\rangle|b\rangle$	subtract a again	(7)
$\neg|b_n\rangle$ **control** $X|t\rangle$	reset t	(8)
$Add \; |c\rangle|a\rangle|b\rangle$	construct final result	(9)

Classically, steps (3) through (6) are all that are needed. In (4) if $M > b$, subtracting M from b causes b_n to become 1. Steps (7) through (9) are needed to reset t. Note that each Add operation internally resets $|c\rangle$ back to its original value.

The condition $0 \le a, b < M$ is necessary, since for values outside that range, an operation that sends $|a, b, M\rangle$ to $|a, b + a \bmod M, M\rangle$ is not reversible and therefore not unitary. If this condition does not hold, for example if $b \ge M$ initially, then the final value of b may still be greater than M, since the algorithm subtracts M at most once.

6.4.5 Modular Multiplication

The *TimesMod* transformation multiplies two n-bit binary numbers a and b modulo another n-bit binary number M. The transformation

$$TimesMod \; |a\rangle|b\rangle|M\rangle|p\rangle \; \rightarrow \; |a\rangle|b\rangle|M\rangle|(p + ba) \bmod M\rangle$$

is defined by the following program that successively adds $b_i 2^i a \bmod M$ to the result register p. It is assumed that $a < M$, but b can be arbitrary. Both a and p are $(n+1)$-qubit registers; the additional high-order bit is needed for intermediate results. The operation *Shift* simply cyclically shifts all bits by 1, which can easily be done by swapping bits a_{i+1} with a_i for all i, starting with the high-order bits. *Shift* acts as multiplication by 2, since the high-order bit of a will be 0.

define *TimesMod* $|a[n+1]\rangle|b[k]\rangle|M[n]\rangle|p[n+1]\rangle =$

qubit $t[k]$	use k temporary bits	(1)
qubit $c[n]$	carry register for addition	(2)
for $i \in [0 \ldots k-1]$	iterate through bits of b	(3)
$\quad Add^{-1}\;|c\rangle|M\rangle|a\rangle$	subtract M from a	(4)
$\quad |a_n\rangle$ **control** $X|t_i\rangle$	$t_i = 1$ if $M > a$	(5)
$\quad |t_i\rangle$ **control** $Add \; |c\rangle|M\rangle|a\rangle$	add M to a if t_i is set	(6)

$|b_i\rangle$ **control**

$$
\begin{array}{lll}
AddMod \; |a_{n-1} \cdots a_0\rangle |p\rangle |M\rangle & \text{add } a \text{ to } p \text{ if } b_i \text{ is set} & (7) \\
Shift \; |a\rangle & \text{multiply } a \text{ by 2} & (8) \\
\textbf{for } i \in [k-1\ldots 0] & \text{clear } t \text{ and restore } a & (9) \\
Shift^{-1} \; |a\rangle & \text{divide } a \text{ by 2} & (10) \\
|t_i\rangle \textbf{ control } Add^{-1} \; |c\rangle |M\rangle |a\rangle & \text{perform all steps in reverse} & (11) \\
|a_n\rangle \textbf{ control } X|t_i\rangle & \text{clear } i\text{th bit of } t & (12) \\
Add \; |c\rangle |M\rangle |a\rangle & \text{add } M \text{ to } a & (13)
\end{array}
$$

Lines (4)–(6) compute the $a \bmod M$. The second loop, (9)–(13), undoes all the steps of the first one, (3)–(8), except the conditional addition to the output p (line 8).

Note that modular multiplication cannot be defined as an in-place operation because the transformation that sends $|a, b, M\rangle$ to $|a, ab \bmod M, M\rangle$ is not unitary: both $|2, 1, 4\rangle$ and $|2, 3, 4\rangle$ would be mapped to the same state $|2, 2, 4\rangle$.

6.4.6   Modular Exponentiation

We implement modular exponentiation,

$$ExpMod \; |a\rangle |b\rangle |M\rangle |0\rangle \rightarrow |a\rangle |b\rangle |M\rangle |a^b \bmod M\rangle$$

using $O(n^2)$ temporary qubits where a, b, and M are n-qubit registers.

First, we define two transformations we will use in our implementation of *ExpMod*, an n-bit copy and an n-bit modular squaring function. The *Copy* transformation

$$Copy : |a\rangle |b\rangle \rightarrow |a\rangle |a \oplus b\rangle$$

copies the contents of an n-bit register a to another n-bit register b whenever the register b is initialized to 0. The operation *Copy* can be implemented as bitwise XOR operations between the corresponding bits in registers a and b.

define *Copy* $|a[n]\rangle |b[n]\rangle =$
    for $i \in [0..n-1]$ bit-wise
      $|a_i\rangle$ **control** $X|b_i\rangle$ XOR a with b

The modular squaring operation *SquareMod*

$$SquareMod : |a\rangle |M\rangle |s\rangle \rightarrow |a\rangle |M\rangle |(s + a^2) \bmod M\rangle$$

places the result of squaring the contents of register a, modulo the contents of register M, in the register s.

define *SquareMod* $|a[n+1]\rangle |M[n]\rangle |s[n+1]\rangle =$

$$
\begin{array}{lll}
\textbf{qubit } t[n] & \text{use } n \text{ temporary bits} & (1) \\
Copy \; |a_{n-1} \cdots a_0\rangle |t\rangle & \text{copy } n \text{ bits of } a \text{ to } t & (2) \\
TimesMod \; |a\rangle |t\rangle |M\rangle |s\rangle & \text{compute } a^2 \bmod M. & (3) \\
Copy^{-1} \; |a_{n-1} \cdots a_0\rangle |t\rangle & \text{clear } t & (4)
\end{array}
$$

Finally, we can give a recursive definition of modular exponentiation with the signature

$$ExpMod : |a\rangle|b\rangle|M\rangle|p\rangle|e\rangle \rightarrow |a\rangle|b\rangle|M\rangle|p\rangle|e \oplus (pa^b) \bmod M\rangle$$

define $ExpMod\ |a[n+1]\rangle|b[1]\rangle|M[n]\rangle|p[n+1]\rangle|e[n+1]\rangle = \qquad$ base case

$\neg|b_0\rangle$ **control** $Copy\ |p\rangle\ |e\rangle$	result is p	(1)
$|b_0\rangle$ **control** $TimesMod\ |a\rangle|p\rangle|M\rangle|e\rangle$	result is $pa^1 \bmod M$	(2)

define $ExpMod\ |a[n+1]\rangle|b[k]\rangle|M[n]\rangle|e[n+1]\rangle = \qquad$ general case $k > 1$

qubit $u[n+1]$	for $a^2 \bmod M$	(3)
qubit $v[n+1]$	for $(p * a^{b_0}) \bmod M$	(4)
$\neg|b_0\rangle$ **control** $Copy\ |p\rangle\ |v\rangle$	$v = pa^0 \bmod M$	(5)
$|b_0\rangle$ **control** $TimesMod\ |a\rangle|p\rangle|M\rangle|e\rangle$	$e = pa^1 \bmod M$	(6)
$SquareMod\ |a\rangle|M\rangle|u\rangle$	compute $a^2 \bmod M$ in u	(7)
$ExpMod\ |u\rangle|b_{k-1}\cdots b_1\rangle|M\rangle|v\rangle|e\rangle$	compute $v(a^2)^{b/2} \bmod M$	(8)
$SquareMod^{-1}\ |a\rangle|M\rangle|u\rangle$	uncompute u	(9)
$|b_0\rangle$ **control** $TimesMod^{-1}\ |a\rangle|p\rangle|M\rangle|e\rangle$	uncompute e	(10)
$\neg|b_0\rangle$ **control** $Copy^{-1}\ |p\rangle\ |v\rangle$	uncompute v	(11)

The program unfolds recursively k times, once for each bit of b. Steps (5)–(8) and the base case (1) and (2) perform the classical computation. The division $b/2$ in step (8) is integer division. Each recursive step requires two temporary registers of size $n + 1$ that are reset at the end in steps (9) and (11). Thus, the algorithm requires a total of $2(k - 1)(n + 1)$ temporary qubits.

The algorithm for modular multiplication given in 6.4.5 requires $O(n^2)$ steps to multiply two n-bit numbers. Thus, the modular exponentiation requires $O(kn^2)$ steps. But more efficient multiplication algorithms are possible and this complexity can be reduced to $O(kn \log n \log \log n)$ using the Schönhage-Strassen multiplication algorithm.

6.5 References

See Feynman's *Lectures on Computation* [121] for an account of reversible computation and its relation to the energy of computation and information.

In his 1980 paper [270], Tommaso Toffoli shows that any (classical) function with finite domain and range can be realized as a reversible function using additional bits. To prove this theorem, he introduces a family of controlled gates $\theta^{(n)}$ that we write as $\bigwedge_{n-1} X$. The instance $\bigwedge_2 X$ is generally known as the Toffoli gate. The Fredkin gate was first described as a billiard-ball gate in [124].

Reversible classical computations were first discussed by Bennett in [39], where he constructs reversible Turing machines from nonreversible ones. In [40] Bennett discusses the recursive decomposition presented in section 6.2. Bennett's argument uses multitape Turing machines instead of registers.

Deutsch [99] shows how to construct reversible quantum gates for any classically computable function. Deutsch defines, and Yao [287] and Bernstein and Vazirani [48] refine, the definition of a universal quantum Turing machine. This construction assumes a sufficient supply of qubits that correspond to the finite but unbounded tape of a Turing machine. Section 7.2 discusses quantum Turing machines briefly.

The implementations of the m-way AND and $Conditional(z, Q)$ are due to Barenco et al. [31], who also describe a $O(n^2)$-gate circuit for $Conditional(z, Q)$ that uses no additional qubits. Vedral, Barenco, and Ekert [275] give a comprehensive definition of quantum circuits for arithmetic operations. In particular, they show how modular exponentiation $a^x \bmod M$ can be done with fewer temporary qubits than the version presented here for the case where a and M are classical and relative prime. Fast multiplication was first described in Schönhage and Strassen's paper [245]. Descriptions in English can be found in most books on algorithms such as [182].

6.6 Exercises

Exercise 6.1. Show that it is impossible to perform a reversible AND operation with only two bits.

Exercise 6.2.

a. Construct a classical Boolean circuit with three input bits and two output bits that computes as a two-bit binary number the number of 1 bits in the input.

b. Convert your circuit into a classical reversible one.

c. Give an equivalent quantum circuit.

Exercise 6.3. Given two-qubit registers $|c\rangle$ and $|a\rangle$ and three-qubit register $|b\rangle$, construct the quantum circuit that computes $Add\ |c\rangle\ |a\rangle\ |b\rangle$.

Exercise 6.4.

a. Define a quantum algorithm that computes the maximum of two n-qubit registers.

b. Explain why such an algorithm requires one additional qubit that cannot be reused, that is, the algorithm will have to have $2n + 1$ input and output qubits.

Exercise 6.5. Show how to construct an efficient reversible circuit for every classical circuit along the lines of the construction of section 6.2.2, but without the assumption that t is a power of 2. Give the time and space bounds for your construction.

II QUANTUM ALGORITHMS

7 Introduction to Quantum Algorithms

The previous chapter used quantum computers in an essentially classical manner; in each of the algorithms of part I, if the quantum computer starts in a standard basis state, the state after every step of the computation is also a standard basis vector, not a superposition, so the computational state always has an obvious interpretation as a classical state. These algorithms do not make use of the ability of qubits to be in superposition or of sets of qubits to be entangled. In part I, we showed that quantum computation is at least as powerful as classical computation: for any classical circuit, there exists a quantum circuit that performs the same computation with similar efficiency. We now turn our attention to showing that quantum computation is more powerful than classical computation. Part II is concerned with truly quantum algorithms, quantum computations that outperform classical ones.

The algorithms in this part make use of the simple gates used in the quantum analogs of classical computations of chapter 6, and they also use more general unitary transformations that have no classical counterpart. Geometrically, all quantum state transformations on n qubits are rotations of 2^n-dimensional complex state space. Nonclassical quantum computations involve rotations to nonstandard bases, whereas, as explained in section 6.1, the steps of any classical computation merely permute the standard basis elements. Section 5.4.4 showed how any quantum transformation can be implemented in terms of simple gates. We now concentrate on quantum transformations that can be implemented *efficiently* and how such transformations can be used to speed up certain types of computation. The key to designing a truly quantum algorithm is figuring out how to use these nonclassical basic unitary gates to perform a computation more efficiently.

In this and the next few chapters, all discussion is in terms of the standard circuit model of quantum computation we described in section 5.6. We use the language introduced in 6.3 to specify general sequences of simple quantum gates as we did when we discussed quantum analogs of classical computations in chapter 6, but now we allow basic unitary transformations that have no classical counterpart. The way efficiency is computed in the quantum circuit model resembles the way it is computed classically, which makes it easy to compare the efficiency of quantum and classical algorithms. Early quantum algorithms were designed in the circuit model, but it is not the only, or necessarily the best, model to use for quantum algorithm design. Other models of quantum computation exist, and algorithms in these models have a different flavor. In chapter 13 we describe alternative models that have been shown to be equivalent in terms of computational power to the standard circuit model of quantum computation. In addition to having led to new types of quantum algorithms, these models underlie some promising efforts to build quantum computers.

In the standard circuit model of quantum computation, the efficiency of a quantum algorithm is computed in terms of the circuit complexity, the number of basic gates together with the number of qubits used, of the circuits used to implement the algorithm. Sometimes we are interested in the efficient use of other resources, so we will measure, say, the number of bits or qubits transmitted between two parties to carry out a task, or the number of calls to a (usually expensive to compute) function. Such functions are often called *black box* or *oracle* functions, since it is assumed that one does not have access to the inner workings of the computation of this function, only to the result of its application. These various notions of complexity are discussed in section 7.2.

Section 7.1 begins the chapter with a general discussion of computing with superpositions, including the notion of *quantum parallelism*. Section 7.2 describes various notions of complexity including circuit complexity, query complexity, and communication complexity. Deutsch's algorithm of section 7.3.1 provides the first example of a truly quantum algorithm, one for which there is no classical analog. The quantum subroutines of section 7.4 pave the way for the description in section 7.5 of four simple quantum algorithms, including Simon's algorithm, which inspired Shor's factoring algorithm. While the problems these algorithms solve are not so interesting, a study of the techniques they use will aid in understanding Grover's algorithm and Shor's algorithm. Section 7.7 defines quantum complexity and describes relations between quantum complexity classes and classical complexity classes. The final section of the chapter, section 7.8, discusses quantum Fourier transforms, which, in one form or another, are used in most of the algorithms described in this book.

7.1 Computing with Superpositions

Many quantum algorithms use quantum analogs of classical computation as at least part of their computation. Quantum algorithms often start by creating a quantum superposition and then feeding it into a quantum version U_f of a classical circuit that computes a function f. This setup, called *quantum parallelism*, accomplishes nothing by itself—any algorithm that stopped at this point would have no advantage over a classical algorithm—but this construction leaves the system in a state that quantum algorithm designers have found a useful starting point. Both Shor's algorithm and Grover's algorithm begin with the quantum parallelism setup.

7.1.1 The Walsh-Hadamard Transformation

Quantum parallelism, the first step of many quantum algorithms, starts by using the Walsh-Hadamard transformation, a generalization of the Hadamard transformation, to create a superposition of all input values. Recall from section 5.2.2 that the Hadamard transformation H applied to $|0\rangle$ creates a superposition state $\frac{1}{\sqrt{2}}(|0\rangle + |1\rangle)$. Applied to n qubits individually, all in state $|0\rangle$, H generates a superposition of all 2^n standard basis vectors, which can be viewed as the binary representation of the numbers from 0 to $2^n - 1$:

$$(H \otimes H \otimes \cdots \otimes H)|00\ldots 0\rangle$$

$$= \frac{1}{\sqrt{2^n}} \left((|0\rangle + |1\rangle) \otimes (|0\rangle + |1\rangle) \otimes \cdots \otimes (|0\rangle + |1\rangle)\right)$$

$$= \frac{1}{\sqrt{2^n}} \left(|0\ldots 00\rangle + |0\ldots 01\rangle + |0\ldots 10\rangle + \cdots + |1\ldots 11\rangle\right)$$

$$= \frac{1}{\sqrt{2^n}} \sum_{x=0}^{2^n-1} |x\rangle.$$

Box 7.1
Hamming Weight and Hamming Distance

The Hamming distance $d_H(x, y)$ between two bit strings x and y is the number of bits in which the two strings differ. The Hamming weight $d_H(x)$ of a bit string x is the number of 1-bits in x, which is equal to the Hamming distance between x and the bit string consisting of all zeros: $d_H(x) = d_H(x, 0)$.

For two bit strings x and y, $x \cdot y$ is the number of common 1 bits in x and y, $x \oplus y$ is the bitwise exclusive-OR, and $x \wedge y$ is the bitwise AND of x and y. The bitwise exclusive-OR $\oplus$ can also be viewed as bitwise modular addition of the strings x and y, viewed as elements of $\mathbf{Z}_2^n$. We use $\neg x$ to denote the bit string that flips 0 and 1 throughout bit string x, so $\neg x = x \oplus 11\ldots 1$.

The following identities hold:

$$x \cdot y = d_H(x \wedge y)$$

$$(x \cdot y \bmod 2) = \frac{1}{2}\left(1 - (-1)^{x \cdot y}\right)$$

$$x \cdot y + x \cdot z =_2 x \cdot (y \oplus z)$$

$$d_H(x \oplus y) =_2 d_H(x) + d_H(y)$$

where the notation $x =_2 y$ means equality modulo 2; it is shorthand for $x \bmod 2 = y \bmod 2$. Note that

$$\sum_{x=0}^{2^n-1} (-1)^{x \cdot x} = 0$$

since the successive ($2i$ and $2i + 1$) terms cancel.

Finally, we note that

$$\sum_{x=0}^{2^n-1} (-1)^{x \cdot y} = \begin{cases} 2^n & \text{if } y = 0 \\ 0 & \text{otherwise.} \end{cases} \tag{7.1}$$

The transformation $W = H \otimes H \otimes \cdots \otimes H$, which applies H to each of the qubits in an n-qubit state, is called the Walsh, or Walsh-Hadamard, transformation. Using $N = 2^n$, we may write

$$W|0\rangle = \frac{1}{\sqrt{N}} \sum_{x=0}^{N-1} |x\rangle.$$

Another way of writing W is useful for understanding the effect of W in quantum algorithms. In the standard basis, the matrix for the n-qubit Walsh-Hadamard transformation is a $2^n \times 2^n$ matrix W with entries W_{rs}, such that

$$W_{sr} = W_{rs} = \frac{1}{\sqrt{2^n}}(-1)^{r \cdot s},$$

where $r \cdot s$ is the number of common one-bits in s and r (see box 7.1) and both r and s range from 0 to $2^n - 1$. To see this equality, note that

$$W(|r\rangle) = \sum_s W_{rs}|s\rangle.$$

Let $r_{n-1} \ldots r_0$ be the binary representation of r, and $s_{n-1} \ldots s_0$ be the binary representation of s.

$$W(|r\rangle) = (H \otimes \cdots \otimes H)(|r_{n-1}\rangle \otimes \cdots \otimes |r_0\rangle)$$

$$= \frac{1}{\sqrt{2^n}}(|0\rangle + (-1)^{r_{n-1}}|1\rangle) \otimes \cdots \otimes (|0\rangle + (-1)^{r_0}|1\rangle)$$

$$= \frac{1}{\sqrt{2^n}} \sum_{s=0}^{2^n-1} (-1)^{s_{n-1}r_{n-1}}|s_{n-1}\rangle \otimes \cdots \otimes (-1)^{s_0 r_0}|s_0\rangle$$

$$= \frac{1}{\sqrt{2^n}} \sum_{s=0}^{2^n-1} (-1)^{s \cdot r}|s\rangle.$$

7.1.2 Quantum Parallelism

Any transformation of the form $U_f = |x, y\rangle \rightarrow |x, y \oplus f(x)\rangle$ from section 6.1 is linear and therefore acts on a superposition $\sum a_x|x\rangle$ of input values as follows:

$$U_f : \sum_x a_x|x, 0\rangle \rightarrow \sum_x a_x|x, f(x)\rangle.$$

Consider the effect of applying U_f to the superposition of values from 0 to $2^n - 1$ obtained from the Walsh transformation:

$$U_f : (W|0\rangle) \otimes |0\rangle = \frac{1}{\sqrt{N}} \sum_{x=0}^{N-1} |x\rangle|0\rangle \rightarrow \frac{1}{\sqrt{N}} \sum_{x=0}^{N-1} |x\rangle|f(x)\rangle.$$

After only one application of U_f, the superposition now contains all of the 2^n function values $f(x)$ entangled with their corresponding input value x. This effect is called quantum parallelism. Since n qubits enable us to work simultaneously with 2^n values, quantum parallelism in some sense circumvents the time/space trade-off of classical parallelism through its ability to hold exponentially many computed values in a linear amount of physical space. However, this effect is less powerful than it may initially appear.

To begin with, it is possible to gain only limited information from this superposition: these 2^n values of f are not independently accessible. We can gain information only from measuring the states, but measuring in the standard basis will project the final state onto a single input/output pair $|x, f(x)\rangle$, and a random one at that. The following simple example uses the basic setup of quantum parallelism and illustrates how useless the raw superposition arising from quantum parallelism is on its own, without performing any additional transformations.

Example 7.1.1 The controlled-controlled-NOT (Toffoli) gate, T, of section 5.4.3 computes the conjunction of two values:

$$
\begin{array}{lll}
|x\rangle & \text{———}\!\!\!\!\!\bullet\text{———} & |x\rangle \\
|y\rangle & \text{———}\!\!\!\!\!\bullet\text{———} & |y\rangle \\
|0\rangle & \text{———}\!\!\!\!\!\times\text{———} & |x \wedge y\rangle
\end{array}
$$

Take as input the superposition of all possible bit combinations of x and y together with a single-qubit register, initially set to $|0\rangle$, to contain the output. We use quantum parallelism to construct this input state in the standard way:

$$
W(|00\rangle) \otimes |0\rangle = \frac{1}{\sqrt{2}}(|0\rangle + |1\rangle) \otimes \frac{1}{\sqrt{2}}(|0\rangle + |1\rangle) \otimes |0\rangle
$$

$$
= \frac{1}{2}(|000\rangle + |010\rangle + |100\rangle + |110\rangle).
$$

Applying the Toffoli gate T to this superposition of inputs yields

$$
T(W|00\rangle \otimes |0\rangle) = \frac{1}{2}(|000\rangle + |010\rangle + |100\rangle + |111\rangle).
$$

This superposition can be viewed as a truth table for conjunction. The values of x, y, and $x \wedge y$ are entangled in such a way that measuring in the standard basis will give one line of the truth table. Computing the AND using quantum parallelism, and then measuring in the standard basis, gives no advantage over classical parallelism: only one result is obtained and, worse still, we cannot even choose which result we get.

7.2 Notions of Complexity

Complexity theory analyzes the amount of resources, most often time or space, asymptotically required to perform a computation. Turing machines provide a formal model of computation often used for reasoning about computational complexity. Early work by Benioff, improved by David Deutsch, then Andrew Yao, then Ethan Bernstein and Umesh Vazirani, defined quantum Turing machines and enabled the formalization of quantum complexity and comparison with classical results. In both quantum and classical settings, other methods, such as the circuit model, provide alternative means for formalizing complexity notions. Because most research on quantum algorithms discusses complexity in terms of *quantum circuit complexity*, we have chosen to take that approach in this book. Another common complexity measure used in the analysis of quantum algorithms is *quantum query complexity*, which will be discussed in section 7.2.1. Furthermore, there are a number of complexity measures used for analyzing quantum communication protocols. Communication complexity will be discussed in section 7.2.2.

A circuit family $C = \{C_n\}$ consists of circuits C_n indexed by the maximum input size for that circuit; the circuit C_n handles input of size n (bits or qubits). The *complexity* of a circuit C is defined to be the number of simple gates in the circuit, where the set of simple gates under consideration must be specified. Any of the finite sets of gates discussed in section 5.5 may be used, or the infinite set consisting of all single qubit operations together with the C_{not} may be used. The *circuit complexity*, or *time complexity*, of a family of circuits $C = \{C_n\}$ is the asymptotic number of simple gates in the circuits expressed as a function of the input size; the circuit complexity for a circuit family $C = \{C_n\}$ is $O(f(n))$ if the size of the circuit is bounded by $O(f(n))$: the function $t(n) = |C_n|$ satisfies $t(n) \in O(f(n))$. Any of the simple gate sets mentioned earlier give the same asymptotic circuit complexity.

Circuit complexity models are nonuniform in that different, larger circuits are required to handle larger input sizes. Both quantum and classical Turing machines, by contrast, propose a single machine that can handle arbitrarily large input. The nonuniformity of circuit models makes circuit complexity more complicated to define than Turing machine models because of the following issue: complexity can be *hidden* in the complexity of constructing the circuits C_n themselves, even if the size of the circuits C_n is asymptotically bounded. To get sensible notions of complexity, in particular to obtain circuit complexity measures similar to Turing machine based ones, a separate *uniformity condition* must be imposed. Both quantum and classical circuit complexity use similar uniformity conditions.

In addition to uniformity, a requirement that the behavior of the circuits C_n in a circuit family C behave in a consistent manner is usually imposed as well. This consistency condition is usually phrased in terms of a function $g(x)$, and says that all circuits $C_n \in C$ that can take x as input give $g(x)$ as output. This condition is sometimes misunderstood to include restrictions on the sorts of functions $g(x)$ a consistent circuit family can compute. For this reason, and to generalize easily to the quantum case, we phrase this same consistency condition without explicit reference to a function $g(x)$.

Consistency Condition A quantum or classical circuit family C is *consistent* if its circuits C_n give consistent results: for all $m < n$: applying circuit C_n to input x of size m must give the same result as applying C_m to that input.

The most common uniformity condition, and the one we impose here, is the polynomial uniformity condition.

Uniformity condition A quantum or classical circuit family $C = \{C_n\}$ is *polynomially uniform* if there exists a polynomial-time classical algorithm that generates the circuits. In other words, C is polynomially uniform if there exists a polynomial $f(n)$ and a classical program that, given n, constructs the circuit C_n in at most $O(f(n))$ steps.

The uniformity condition means that the circuit construction cannot be arbitrarily complex.

The relation between the circuit complexity of polynomially uniform, consistent circuit families and the Turing machine complexity is understood for both the classical and quantum case. In the classical case, for any classical function $g(x)$ computable on a Turing machine in time $O(f(n))$, there is a polynomially uniform, consistent classical circuit family that computes $g(x)$ in time $O(f(n) \log f(n))$. Conversely, a polynomially uniform, consistent family of Boolean circuits can be simulated efficiently by a Turing machine. In the quantum case, Yao has shown that any polynomial time computation on a quantum Turing machine can be computed by a polynomially uniform, consistent family of polynomially sized quantum circuits. As in the classical case, demonstrating that any polynomially uniform, consistent family of quantum circuit can be simulated by a quantum Turing machine is straightforward. Since we are not concerned with sublinear complexity differences, asymptotic differences of at most a polynomial in $\log(f(n))$, we discuss quantum complexity in terms of circuit complexity with the polynomial uniformity condition instead of using quantum Turing machines.

7.2.1 Query Complexity

The earliest quantum algorithms solve *black box*, or *oracle*, problems. A *classical black box* outputs $f(x)$ upon input of x. A *quantum black box* behaves like U_f, outputting $\sum_x \alpha_x |x, f(x) \oplus y\rangle$ upon input of $\sum_x \alpha_x |x\rangle|y\rangle$. Black boxes are theoretical constructs; they may or may not have an efficient implementation. For this reason, they are often called *oracles*. The black box terminology emphasizes that only the output of a black box can be used to solve the problem, not anything about its implementation or any of the intermediate values computed along the way; we cannot see inside it. The most common type of complexity discussed with respect to black box problems is *query complexity*: how many calls to the oracle are required to solve the problem.

Black box algorithms of low query complexity, algorithms that solve a black box problem with few calls to the oracle, are only of practical use if the black box has an efficient implementation. The black box approach is very useful, however, in establishing *lower bounds* on the circuit complexity of a problem. If the query complexity is $\Omega(N)$—in other words, at least $\Omega(N)$ calls to the oracle are required—then the circuit complexity must be at least $\Omega(N)$.

Black boxes have been used to establish lower bounds on the circuit complexity for quantum algorithms, but their first use in quantum computation was to show that the quantum query complexity of certain black box problems was strictly less than the classical query complexity: the number of calls to a quantum oracle needed to solve certain problems is strictly less than the required number of calls to a classical oracle to solve the same problem.

The first few quantum algorithms solve black box problems: Deutsch's problem (section 7.3.1), the Deutsch-Jozsa problem (section 7.5.1), the Bernstein-Vazirani problem (section 7.5.2), and Simon's problem (section 7.5.3). The most famous query complexity result is Grover's: that it takes only $O(\sqrt{N})$ calls to a quantum black box to solve an unstructured search problem over N elements, where as the classical query complexity of unstructured search is $\Omega(N)$. Grover's algorithm, and the extent to which its superior query complexity provide practical benefit, are discussed in chapter 9.

7.2.2 Communication Complexity

For communication protocols, common complexity measures include the minimum number of bits, or the minimum number of qubits, that must be transmitted to accomplish a task. Bounds on other resources, such as the number of bits of shared randomness or, in the quantum case, the number of shared EPR pairs, may or may not be of interest as well. Various notions of communication complexities exist, depending on whether the task requires quantum or classical information to be transmitted, whether qubits or bits can be sent, and what entanglement resources can be used.

We have already seen some examples of communication complexity results. The complexity notion of interest in dense coding is the number of qubits that must be sent in order to communicate n bits of information. While classical protocols require the transmission of n bits, only $n/2$ qubits need to be sent in order to communicate n bits of information. The other resource used in dense coding, the number of EPR pairs, sometimes called ebits in the communication protocol context, required in the setup is also $n/2$. Teleportation, by contrast, aims to transmit quantum information using a classical channel that can only send bits not qubits. The relevant complexity notion is the number of bits needed to transmit n qubits worth of quantum information. Using quantum teleportation, $2n$ bits can be used to transmit the state of n qubits. The number of ebits used to teleport n qubits is n.

The distributed computation protocol described in section 7.5.4 does not require the transmission of any bits or qubits, but it requires n ebits in order to accomplish a task concerning exponentially large bit strings, bit strings of length $N = 2^n$. A classical solution to this problem requires a minimum of $N/2$ bits to be transmitted. Since this book is concerned primarily with quantum computation not quantum communication, we will not discuss quantum communication complexity again except briefly in section 13.5.

7.3 A Simple Quantum Algorithm

We are now in a position to describe our first truly quantum algorithm. This algorithm, due to David Deutsch in 1985, was the first result that showed that quantum computation could outperform classical computation. The problem Deutsch's algorithm solves is a *black box* problem. Deutsch showed that his quantum algorithm has better query complexity than any possible classical algorithm: it can solve the problem with fewer calls to the black box than is possible classically. While the problem it solves is too trivial to be of practical interest, the algorithm contains simple versions of a number of key elements of intrinsically quantum computation, including the use of nonstandard bases and quantum analogs of classical functions applied to superpositions, that will recur in more complex quantum algorithms.

7.3.1 Deutsch's Problem

Deutsch's Problem   Given a Boolean function $f : \mathbf{Z}_2 \to \mathbf{Z}_2$, determine whether f is constant.

Deutsch's quantum algorithm, described in this section, requires only a single call to a black box for U_f to solve the problem. Any classical algorithm requires two calls to a classical black box for C_f, one for each input value. The key to Deutsch's algorithm is the nonclassical ability to place the second qubit of the input to the black box in a superposition. The subroutine of section 7.4.2 generalizes this trick.

Recall from 6.1 that U_f for a single bit function f takes two qubits of input and produces two qubits of output. On input $|x\rangle|y\rangle$, U_f produces $|x\rangle|f(x) \oplus y\rangle$, so when $|y\rangle = |0\rangle$, the result of applying U_f is $|x\rangle|f(x)\rangle$. The algorithm applies U_f to the two-qubit state $|+\rangle|-\rangle$, where the first qubit is a superposition of the two values in the domain of f, and the third qubit is in the superposition $|-\rangle = \frac{1}{\sqrt{2}}(|0\rangle - |1\rangle)$. We obtain

$$U_f (|+\rangle|-\rangle) = U_f \left(\frac{1}{2}(|0\rangle + |1\rangle)(|0\rangle - |1\rangle) \right)$$

$$= \frac{1}{2} (|0\rangle(|0 \oplus f(0)\rangle - |1 \oplus f(0)\rangle) + |1\rangle(|0 \oplus f(1)\rangle - |1 \oplus f(1)\rangle)).$$

In other words,

$$U_f (|+\rangle|-\rangle) = \frac{1}{2} \sum_{x=0}^{1} |x\rangle(|0 \oplus f(x)\rangle - |1 \oplus f(x)\rangle).$$

When $f(x) = 0$, $\frac{1}{\sqrt{2}}(|0 \oplus f(x)\rangle - |1 \oplus f(x)\rangle)$ becomes $\frac{1}{\sqrt{2}}(|0\rangle - |1\rangle) = |-\rangle$. When $f(x) = 1$, $\frac{1}{\sqrt{2}}(|0 \oplus f(x)\rangle - |1 \oplus f(x)\rangle)$ becomes $\frac{1}{\sqrt{2}}(|1\rangle - |0\rangle) = -|-\rangle$. Therefore

$$U_f \left(\frac{1}{\sqrt{2}} \sum_{x=0}^{1} |x\rangle|-\rangle \right) = \frac{1}{\sqrt{2}} \sum_{x=0}^{1} (-1)^{f(x)} |x\rangle|-\rangle.$$

For f constant, $(-1)^{f(x)}$ is just a physically meaningless global phase, so the state is simply $|+\rangle|-\rangle$. For f not constant, the term $(-1)^{f(x)}$ negates exactly one of the terms in the superposition so, up to a global phase, the state is $|-\rangle|-\rangle$. If we apply the Hadamard transformation H to the first qubit and then measure it, with certainty we obtain $|0\rangle$ in the first case and $|1\rangle$ in the second case. Thus with a single call to U_f we can determine, with certainty, whether f is constant or not. We now have our first example of a quantum algorithm that outperforms any classical algorithm!

It may surprise readers that this algorithm succeeds with certainty; the most commonly remembered aspect of quantum mechanics is its probabilistic nature, so people often naively expect that anything done with quantum means must be probabilistic, or at least that anything that exhibits peculiarly quantum properties must be probabilistic. We already know, from our study of quantum analogs to classical computations, that the first of these expectations does not hold. The algorithm for Deutsch's problem shows that even inherently quantum processes do not have to be probabilistic.

7.4 Quantum Subroutines

We now look at some useful nonclassical operations that can be performed on a quantum computer. The first subroutine, discussed in section 7.4.2, is commonly used; in particular, it is part of Grover's search algorithm as well as being used in most of the simpler quantum algorithms of section 7.5, including the Deutsch-Jozsa problem, a multiple bit generalization of Deutsch's problem. To illustrate further how to work with quantum superpositions, we describe a couple of other subroutines, though these subroutines are not used elsewhere in the book.

7.4.1 The Importance of Unentangling Temporary Qubits in Quantum Subroutines

Chapter 6, when describing the constructions of section 6.2, emphasized the importance of uncomputing temporarily used bits to conserve space in classical computations. In quantum computation, uncomputing qubits used temporarily as part of subroutines is crucial even when conserving space and reusing qubits is not an issue; failing to uncompute temporary qubits can result in entanglement between the computational qubits and the temporary qubits, and in this way can destroy the calculation. More specifically, if a subroutine claims to compute state $\sum_i \alpha_i |x_i\rangle$, it is not okay if it actually computes $\sum_i \alpha_i |x_i\rangle |y_i\rangle$ and throws away the qubits storing $|y_i\rangle$ unless there is no entanglement between the two registers. There is no entanglement if $\sum_i \alpha_i |x_i\rangle |y_i\rangle = \left(\sum_i \alpha_i |x_i\rangle \right) \otimes |y_i\rangle$, which can happen only if $|y_i\rangle = |y_j\rangle$ for all i and j. In general, the states $\sum_i \alpha_i |x_i\rangle$ and $\sum_i \alpha_i |x_i\rangle |y_i\rangle$ behave quite differently, even if we have access only to the first register of the second state. Chapter 10, which discusses quantum subsystems, provides the means of talking about the differences between these two situations without looking at the consequences for computation. In this section, we illustrate the difference by showing how using the first state when expecting the second can mess up computation. In particular, we show that if we replace the black box U_f used in Deutsch's problem with the black box for V_f that outputs

$$V_f : |x, t, y\rangle \rightarrow |x, t \oplus x, y \oplus f(x)\rangle,$$

Deutsch's algorithm no longer works.

Begin with qubit $|t\rangle$ in the state $|0\rangle$ and, as before, the first qubit in the state $|+\rangle$, and the third qubit in the state $|-\rangle$. Apply V_f to obtain

$$V_f\left(|+\rangle|0\rangle|-\rangle\right) = V_f\left(\frac{1}{\sqrt{2}}\sum_{x=0}^{1}|x\rangle|0\rangle|-\rangle\right) = \frac{1}{\sqrt{2}}\sum_{x=0}^{1}(-1)^{f(x)}|x\rangle|x\rangle|-\rangle.$$

The first qubit is now entangled with the second qubit. Because of this entanglement, applying H to the first qubit and then measuring it no longer has the desired effect. For example, when f is constant, the state is $(|00\rangle + |11\rangle)|-\rangle$, and applying $H \otimes I \otimes I$ results in the state

$$\frac{1}{2}(|00\rangle + |10\rangle + |01\rangle - |11\rangle)|-\rangle.$$

The second and fourth terms canceled before, but they do so no longer. Now there is an equal chance of measuring the first qubit as $|0\rangle$ or $|1\rangle$. A similar calculation shows that when the function is not constant, there is also an equal chance of measuring the first qubit as $|0\rangle$ or $|1\rangle$. Thus, we can no longer distinguish the two cases. Entanglement with the qubit $|t\rangle$ has destroyed the quantum computation.

Had V_f properly uncomputed t so that at the end of the calculation it was in state $|0\rangle$, the algorithm would still work properly. For example, for f constant, we would have state

$$\frac{1}{2}(|00\rangle + |10\rangle + |00\rangle - |10\rangle)|-\rangle$$

in which case the appropriate terms would cancel to yield

$$(|00\rangle)|-\rangle.$$

If a quantum subroutine claims to produce a state $|\psi\rangle$, it must not produce a state that *looks like* $|\psi\rangle$ but is entangled with other qubits. In particular, if a subroutine makes use of other qubits, by the end of the subroutine these qubits must not be entangled with the other qubits. For this reason, the following quantum subroutines are careful to *uncompute* any auxiliary qubits so that at the end of the algorithm they are always in state $|0\rangle$.

7.4.2 Phase Change for a Subset of Basis Vectors

Aim   Change the phase of terms in a superposition $|\psi\rangle = \sum a_i|i\rangle$ depending on whether i is in a subset X of $\{0, 1, \ldots, N-1\}$ or not. More specifically, we wish to find an efficient implementation of the quantum transformation

$$S_X^\phi : \sum_{x=0}^{N-1} a_x|x\rangle \to \sum_{x\in X} a_x e^{i\phi}|x\rangle + \sum_{x\notin X} a_x|x\rangle.$$

Section 5.4 explained how to realize an arbitrary unitary transformation without regard to efficiency. Applying that algorithm blindly would give an implementation of S_X^ϕ using more than $N = 2^n$ simple gates. This section shows how, for any efficiently computable subset X, the transformation S_X^ϕ can be implemented efficiently. An efficiently implementable S_X^ϕ is used in some of the quantum algorithms we describe later that outperform classical ones.

We can hope to implement S_X^ϕ efficiently only if there is an efficient algorithm for computing membership in X: the Boolean function $f : \mathbf{Z}_{2^n} \to \mathbf{Z}_2$, where

$$f(x) = \begin{cases} 1 & \text{if } x \in X \\ 0 & \text{otherwise} \end{cases}$$

must be efficiently computable, say polynomial in n. Most subsets X do not have this property. For subsets X with this property, the main result of chapter 6 implies that there is an efficient quantum circuit for U_f. Given such an implementation for U_f, we can compute S_X^ϕ using a few additional steps. We use U_f to compute f in a temporary qubit, use the value in that qubit to effect the phase change, and then uncompute f in order to remove any entanglement between the temporary qubit and the rest of the state.

define $Phase_f(\phi)|x[k]\rangle =$

     qubit $a[1]$ a temporary bit   (1)
     $U_f|x, a\rangle$ compute f in a (2)
     $K(\phi/2)|a\rangle$ (3)
     $T(-\phi/2)|a\rangle$ (4)
     $U_f^{-1}|x, a\rangle$ uncompute f (5)

Since

$$T(-\phi/2)K(\phi/2) = \begin{pmatrix} 1 & 0 \\ 0 & e^{i\phi} \end{pmatrix},$$

where K and T are the single-qubit operations introduced in 5.4.1, together steps (3) and (4) shift the phase by $e^{i\phi}$ if and only if bit a is one. Strictly speaking, we do not need to do step (3) at all since it is a physically meaningless global phase shift: performing step (3) merely makes it easier to see that we get the desired result. Alternatively, we could replace steps (3) and (4) by a single step $\bigwedge_1 K(\phi)|a\rangle|x_i\rangle$, where i can be any of the qubits in register x, since placing a phase in any term of the tensor product is the same as placing it in any other term. We need to uncompute U_f in step (5) to remove the entanglement between register $|x\rangle$ and the temporary qubit so that $|x\rangle$ ends up in the desired state, no longer entangled with the temporary qubits.

Special case $\phi = \pi$ The important special case $\phi = \pi$ has an alternative, surprisingly simple implementation that generalizes the trick used in the algorithm for Deutsch's problem. Given U_f as above, the transformation S_X^π can be implemented by initializing a temporary qubit b to $|-\rangle = \frac{1}{\sqrt{2}}(|0\rangle - |1\rangle)$, and then using U_f to compute into this register: consider $|\psi\rangle = \sum_{x \in X} a_x|x\rangle + \sum_{x \notin X} a_x|x\rangle$, and compute

$$U_f(|\psi\rangle \otimes |-\rangle) = U_f\left(\sum_{x \in X} a_x |x\rangle \otimes |-\rangle\right) + U_f\left(\sum_{x \notin X} a_x |x\rangle \otimes |-\rangle\right)$$

$$= -\left(\sum_{x \in X} a_x |x\rangle \otimes |-\rangle\right) + \left(\sum_{x \notin X} a_x |x\rangle \otimes |-\rangle\right)$$

$$= (S_X^\pi |\psi\rangle) \otimes |-\rangle.$$

In particular, the following circuit, acting on the n-qubit state $|0\rangle$ together with an ancilla qubit in state $|1\rangle$ creates the superposition $|\psi_X\rangle = \sum(-1)^{f(x)}|x\rangle$:

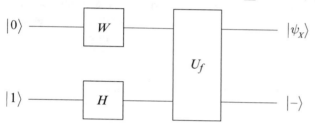

For elegance, and to be able to reuse the ancilla qubit, we may want to apply a final Hadamard transformation to the ancilla qubit, in which case the circuit is

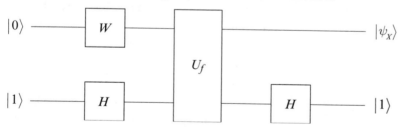

Geometrically, when acting on the N-dimensional vector space associated with the quantum system, the transformation S_X^π is a reflection about the $N - k$-dimensional hyperplane perpendicular to the k-dimensional hyperplane spanned by $\{|x\rangle | x \in X\}$: a reflection in a hyperplane sends any vector $|v\rangle$ perpendicular to the hyperplane to its negative $-|v\rangle$. For any unitary transformation U, the transformation

$$U S_X^\pi U^{-1}$$

is a reflection in the hyperplane perpendicular to the hyperplane spanned by the vectors $\{U|x\rangle | x \in X\}$. Section 9.2.1 uses this geometric view of S_X^π to build intuition for Grover's algorithm.

We can write the result of applying S_X^π to the superposition $W|0\rangle$ as

$$\frac{1}{\sqrt{N}} \sum (-1)^{f(x)} |x\rangle$$

where f is the Boolean function for membership in X,

$$f(x) = \begin{cases} 1 & \text{if } x \in X \\ 0 & \text{otherwise.} \end{cases}$$

Conversely, given a Boolean function f, we define S_f^π to be S_X^π where $X = \{x | f(x) = 1\}$.

7.4.3 State-Dependent Phase Shifts

Section 7.4.2 explained how to change efficiently the phase of all terms in a superposition corresponding to certain subsets of basis elements, but that construction performs the same phase change on all of those terms. This section considers the problem of implementing different phase shifts in different terms; we wish to implement transformations in which the amount of the phase shift depends on the quantum state.

Aim Efficiently approximate to accuracy s the transformation on n-qubits that changes the phase of the basis elements by

$$|x\rangle \rightarrow e^{i\phi(x)} |x\rangle$$

where the function $\phi(x)$ that describes the desired phase shift angle ϕ for each term x has an associated function $f : \mathbf{Z}_n \rightarrow \mathbf{Z}_s$ that is efficiently computable, and the value of the ith bit of $f(x)$ is the ith term in the following binary expansion for $\phi(x)$:

$$\phi(x) \approx 2\pi \frac{f(x)}{2^s}.$$

The implementation can be only as efficient as the function f. Given a quantum circuit that efficiently implements U_f, we can perform the state-dependent phase shift in $O(s)$ steps in addition to 2 uses of U_f. The ability to compute f efficiently is a strong one: most functions do not have this property.

This paragraph shows how to implement efficiently the subprogram that changes the phase of an s-qubit standard basis state $|x\rangle$ by the angle $\phi(x) = \frac{2\pi x}{2^s}$. Let

$$P(\phi) = T(-\phi/2) K(\phi/2) = \begin{pmatrix} 1 & 0 \\ 0 & e^{i\phi} \end{pmatrix}$$

be the transformation that shifts the phase in a qubit if that bit is 1 but does nothing if that bit is 0. The program

define *Phase* $|a[s]\rangle =$
    for $i \in [0 \ldots s-1]$
        $P(\frac{2\pi}{2^i})|a_i\rangle$

performs the s-qubit transformation *Phase* : $|a\rangle \rightarrow \exp(i2\pi \frac{a}{2^s})|a\rangle$.

The *Phase* program is used as a subroutine in the following program that implements the n-qubit transformation $Phase_f : |x\rangle \rightarrow \exp(2\pi i \frac{f(x)}{2^s})|x\rangle$:

define $Phase_f|x[k]\rangle =$

qubit $a[s]$	an s-bit temporary register	(1)
$U_f|x\rangle|a\rangle$	compute f in a	(2)
$Phase \ |a\rangle$	perform phase shift by $2\pi a/2^s$	(3)
$U_f^{-1}|x\rangle|a\rangle$	uncompute f	(4)

After step (2), register a is entangled with x and contains the binary expansion of the angle $\phi(x)$ for the desired phase shift for the basis vector $|x\rangle$. Since registers a and x are entangled, changing the phase in register a during step (3) is equivalent to changing the phase in register x. Step (4) uncomputes U_f to remove this entanglement so that the contents of register x end up in the desired state, no longer entangled with the temporary qubits.

7.4.4 State-Dependent Single-Qubit Amplitude Shifts

Aim Efficiently approximate, to accuracy s, rotating each term in a superposition by a single-qubit rotation $R(\beta(x))$ (see Section 5.4.1), where the angle $\beta(x)$ depends on the quantum state in another register. More specifically, we wish to implement a transformation that takes

$$|x\rangle \otimes |b\rangle \rightarrow |x\rangle \otimes (R(\beta(x))|b\rangle),$$

where $\beta(x) \approx f(x)\frac{2\pi}{2^s}$ and the approximating function $f : \mathbf{Z}_n \rightarrow \mathbf{Z}_s$ is efficiently computable.

From an efficient implementation of U_f, we can implement this transformation in $O(s)$ steps plus two calls to U_f. The subroutine uses an auxiliary transformation Rot that shifts the amplitude in qubit b by the amount specified in register a; $|a\rangle \otimes |b\rangle \rightarrow |a\rangle \otimes \left(R(a\frac{2\pi}{2^s})|b\rangle\right)$ where the contents of the s-qubit register a give the angle by which to rotate up to accuracy 2^{-s}. Figure 7.1 shows a circuit that implements Rot. Using our program notation, this transformation can be described more concisely by

define $Rot \ |a[s]\rangle|b[1]\rangle =$
    for $i \in [0 \ldots s-1]$
        $|a_i\rangle$ **control** $R(\frac{2\pi}{2^i})|b\rangle.$

The desired rotation specified by the function f can be achieved by the program

define $Rot_f|x[k]\rangle|b[1]\rangle =$

qubit $a[s]$	an s-bit temporary register
$U_f|x\rangle|a\rangle$	compute f in a
$Rot \ |a, b\rangle$	perform rotation by $2\pi a/2^s$
$U_f^{-1}|x\rangle|a\rangle$	uncompute f

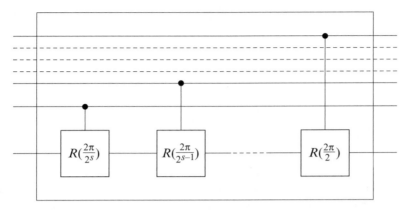

Figure 7.1
Circuit for controlled rotation.

7.5 A Few Simple Quantum Algorithms

This section presents a few simple quantum algorithms. The first three problems are *black box* or *oracle* problems for which the quantum algorithm's query complexity is better than the query complexity of any conceivable classical algorithm. The fourth is a problem for which the communication complexity of a quantum protocol is better than the communication complexity for any possible classical one. Like Deutsch's problem, the problems are a bit artificial, but they have relatively simple quantum algorithms that can be proved to be more efficient than any possible classical approach. Like Deutsch's algorithm, these algorithms solve these problems with certainty.

7.5.1 Deutsch-Jozsa Problem
David Deutsch and Richard Jozsa present a quantum algorithm for the following problem, a multiple bit generalization of Deutsch's problem of section 7.3.1.

Deutsch-Jozsa Problem A function f is *balanced* if an equal number of input values to the function return 0 and 1. Given a function $f : \mathbf{Z}_{2^n} \mapsto \mathbf{Z}_2$ that is known to be either constant or balanced, and a quantum oracle $U_f : |x\rangle|y\rangle \rightarrow |x\rangle|y \oplus f(x)\rangle$ for f, determine whether the function f is constant or balanced.

The algorithm begins by using the phase change subroutine of section 7.4.2 to negate terms of the superposition corresponding to basis vectors $|x\rangle$ with $f(x) = 1$: the subroutine returns the state

$$|\psi\rangle = \frac{1}{\sqrt{N}} \sum_{i=0}^{N-1} (-1)^{f(i)} |i\rangle.$$

(The subroutine uses a temporary qubit in state $|-\rangle$. Just as section 7.4.1 showed for Deutsch's algorithm, it is vital that the subroutine end with that qubit unentangled with any other qubits, so that it can be safely ignored.) Next, apply the Walsh transform W to the resulting state $|\psi\rangle$ to obtain

$$|\phi\rangle = \frac{1}{N}\sum_{i=0}^{N-1}\left((-1)^{f(i)}\sum_{j=0}^{N-1}(-1)^{i\cdot j}|j\rangle\right).$$

For constant f, the $(-1)^{f(i)} = (-1)^{f(0)}$ is simply a global phase and the state $|\phi\rangle$ is simply $|0\rangle$:

$$(-1)^{f(0)}\frac{1}{2^n}\sum_{j\in\mathbf{Z}_2^n}\left(\sum_{i\in\mathbf{Z}_2^n}(-1)^{i\cdot j}\right)|j\rangle = (-1)^{f(0)}\frac{1}{2^n}\sum_{i\in\mathbf{Z}_2^n}(-1)^{i\cdot 0}|0\rangle = (-1)^{f(0)}|0\rangle$$

because, as box 7.1 shows, $\sum_{i\in\mathbf{Z}_2^n}(-1)^{i\cdot j} = 0$ for $j\neq 0$. For f balanced

$$|\phi\rangle = \frac{1}{2^n}\sum_{j\in\mathbf{Z}_2^n}\left(\sum_{i\in X_0}(-1)^{i\cdot j} - \sum_{i\notin X_0}(-1)^{i\cdot j}\right)|j\rangle$$

where $X_0 = \{x | f(x) = 0\}$. This time, for $j = 0$, the amplitude is zero: $\sum_{j\in X_0}(-1)^{i\cdot j} - \sum_{j\notin X_0}(-1)^{i\cdot j} = 0$. Thus, measurement of state $|\phi\rangle$ in the standard basis will return $|0\rangle$ with probability 1 if f is constant and will return a non-zero $|j\rangle$ with probability 1 if f is balanced.

This quantum algorithm solves the Deutsch-Jozsa problem with a single evaluation of U_f, while any classical algorithm must evaluate f at least $2^{n-1} + 1$ times to solve the problem with certainty. Thus, there is an exponential separation between the query complexity of this quantum algorithm and the query complexity for any possible classical algorithm that solves that problem with certainty. There are, however, classical algorithms that solve this problem in fewer evaluations, but only with high probability of success. (See exercise 7.4.)

7.5.2 Bernstein-Vazirani Problem

The problem is to determine the value of an unknown bit string u of length n where one is allowed only queries of the form $q \cdot u$ for some query string q. The best classical algorithm uses $O(n)$ calls to $f_u(q) = q \cdot u \bmod 2$. A quantum algorithm, closely related to the algorithm we just gave for the Deutsch-Jozsa problem, can find u in just a single call to U_{f_u}: on a quantum computer it is possible to determine u exactly with a single query (in superposition). Let $f_u(q) = q \cdot u \bmod 2$ and

$$U_{f_u}: |q\rangle|b\rangle \mapsto |q\rangle|b \oplus f_u(q)\rangle.$$

The following circuit (figure 7.2) solves this problem with certainty using only one call to U_{f_u}. To understand how this circuit works, recall from section 7.4.2 that in the special case $\phi = \pi$, the phase change subroutine can be accomplished by the circuit

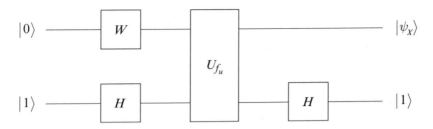

In this case, applying this circuit results in the state

$$|\psi_X\rangle = \frac{1}{\sqrt{2^n}} \sum_q (-1)^{f_u(q)} |q\rangle = \frac{1}{\sqrt{2^n}} \sum_q (-1)^{u \cdot q} |q\rangle$$

in the first register. The next paragraph shows that applying the Walsh-Hadamard transformation W to this state produces the state $|u\rangle$.

Recall that $W|x\rangle = \frac{1}{\sqrt{2^n}} \sum_z (-1)^{x \cdot z} |z\rangle$. Thus

$$W|\psi_X\rangle = W \left(\frac{1}{\sqrt{2^n}} \sum_q (-1)^{u \cdot q} |q\rangle \right)$$

$$= \frac{1}{\sqrt{2^n}} \sum_q (-1)^{u \cdot q} W|q\rangle$$

$$= \frac{1}{2^n} \sum_q (-1)^{u \cdot q} \left(\sum_z (-1)^{q \cdot z} |z\rangle \right).$$

A fact from box 7.1 tells us that $(-1)^{u \cdot q + z \cdot q} = (-1)^{(u \oplus z) \cdot q}$. Furthermore, equation 7.1 tells us that the internal sum is 0 unless $u \oplus z = 0$, which implies that the only term that remains is the $u = z$ term. Thus,

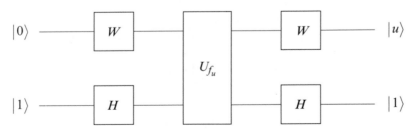

Figure 7.2
Circuit for the Bernstein-Vazirani algorithm.

$$W(|\psi_X\rangle) = \frac{1}{2^n} \sum_z \left(\sum_q (-1)^{u \cdot q + z \cdot q} \right) |z\rangle$$

$$= |u\rangle.$$

Thus measurement in the standard basis gives $|u\rangle$ with certainty.

A Simpler Explanation   Using quantum parallelism to compute on all possible inputs at the same time, and then cleverly manipulate the resulting superposition, is a common explanation for how quantum algorithms work. The description we gave for the Bernstein-Vazirani algorithm fits this framework. There is a question, however, as to whether quantum parallelism is the *right* way of looking at algorithms. To illustrate this point, we give an alternative description, due to Mermin, of exactly this same algorithm.

The key to Mermin's explanation of the algorithm is to look at the circuit in the Hadamard basis. To understand what the quantum black box for U_{f_u} does in the Hadamard basis, recognize that it behaves *as if* it contained a circuit consisting of C_{not} operations from some of the qubits to the ancilla qubit: this circuit contains a C_{not} from qubit i to the ancilla if and only if the ith bit of u is 1 (see figure 7.3). Recall from section 5.2.4 that Hadamard operations reverse the control and target roles of the qubits:

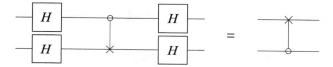

The Bernstein-Vazirani algorithms consists of starting with the state $|0\ldots0\rangle|1\rangle$ and applying Hadamard transformations to every qubit before and after the call to the black box for U_{f_u} (see figure 7.2.) Thus, the Bernstein-Vazirani algorithm behaves as if it were a circuit consisting only of C_{not} operations from the ancilla qubit to the qubits corresponding to 1-bits of u. (See figure

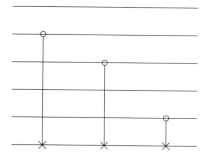

Figure 7.3
For $u = 01101$, the black box for U_{f_u} behaves *as if* it contained this circuit, consisting of C_{not} gates for each 1-bit of u.

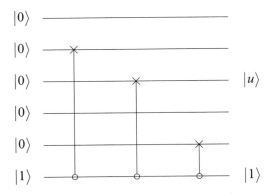

Figure 7.4
For $u = 01101$, the Bernstein-Vazirani algorithm behaves *as if* it were implemented by this simple circuit consisting of a C_{not} for each 1-bit of u.

7.4.) From this view of the circuit, it is immediate that the qubits end up in the state $|u\rangle$, so this much simpler explanation, which does not speak of quantum parallelism or of "computing on all possible inputs," is the *right* way to look at the algorithm.

7.5.3 Simon's Problem

Simon's problem: Given a 2-to-1 function f such that $f(x) = f(x \oplus a)$ for all $x \in \mathbf{Z}_2^n$, find hidden string $a \in \mathbf{Z}_2^n$.

Simon describes a quantum algorithm that can find a in only $O(n)$ calls to U_f, followed by $O(n^2)$ additional steps, whereas the best a classical algorithm that can do is $O(2^{n/2})$ calls to f. Simon's algorithm suggested to Shor an approach to the factoring problem that is now famous as Shor's algorithm. As we will see in chapter 8, there are structural similarities between Shor's algorithm and Simon's algorithm.

To determine a, create the superposition $\sum_x |x\rangle |f(x)\rangle$. Measuring the right part of the register projects the state of the left register to $\frac{1}{\sqrt{2}}(|x_0\rangle + |x_0 \oplus a\rangle)$, where $f(x_0)$ is the measured value. Applying the Walsh-Hadamard transformation W leads to

$$
W\left(\frac{1}{\sqrt{2}}(|x_0\rangle + |x_0 \oplus a\rangle) \right)
$$

$$
= \frac{1}{\sqrt{2}} \left(\frac{1}{\sqrt{2^n}} \sum_y ((-1)^{x_0 \cdot y} + (-1)^{(x_0 \oplus a) \cdot y}) |y\rangle \right)
$$

$$
= \frac{1}{\sqrt{2^{n+1}}} \sum_y (-1)^{x_0 \cdot y} (1 + (-1)^{a \cdot y}) |y\rangle
$$

$$
= \frac{2}{\sqrt{2^{n+1}}} \sum_{y \cdot a \text{ even}} (-1)^{x_0 \cdot y} |y\rangle.
$$

Measurement of this state results in a random y such that $y \cdot a = 0 \bmod 2$, so the unknown bits a_i of a must satisfy the equation $y_0 \cdot a_0 \oplus \cdots \oplus y_{n-1} \cdot a_{n-1} = 0$. This computation is repeated until n linearly independent equations have been found. Each time the computation is repeated, the resulting equation has at least a 50 percentage chance of being linearly independent of the previous equations obtained. After repeating the computation $2n$ times, there is a 50 percentage chance that n linearly independent equations have been found. These equations can be solved to find a in $O(n^2)$ steps. Thus, with high likelihood, the hidden string a will be found with $O(n)$ calls to U_f, followed by $O(n^2)$ steps to solve the resulting set of equations.

7.5.4 Distributed Computation

This section describes a different type of quantum algorithm, one for which communication complexity is the concern. Like dense coding and teleportation, it uses entangled pairs that can be distributed ahead of time, independent of the computation, so these qubits are not counted as qubits transmitted during the solution of the problem (though the exponential savings would remain even if they were counted).

The Problem Let $N = 2^n$. Alice and Bob are each given an N-bit number, u and v respectively. The objective is for Alice to compute an n-bit number a and Bob to compute an n-bit number b such that

$$d_H(u, v) = 0 \rightarrow a = b$$

$$d_H(u, v) = N/2 \rightarrow a \neq b$$

$$\text{else} \rightarrow \text{no condition on } a \text{ and } b$$

where $d_H(u, v)$ is the Hamming distance between u and v. In other words, Alice and Bob need an algorithm that produces a and b from any u and v such that if $u = v$, then $a = b$; if u and v differ in half of their bits, then $a \neq b$; and if the Hamming distance of u and v is anything other than 0 or $N/2$, a and b can be anything.

This problem is nontrivial because u and v are exponentially larger than a and b. Given a sufficient supply of entangled pairs, this problem can be solved without additional communication between Alice and Bob, while a classical solution requires communication of at least $N/2$ bits between the two parties.

Suppose Alice and Bob share n entangled pairs of particles (a_i, b_i), each in state $\frac{1}{\sqrt{2}}(|00\rangle + |11\rangle)$, where Alice can access particles a_i and Bob can access particles b_i. We write the state of the $2n$ particles making up these n entangled pairs in order $a_0, a_1, \ldots, a_{n-1}, b_0, b_1, \ldots, b_{n-1}$, so the entire $2n$-qubit state is written $\frac{1}{\sqrt{N}} \sum_{i=0}^{N-1} |i, i\rangle$, where Alice can manipulate the first n qubits and Bob can manipulate the last n qubits.

The problem can be solved without additional communication as follows. Using the phase change subroutine of section 7.4.2, with $f(i) = u_i$, Alice performs $\sum |i\rangle \rightarrow \sum (-1)^{u_i} |i\rangle$ followed by the Walsh transform W on her n qubits. Bob performs the same computation on his n qubits using $f(i) = v_i$. Together their particles are now in the common global state

$$|\psi\rangle = W\left(\frac{1}{\sqrt{N}}\sum_{i=0}^{N-1}(-1)^{u_i\oplus v_i}|i\rangle|i\rangle\right).$$

Alice and Bob now measure their respective part of the state to obtain results a and b. We need to show that a and b have the desired properties.

The probability that measurement results in $a = x = b$ is $|\langle x, x|\psi\rangle|^2$. We wish to show that this probability is 1 if $u = v$ and 0 if $d_H(u, v) = N/2$. Let us simplify the state as follows, where the superscript in $W^{(l)}$ indicates that W is acting on an l qubit state:

$$|\psi\rangle = W^{(2n)}\frac{1}{\sqrt{N}}\sum_{i=0}^{N-1}(-1)^{u_i\oplus v_i}|i\rangle|i\rangle$$

$$= \frac{1}{\sqrt{N}}\sum_{i=0}^{N-1}(-1)^{u_i\oplus v_i}(W^{(n)}|i\rangle\otimes W^{(n)}|i\rangle)$$

$$= \frac{1}{N\sqrt{N}}\sum_{i=0}^{N-1}\sum_{j=0}^{N-1}\sum_{k=0}^{N-1}(-1)^{u_i\oplus v_i}(-1)^{i\cdot j}(-1)^{i\cdot k}|jk\rangle.$$

Now

$$\langle x, x|\psi\rangle = \frac{1}{N\sqrt{N}}\sum_{i=0}^{N-1}(-1)^{u_i\oplus v_i}(-1)^{i\cdot x}(-1)^{i\cdot x} = \frac{1}{N\sqrt{N}}\sum_{i=0}^{N-1}(-1)^{u_i\oplus v_i}.$$

If $u = v$, then $(-1)^{u_i\oplus v_i} = 1$ and $\langle x, x|\psi\rangle = \frac{1}{\sqrt{N}}$, so the probability $|\langle x, x|\psi\rangle|^2 = \frac{1}{N}$. The probability, summed over the N possible values of x, is 1, so when Alice and Bob measure they obtain, with probability 1, states a and b with $a = b = x$ for some bit string x. For $d_H(u, v) = N/2$, the sum $\langle x, x|\psi\rangle = \frac{1}{N\sqrt{N}}\sum_{i=0}^{N-1}(-1)^{u_i\oplus v_i}$ has an equal number of $+1$ and -1 terms, which cancel to give $\langle x, x|\psi\rangle = 0$. Thus, in this case, Alice and Bob measure the same value with probability 0.

7.6 Comments on Quantum Parallelism

Because quantum parallelism's role in quantum computation has often been misunderstood, we make a few comments to address some common misconceptions. The notation

$$\frac{1}{\sqrt{N}}\sum_{x=0}^{N-1}|x, f(x)\rangle$$

suggests that exponentially more computation is being done by the quantum operation U_f acting on the superposition $\sum_x |x, 0\rangle$ than by a classical computer computing $f(x)$ from x. The next paragraph explains how this view is misleading and how it does not explain the power of quantum computation. Similarly, the exponential size of the n-qubit quantum state space may seem to suggest that an exponential speedup over the classical case can always be obtained using quantum parallelism. This statement is generally incorrect, although in certain special cases quantum computation does provide such speedups. We elaborate briefly on each of these statements.

As explained in section 7.1.2, only one input/output pair can be extracted by measurement in the standard basis from the superposition generated quantum parallelism. It is not possible to extract more input/output pairs in any other way since, as section 4.3.1 explained, only m bits of information can be extracted from an m-qubit state. Thus, while the 2^n values of $f(x)$ appear in the single superposition state, it still takes 2^n computations of U_f to obtain them all, no better than the classical case. This limitation leaves open the possibility that any classical algorithm that takes 2^n steps to obtain n bits of output could be done in a single step on a quantum computer. While some algorithms do give speedups of this magnitude over classical algorithms, the optimality of Grover's algorithm proved in chapter 9.1 shows that there are problems of this form for which it is known that no quantum algorithm can provide an exponential speedup. Furthermore, *lower bound* results exist that show that for many problems quantum computation cannot provide any speedup at all. Thus, quantum parallelism and quantum computation do not, in general, provide the exponential speedup suggested by the notation.

Furthermore, a superposition like $\frac{1}{\sqrt{N}} \sum |x, f(x)\rangle$ is still only a single state of the quantum state space. The n-qubit quantum state space is extremely large, so large that the vast majority of states cannot even be approximated by an efficient quantum algorithm. (The elegant proof goes beyond the scope of this book. A reference is given in section 7.9.) Thus, an efficient quantum algorithm cannot even come close to most states in the state space. For this reason, quantum parallelism does not, and efficient quantum algorithms cannot, make use of the full state space.

As Mermin's explanation of the Bernstein-Vazirani algorithm of section 7.5.2 illustrates, even when quantum parallelism can be used to describe an algorithm, it is not necessarily correct to view it as key to the algorithm. Understanding where the power of quantum computation comes from remains an open research question. The status of entanglement as one of the keys will be discussed in the introduction to chapter 10 and in section 13.9, which addresses this question explicitly.

When algorithms are described in terms of quantum parallelism, the heart of the algorithm is the way in which the algorithm manipulates the state generated by quantum parallelism. This sort of manipulation has no classical analog and requires nontraditional programming techniques. We list a couple of general techniques:

- **Amplify output values of interest.**   The general idea is to transform the state in such a way that values of interest have a larger amplitude and therefore have a higher probability of being measured. Grover's algorithm of chapter 9 exploits this approach, as do the many closely related algorithms.

- **Find properties of the set of all the values of** *f(x)***.**   This idea is exploited in Shor's algorithm of chapter 8, which uses a quantum Fourier transformation to obtain the period of f. The algorithms given in section 7.5 for the Deutsch-Jozsa problem, the Bernstein-Vazirani problem, and Simon's problem all take this approach.

7.7 Machine Models and Complexity Classes

Computational complexity classes are defined in terms of a *language* and *machines* that recognize that language. In this section, the term *machine* refers to any quantum or classical computing device that runs a single algorithm on which we can count the number of computation steps and storage cells used. A *language L* over an alphabet Σ is a subset of the finite strings Σ^* of elements from Σ. A language L is recognized by a machine M if, for each string $x \in \Sigma^*$, the machine M can *determine* if $x \in L$. Exactly what *determine* means depends on the kind of machine we are considering. For example, given input x, a classical deterministic machine may answer *Yes, x ∈ L*, or *No, x ∉ L*, or it may never halt. Probabilistic and quantum machines might answer *Yes* or *No* correctly with certain probabilities. We consider five kinds of classical machines, deterministic (**D**), nondeterministic (**N**), randomized (**R**), probabilistic (**Pr**), and bounded probability of error (**BP**). Each of these types of classical machine has a quantum analog (**EQ, NQ, RQ, PrQ, BQ**). Of particular interest will be quantum deterministic (exact) machines (**EQ**), and quantum bounded probability of error machines (**BQ**). Section 7.7.1 uses these types of machine to define numerous complexity classes of varying resource constraints. We now more rigorously describe exactly how the different kinds of machines recognize a language.

For each kind of machine M, there is a single language L_M that M recognizes. For example, a machine is *deterministic* if whenever it sometimes answers *Yes* on a given input x it always answers *Yes* on that input. A deterministic machine D recognizes the language

$$L_D = \{x \in \Sigma^* | D(x) = Yes\} = \{x | P(D(x) = Yes) = 1\}.$$

By definition of deterministic, for all $x \notin L$, the probability $P(D(x) = Yes)$ is zero. As a second example, a *bounded probability of error* machine, acting on a given input x, either answers *Yes* with probability at least $1/2 + \epsilon$ or with probability no more than $1/2 - \epsilon$. Given a bounded probability of error machine **BP**, $L_{BP} = \{x | P(BP(x) = Yes) \geq 1/2 + \epsilon\}$. For $x \notin L_{BP}$, $P(BP(x) = Yes) \leq 1/2 - \epsilon$.

A machine may not give an answer at all for some inputs. Table 7.1 summarizes the conditions for the various types of machines we consider. The quantum machine types recognize a language with the same probability as their classical counterparts. Figure 7.5 illustrates containment relations between the kinds of machines. Containment means that by definition each **D** machine, for example, is also an **R** machine.

A language is recognized by a kind of machine if there exists a machine of that kind that recognizes it. The set of languages recognized by the types of machines we have defined does not depend on the particular value of ϵ. For example, suppose we are given a **Pr** machine M that answers *Yes* for $x \in L$ with probability $P(x \in L) > \frac{1}{2} + \epsilon$. We can construct a new **Pr** machine M' that runs M three times and answers *Yes* if M answers *Yes* at least two times. Then M' will accept

Table 7.1
Probability for a particular kind of machine to answer *Yes* when given an input x that is or is not and element of language L

Prefix	Kind of machine	$P(x \in L)$	$P(x \notin L)$
Classical			
D	Deterministic	$= 1$	$= 0$
N	Nondeterministic	> 0	$= 0$
R	Randomized (Monte Carlo)	$> \frac{1}{2} + \epsilon$	$= 0$
Pr	Probabilistic	$> \frac{1}{2}$	$\leq \frac{1}{2}$
BP	Bounded probability of error	$> \frac{1}{2} + \epsilon$	$\leq \frac{1}{2} - \epsilon$
Quantum			
EQ	Quantum deterministic (exact)	$= 1$	$= 0$
BQ	Quantum bounded probability of error	$> \frac{1}{2} + \epsilon$	$\leq \frac{1}{2} - \epsilon$

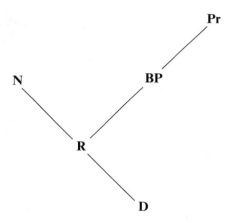

Figure 7.5
Containment relations between kinds of machines. These relations hold for classical and quantum machines, and for time and space complexity.

$x \in L$ with probability $> \frac{1}{2} + \frac{3}{2}\epsilon - \epsilon^3$. Some authors use a fixed value such as $\epsilon = 1/4$. The case $P(x \in L) > 1/2$ is quite different from $P(x \in L) > 1/2 + \epsilon$, however, since in the former case no polynomial number of repetitions can guarantee an increase in the success probability above a given threshold $\frac{1}{2} + \epsilon$.

7.7.1 Complexity Classes

In addition to being concerned about the probability that a machine answer correctly, complexity theory is concerned about quantifying the amount of resources, particularly time and space, that a machine uses to obtain its answers. A machine recognizes a language L in time $O(f)$ if, for any string $x \in \Sigma^*$ of length n, it answers *Yes* or *No* within $t(n)$ steps and $t \in O(f)$. A machine recognizes a language L in space $O(f)$ if, for any string $x \in \Sigma^*$ of length n, it answers *Yes* or *No* using at most $s(n)$ storage units, measured in bits or qubits, where $s \in O(f)$.

A *complexity class* is the set of languages recognized by a particular kind of machine within given resource bounds. Specifically, for $\mathbf{m} \in \{\mathbf{D}, \mathbf{EQ}, \mathbf{N}, \mathbf{R}, \mathbf{Pr}, \mathbf{BP}\}$, we consider the classes $\mathbf{mTime}(f)$ and $\mathbf{mSpace}(f)$. Language L is in complexity class $\mathbf{mTime}(f)$ if there exists a machine M of kind $\mathbf{m}$ that recognizes L in time $O(f)$. Language L is in complexity class $\mathbf{mSpace}(f)$ if there exists a machine M of kind $\mathbf{m}$ that recognizes L in space $O(f)$.

We are particularly interested in machines that use only a polynomial amount of resources, and to a lesser extent in those that use only an exponential amount. For example, we are interested in the class $\mathbf{P} = \mathbf{DTime}(n^k)$ of machines that respond to an input of length n using only $O(n^k)$ time for some k. The following shorthand notations are common:

P	$\mathbf{DTime}(n^k)$
EQP	$\mathbf{EQTime}(n^k)$
NP	$\mathbf{NTime}(n^k)$
R	$\mathbf{RTime}(n^k)$
PP	$\mathbf{PrTime}(n^k)$
BPP	$\mathbf{BPTime}(n^k)$
BQP	$\mathbf{BQTime}(n^k)$
PSpace	$\mathbf{DSpace}(n^k)$
NPSpace	$\mathbf{NSpace}(n^k)$
EXP	$\mathbf{DTime}(k^n)$

For time classes, we can assume that machines always halt because the function f provides an upper bound on the possible runtimes. However, machines in the space complexity classes may never halt on some inputs. Therefore, we define $\mathbf{m}_H \mathbf{Space}(f)$ to be the class of languages that are recognized by a halting machine of type $\mathbf{m}$ in space $O(f)$. Obviously, $\mathbf{m}_H \mathbf{Space}(f) \subseteq \mathbf{mSpace}(f)$. Note that in the circuit model all computations will terminate. Analysis of the complexity of nonhalting space classes requires a different model of computation, such as quantum Turing machines.

7.7.2 Complexity: Known Results
We give informal arguments for some of the containment relations involving quantum complexity classes. Figure 7.6 depicts the known containment relation involving classical and quantum time complexity classes. Nothing is as yet known about the relation between **BQP** and **NP** or **PP**.

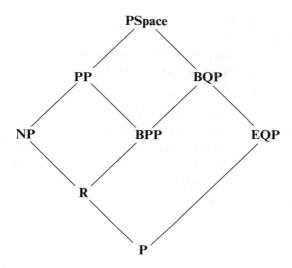

Figure 7.6
Containment relation involving classical and quantum complexity classes.

P ⊆ EQP Any classical polynomial time computation can be performed by a polynomial size circuit family. This inclusion follows from the main result of chapter 6: all classical circuits can be done reversibly with only a slight increase in time and space, and any reversible polynomial time algorithm can be turned into a polynomial time exact quantum algorithm.

EQP ⊆ BQP This containment is trivial since every exact quantum algorithm has bounded probability of error.

BPP ⊆ BQP Any computation performed by a machine M in **BPP** can be approximated arbitrarily closely by an machine $\tilde{M}$ that makes a single equiprobable binary decision at each step. Furthermore, this decision tree is of polynomial depth, so a sequence of choices can be encoded by a polynomial size bit string c. From $\tilde{M}$ one can construct a deterministic machine $\tilde{M}_d$ that, when applied to c and x, will perform the same computation as $\tilde{M}$ applied to x making the random choices c. For the deterministic machine $\tilde{M}_d$ there is a polynomial time quantum machine $\tilde{M}_q$ that can be applied to the superposition of all possible random choices c applied to x, $\sum_c |c, x, 0\rangle$, producing $\sum_c |c, x, \tilde{M}_d(c, x)\rangle$. In effect, $\tilde{M}_q$ performs all possible computations of $\tilde{M}$ on x in parallel. The probability of reading an accepting answer from $\tilde{M}_q$ is the same as the probability that $\tilde{M}$ would accept x.

It is not known whether **BPP ⊆ BQP** is a proper inclusion. In fact, showing **BPP ≠ BQP** would answer the open question as to whether **BPP = PSpace**.

BQP ⊆ PSpace Consider a machine in **BQP** acting on an input of size n that starts from a known state $|\psi_0\rangle = |0\rangle$ and proceeds for k steps followed by a measurement. We show that such a machine can be approximated arbitrarily closely, in the sense of computing any amplitude of the final state to a specified precision, in polynomial space. Let $|\psi_i\rangle = \sum_j a_{ij}|j\rangle$ denote the state after step i. Each state $|\psi_i\rangle$, $i \neq 0$ may be a superposition of an exponential (in n) number of basis vectors. Yet, using only space polynomial in n, it is possible to compute the amplitude a_{kj} of an arbitrary basis vector in the final superposition $|\psi_k\rangle$.

We may assume each step corresponds to a primitive quantum gate U_i that operates on at most $d \leq 3$ quantum bits. For these transformations, we show that the amplitude $a_{i+1,j}$ of basis vector $|j\rangle$ in state $|\psi_{i+1}\rangle$ depends only on the amplitudes $a_{i,j}$ of the small number ($2^d \leq 8$) of basis vectors of the preceding state $|\psi_i\rangle$ that differ from $|j\rangle$ only in the bits that are being operated on by the gate. Without loss of generality, assume that $U = U_{i+1}$ operates on the last d quantum bits. We will use the the shorthand $x \circ y$ to stand for $2^d x + y$ and let $u_{qr} = \langle r|U|q\rangle$ for basis elements $|r\rangle$ and $|q\rangle$ in the standard basis for a 2^d-dimensional space.

$$|\psi_{i+1}\rangle = (I^{n-d} \otimes U)|\psi_i\rangle$$

$$= \sum_j a_{ij}(I^{n-d} \otimes U)|j\rangle$$

$$= \sum_{p=0}^{2^{n-d}-1} \sum_{q=0}^{2^d-1} a_{i,p\circ q}|p\rangle \otimes U|q\rangle$$

$$= \sum_p \sum_q a_{i,p\circ q}|p\rangle \otimes \sum_{r=0}^{2^d-1} u_{qr}|r\rangle$$

$$= \sum_p \sum_r \left(\sum_{q+0}^{2^d-1} u_{qr} a_{i,p\circ q} \right) |p\rangle|r\rangle.$$

It follows that each amplitude $a_{i+1,p\circ r} = \sum_{q=0}^{2^d-1} u_{qr}a_{i,p\circ q}$ depends only on 2^d amplitudes $a_{i,p\circ q}$ of the preceding state.

By induction, we argue that it requires storage of $i2^d$ amplitudes to compute a single amplitude of state $|\psi_i\rangle$. Since we know $|\psi_0\rangle$, it takes no space to compute the amplitude $\langle j|\psi_0\rangle$ for any j. As we have just seen, the amplitude $a_{i+1,j}$ can be computed from 2^d amplitudes of $|\psi_i\rangle$. We can do this by computing each of these amplitudes in turn, which requires storing at most $i2^d$ amplitude values, storing the resulting 2^d amplitudes, and computing $a_{i+1,j}$. Overall, this process requires storage of $(i + 1)2^d$ amplitude values.

We take M to be the maximum precision required at any point in the computation to obtain the desired precision at the end. The total accumulated error is no larger than the sum of the errors of individual steps. Thus, the number M grows only linearly in the number of steps needed, and any one amplitude value can be stored in space M and the amplitude of any basis vector of the final superposition after k steps can be computed in $k2^d M$ space. Since by assumption k is polynomial in n, d is a constant no more than 3, and M only grows linearly with k, it takes only polynomial space to compute a single amplitude of the final state $|\psi_k\rangle$.

To simulate the algorithm, choose a basis vector $|j\rangle$ randomly (or, if you prefer, in a specified order) and calculate the amplitude a_{kj}. Generate a random number between 0 and 1 and see if it is less than $|a_{kj}|$. If so, return $|j\rangle$. Otherwise free all the space, choose another basis vector, and repeat. Repeat as often as necessary until a basis vector is returned (time is not an issue!). Thus, any computation in **BQP** can be simulated classically in polynomial space.

7.8   Quantum Fourier Transformations

The quantum Fourier transformation (QFT) is the single most important quantum subroutine. It and its generalizations are used in many quantum algorithms that achieve a speedup over classical algorithms. Appendix B.2.2 discusses generalizations of quantum Fourier transforms and shows that the Walsh-Hadamard transformation is a generalized quantum Fourier transform. The quantum Fourier transformation (QFT) is based on the classical discrete Fourier transformation (DFT) and its efficient implementation, the fast Fourier transform (FFT). We briefly describe the classical discrete Fourier transform (DFT) and the fast Fourier transform (FFT) before describing the quantum Fourier transform (QFT) and its surprisingly efficient quantum implementation.

7.8.1   The Classical Fourier Transform

Discrete Fourier Transform   The *discrete Fourier transform* (DFT) operates on a discrete complex-valued function to produce another discrete complex-valued function. Given a function $a : [0, \ldots, N-1] \rightarrow \mathbf{C}$, the discrete Fourier transform produces a function $A : [0, \ldots, N-1] \rightarrow \mathbf{C}$ defined by

$$A(x) = \frac{1}{\sqrt{N}} \sum_{k=0}^{N-1} a(k) \exp\left(2\pi \mathbf{i} \frac{kx}{N}\right).$$

The discrete Fourier transform can be viewed as a linear transformation taking column vector $(a(0), \ldots, a(N-1))^T$ to $(A(0), \ldots, A(N-1))^T$ with matrix representation F with entries $F_{xk} = \frac{1}{\sqrt{N}} \exp(2\pi \mathbf{i} \frac{kx}{N})$. The values $A(0), \ldots, A(N-1)$ are called the *Fourier coefficients* of the function a.

Example 7.8.1   Let $a : [0, \ldots, N-1] \rightarrow \mathbf{C}$ be the periodic function $a(x) = \exp(-2\pi \mathbf{i} \frac{ux}{N})$ for some frequency u evenly dividing N. We assume that the function is not constant: $0 < u < N$. The Fourier coefficients for this function are

$$A(x) = \frac{1}{\sqrt{N}} \sum_{k=0}^{N-1} a(k) \exp\left(2\pi i \frac{kx}{N}\right)$$

$$= \frac{1}{\sqrt{N}} \sum_{k=0}^{N-1} \exp\left(-2\pi i \frac{uk}{N}\right) \exp\left(2\pi i \frac{kx}{N}\right)$$

$$= \frac{1}{\sqrt{N}} \sum_{k=0}^{N-1} \exp\left(2\pi i \frac{k(x-u)}{N}\right).$$

It is a well-known fact that sums of the form $\sum_{k=0}^{N-1} \exp(2\pi i k \frac{r}{N})$ vanish unless $r = 0 \bmod N$. (We prove a more general fact in appendix B.) Since $u < N$, $A(x) = 0$ unless $x - u = 0$: only $A(u)$ will be non-zero.

Any periodic complex-valued function a with period r and frequency $u = N/r$ can be approximated, using its Fourier series, as the sum of exponential functions whose frequencies are multiples of u. Since the Fourier transform is linear, the Fourier coefficients $A(x)$ of any periodic function will be the sum of the Fourier coefficients of the component functions. If r divides N evenly, the Fourier coefficients $A(x)$ will be non-zero only for those x that are multiples of $u = N/r$. If r does not divide N evenly, the result only approximates this behavior, with the highest values at the integers closest to multiples of $u = N/r$ and low values at integers far from these multiples.

Fast Fourier Transform   The fast Fourier transform (FFT) is an efficient implementation of the discrete Fourier transform (DFT) when N is a power of two: $N = 2^n$. The key to the implementation is that $F^{(n)}$ can be recursively decomposed in terms of Fourier transforms for lower powers of 2.

Let $\omega_{(n)}$ be the Nth root of unity, $\omega_{(n)} = \exp(\frac{2\pi i}{N})$. The entries of the $N \times N$ matrix $F^{(n)}$ for the $N = 2^n$ dimensional Fourier transform are simply

$$F_{ij}^{(n)} = \omega_{(n)}^{ij},$$

where we index the entries of all $N \times N$ matrices by $i \in \{0, \dots, N-1\}$ and $j \in \{0, \dots, N-1\}$. Let $F^{(k)}$ be the $2^k \times 2^k$ matrix for the 2^k-dimensional Fourier transform.

Let $I^{(k)}$ be the $2^k \times 2^k$ identity matrix. Let $D^{(k)}$ be the $2^k \times 2^k$ diagonal matrix with elements $\omega_{(k+1)}^0, \dots \omega_{(k+1)}^{2^k-1}$. Let $R^{(k)}$ be the permutation shown in figure 7.7 that maps the vector entries at index $2i$ to position i and at index $2i + 1$ to position $i + 2^{k-1}$. The entries of the $2^k \times 2^k$ matrix for $R^{(k)}$ are given by

$$R_{ij}^{(k)} = \begin{cases} 1 & \text{if } 2i = j \\ 1 & \text{if } 2(i - 2^k) + 1 = j \\ 0 & \text{otherwise.} \end{cases}$$

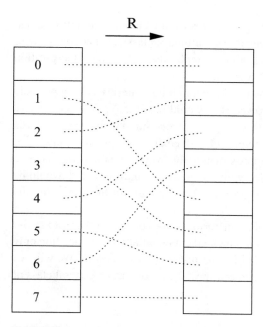

Figure 7.7
An example of the shuffle transform R.

The reader may verify (see exercise 7.7) that

$$F^{(k)} = \frac{1}{\sqrt{2}} \begin{pmatrix} I^{(k-1)} & D^{(k-1)} \\ I^{(k-1)} & -D^{(k-1)} \end{pmatrix} \begin{pmatrix} F^{(k-1)} & 0 \\ 0 & F^{(k-1)} \end{pmatrix} R^{(k)}.$$

The reader may consult any standard reference on the fast Fourier transform for an implementation based on this recursive decomposition that uses only $O(nN)$ steps.

7.8.2 The Quantum Fourier Transform

The quantum Fourier transform (QFT) is a variant of the discrete Fourier transform, which, like the fast Fourier transform (FFT), assumes that $N = 2^n$. The amplitudes a_x of any quantum state $\sum_x a_x |x\rangle$ can be viewed as a function of x, which we will denote by $a(x)$. The quantum Fourier transform operates on a quantum state by sending

$$\sum_x a(x)|x\rangle \rightarrow \sum_x A(x)|x\rangle$$

where the $A(x)$ are the Fourier coefficients of the the discrete Fourier transform of $a(x)$, and x ranges over the integers between 0 and $N-1$. If the state were measured in the standard basis right after the Fourier transform was performed, the probability that the resulting state would be $|x\rangle$

would be $|A(x)|^2$. The quantum Fourier transform generalizes from a classical complex-valued function in quite a different way from how U_f generalizes a binary classical function f; here the output of the classical function is placed in the complex amplitudes of the final superposition state, and there is no need for an additional output register.

Applying the quantum Fourier transform to a state whose amplitudes are given by a periodic function $a(x) = a_x$ with period r, where r is a power of 2, would result in $\sum_x A(x)|x\rangle$, where $A(x)$ is zero except when x is a multiple of $\frac{N}{r}$. Thus, were the state measured in the standard basis at this point, the result would be one of the basis vectors $|x\rangle$ with label a multiple of $\frac{N}{r}$, say $|j\frac{N}{r}\rangle$. The quantum Fourier transform behaves in only approximately this way when the period is not a power of 2 (does not divide $N = 2^n$): states labeled with integers near multiples of $\frac{N}{r}$ would be measured with high probability. The larger the power of 2 used as a base for the transform, the closer the approximation.

While the implementation of the quantum Fourier transform is based on that of the fast Fourier transform, the quantum Fourier transform can be implemented exponentially faster, needing only $O(n^2)$ operations, not the $O(nN)$ operations needed for the fast Fourier transform. We will see in appendix B.2.2 that the quantum Fourier transform is a special case of a more general class of efficiently implementable quantum transformations.

7.8.3   A Quantum Circuit for Fast Fourier Transform

We show how to implement efficiently the quantum Fourier transform $U_F^{(n)}$ for $N = 2^n$, defined by

$$U_F^{(n)} : |k\rangle \rightarrow \frac{1}{\sqrt{N}} \sum_{x=0}^{N-1} \exp(\frac{2\pi \mathbf{i} kx}{N})|x\rangle.$$

The quantum Fourier transform for $N = 2$ is the familiar Hadamard transformation:

$$U_F^{(1)} : |0\rangle \rightarrow \frac{1}{\sqrt{2}} \sum_{x=0}^{1} e^0 |x\rangle = \frac{1}{\sqrt{2}}(|0\rangle + |1\rangle),$$

$$|1\rangle \rightarrow \frac{1}{\sqrt{2}} \sum_{x=0}^{1} e^{\pi \mathbf{i} x} |x\rangle = \frac{1}{\sqrt{2}}(|0\rangle - |1\rangle).$$

Using the recursive decomposition of section 7.8.1,

$$U_F^{(k+1)} = \frac{1}{\sqrt{2}} \begin{pmatrix} I^{(k)} & D^{(k)} \\ I^{(k)} & -D^{(k)} \end{pmatrix} \begin{pmatrix} U_F^{(k)} & 0 \\ 0 & U_F^{(k)} \end{pmatrix} R^{(k+1)},$$

we can compute $U_F^{(n)}$. All of the component matrices are unitary (the multiplicative factor in front goes with the first matrix). It remains to be shown how these components can be efficiently realized on a quantum computer.

We proceed as follows:

1. We can write the rotation $R^{(k+1)}$ as

$$R^{(k+1)} = \sum_{i=0}^{2^k-1} |i\rangle\langle 2i| + |i + 2^k\rangle\langle 2i + 1|.$$

It can be accomplished by a simple permutation of the $k + 1$ qubits: qubit 0 becomes qubit k and qubits 1 through k become qubits 0 through $k - 1$. This permutation can be implemented using $k - 1$ swap operations of section 5.2.4.

2. The transformation

$$\begin{pmatrix} U_F{}^{(k)} & 0 \\ 0 & U_F{}^{(k)} \end{pmatrix} = I \otimes U_F{}^{(k)}$$

can be implemented by recursively applying the quantum Fourier transform to qubits 0 through k.

3. For $k \geq 1$, the $2^k \times 2^k$-diagonal matrix of phase shifts $D^{(k)}$ can be recursively decomposed as

$$D^{(k)} = D^{(k-1)} \otimes \begin{pmatrix} 1 & 0 \\ 0 & \omega_{(k+1)} \end{pmatrix}.$$

Recursively decomposing $D^{(k)}$ in this way, the transformation $D^{(k)}$ can be implemented by applying $\begin{pmatrix} 1 & 0 \\ 0 & \omega_{(i+1)} \end{pmatrix}$ to qubit i for $1 \leq i \leq k$. Thus altogether $D^{(k-1)}$ can be implemented using k single-qubit gates.

4. Given this implementation of $D^{(k)}$, then

$$\frac{1}{\sqrt{2}} \begin{pmatrix} I^{(k)} & D^{(k)} \\ I^{(k)} & -D^{(k)} \end{pmatrix}$$

can be implemented with only k gates.

$$\frac{1}{\sqrt{2}} \begin{pmatrix} I^{(k)} & D^{(k)} \\ I^{(k)} & -D^{(k)} \end{pmatrix} = \frac{1}{\sqrt{2}}(|0\rangle + |1\rangle)\langle 0| \otimes I^{(k)} + \frac{1}{\sqrt{2}}(|0\rangle - |1\rangle)\langle 1| \otimes D^{(k)}$$

$$= (H|0\rangle\langle 0|) \otimes I^{(k)} + (H|1\rangle\langle 1|) \otimes D^{(k)}$$

$$= (H \otimes I^{(k)})(|0\rangle\langle 0| \otimes I^{(k)} + |1\rangle\langle 1| \otimes D^{(k)}).$$

The transformation $(|0\rangle\langle 0| \otimes I^{(k)} + |1\rangle\langle 1| \otimes D^{(k)})$ applies $D^{(k)}$ to the low-order bits controlled by the high-order bit: it applies $D^{(k)}$ to bits 0 through $k - 1$ if bit k is one. This controlled version of $D^{(k)}$ can be implemented as a sequence of k two-qubit controlled gates that apply each of the single-qubit operations making up $D^{(k)}$ to bit i controlled by bit k.

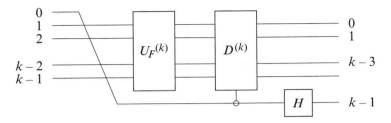

Figure 7.8
A recursive quantum circuit for Fourier transform.

Since $D^{(k)}$ and $R^{(k)}$ can both be implemented with $O(k)$ operations, the kth step in the recursion adds $O(k)$ steps to the implementation of $U_F^{(n)}$. Overall, $U_F^{(n)}$ takes $O(n^2)$ gates to implement, which is exponentially faster than the $O(n2^n)$ steps required for classical fast Fourier transform.

A circuit for this implementation of the quantum Fourier transform is shown in figure 7.8.

A recursive program for this implementation would be

define $QFT\,|x[1]\rangle = H|x\rangle$
$QFT\,|x[n]\rangle =$
    $Swap\,|x_0\rangle|x_1\cdots x_{n-1}\rangle$
    $QFT\,|x_0\cdots x_{n-2}\rangle$
    $|x_{n-1}\rangle$ **control** $D^{(n-1)}|x_0\cdots x_{n-2}\rangle$
    $H|x_{n-1}\rangle$.

7.9 References

Classical circuit complexity is discussed in Goldreich [131] and Vollmer [279]. Watrous [281] provides an excellent and extensive survey of quantum complexity theory. An older survey by Cleve [85], unlike Watrous, discusses quantum communication complexity as well as quantum computational complexity. Brassard [59] and de Wolf [97] both survey quantum communication complexity.

Deutsch described the solution to the 1-qubit version of his problem in [99]. The three subroutines were discussed in Hogg, Mochon, Polak, and Rieffel [154]. Deutsch and Jozsa presented the n-qubit version and its solution in [102]. Simon's problem with solution appeared in [256]. The Bernstein-Vazirani problem first appears in Bernstein and Vazirani [49] as part of a more complex algorithm. The simpler explanation of the algorithm appears in Mermin [209]. Both Grover [144] and Terhal and Smolin [269] independently rediscovered the problem and quantum algorithms for its solution. The latter reference contains a proof of the complexity of the best possible classical algorithm.

The example of section 7.5.4 was presented by Brassard, Cleve, and Tapp [60] in the context of their study of quantum communication complexity. Various notions of communication complexity are discussed in [155, 96, 74].

Beals et al. [35] proved that for a broad class of problems quantum computation cannot provide any speedup. Their methods were used by others to provide lower bounds for other types of problems. Ambainis [21] found another powerful method for establishing lower bounds.

Bernstein and Vazirani [49] analyze the accumulation of errors, and their result implies that the needed precision for simulating a quantum computation grows only linear with the number steps. Bennett et al. [41] provide another, more accessible, account. Yao [287] shows that any function computable in polynomial time on a quantum Turing machine is computable by a polynomial quantum circuit. The same is true for classical Turing machines and Boolean circuits, and proofs can be found in the paper by Pippenger and Fischer [229] or Papadimitriou's book [223]. Papadimitriou's book also contains a comprehensive definition of classical complexity classes, as does Johnson [164]. Boppana and M. Sipser [55] discuss classical complexity for Boolean circuits. Formal proofs of the complexity results given here can be found in the papers of Berthiaume-Brassard [50] and Bernstein-Vazirani [49].

The idea of Fourier transformation goes back to Joseph Fourier's 1822 book *The Analytical Theory of Heat* [123]. The algorithm for fast Fourier transformation was proposed by Cooley and Tukey [88]; more comprehensive treatments can be found in Brigham [66], Cormen et al. [90], Knuth [182], and Strang [264] .

The quantum Fourier transform was developed independently by Shor [250] and Coppersmith [89], and by Deutsch in an unpublished paper. Ekert and Jozsa [112] provide an attractive presentation of quantum Fourier transforms, including some of the circuit diagrams we give here. Approximate implementations of the quantum Fourier transform are analyzed in Barenco et al. [32]. For instance, it is shown that for some applications approximate computations may lead to better performance.

Aharonov, Landau, and Makowsky [12], Yoran and Short [288], and Browne [67] show that quantum Fourier transforms can be simulated efficiently on a classical computer in the sense that there exist efficient classical algorithms that provide a means of sampling from a distribution identical to that obtained by measuring the output of the quantum Fourier transform when the input is a product state. Browne exhibits a method for efficient classical simulation of the quantum Fourier transform applied to a broader class of input states. It is not known how to simulate efficiently the output distribution of the quantum Fourier transform for certain other input states. One such state is the output of the modular exponentiation circuit of section 6.4.6 when applied to a superposition of all inputs. Were it possible classically and efficiently to simulate sampling from such a distribution, Shor's algorithm, described in chapter 8, would be classically simulatable, yielding an efficient classical solution to the factoring problem. For this reason, it is suspected that such simulation is impossible.

7.10   Exercises

Exercise 7.1.   In the standard circuit model of section 5.6, the computation takes place by applying quantum gates. Only at the end are measurements performed. Imagine a computation that proceeds instead as follows. Gates $G_0, G_1, \ldots, G_n$ are applied, then qubit i is measured in the standard basis and never used again. If the result of the measurement is 0, the gates $G_{01}, G_{02}, \ldots, G_{0k}$ are applied. If the result is 1, then gates $G_{11}, G_{12}, \ldots, G_{1l}$ are applied. Find a single quantum circuit in the standard circuit model, with only measurement at the very end, that carries out this computation.

Exercise 7.2. Prove equation 7.1:

$$\sum_{x=0}^{2^n-1} (-1)^{x \cdot y} = \begin{cases} 2^n & \text{if } y = 0 \\ 0 & \text{otherwise.} \end{cases}$$

Exercise 7.3. Let f and g be functions from the space of n-bit strings to the space of m-bit strings. Design a quantum subroutine that changes the sign of exactly the basis states $|x\rangle$ such that $f(x) = g(x)$, and which is efficient if f and g have efficient implementations.

Exercise 7.4.

a. Prove that any classical algorithm requires at least two calls to C_f to solve Deutsch's problem.

b. Prove that any classical algorithm requires $2^{n-1} + 1$ calls to C_f to solve the Deutsch-Jozsa problem with certainty.

c. Describe a classical approach to the Deutsch-Jozsa problem that solves it with high probability using fewer than $2^{n-1} + 1$ calls. Calculate the success probability of your approach as a function of the number of calls.

Exercise 7.5. Show that a classical solution to Simon's problem requires $O(2^{n/2})$ calls to the black box, and describe such a classical algorithm.

Exercise 7.6. Show directly that, in the distributed computation algorithm of section 7.5.4, when $u = v$, $|\langle x, y | \psi \rangle|^2 = 0$ for all $x \neq y$.

Exercise 7.7. *Fast Fourier transform decomposition.*

a. For $k < l$, write the entries $F_{ij}^{(k)}$ of the $2^k \times 2^k$ matrix for the Fourier transform $U_F^{(k)}$ in terms of $\omega_{(l)}$.

b. Find m in terms of k such that $-\omega_{(k)}^i = \omega_{(k)}^{m+i}$ for all $i \in \mathbf{Z}$.

c. Compute the product

$$\begin{pmatrix} I^{(k-1)} & D^{(k-1)} \\ I^{(k-1)} & -D^{(k-1)} \end{pmatrix} \begin{pmatrix} U_F^{(k-1)} & 0 \\ 0 & U_F^{(k-1)} \end{pmatrix},$$

ultimately writing each entry as a power of $\omega_{(k)}$.

d. Let A be any $2^k \times 2^k$ matrix with columns A_j. The product matrix $AR^{(k)}$ is just a permutation of the columns. Where does column A_j end up in the product $AR^{(k)}$?

e. Verify that

$$U_F{}^{(k)} = \frac{1}{\sqrt{2}} \begin{pmatrix} I^{(k-1)} & D^{(k-1)} \\ I^{(k-1)} & -D^{(k-1)} \end{pmatrix} \begin{pmatrix} U_F{}^{(k-1)} & 0 \\ 0 & U_F{}^{(k-1)} \end{pmatrix} R^{(k)}.$$

Exercise 7.8. Even though we know little about quantum hardware, it makes sense that we may not want to require multiple qubit transformations that involve physically distant qubits, since these may be difficult to implement. To avoid such transformations, we can modify the implementation we gave very slightly.

a. Give a quantum circuit like that of figure 7.8 for the Fourier transform that does not swap qubits but changes the order of the output qubits instead.

b. Give a complete quantum circuit for the Fourier transform $U_F{}^{(3)}$ that contains only single-qubit transformations and two-qubit transformations on adjacent qubits. You may want to use the two-qubit swap operator defined in section 5.2.4.

8 Shor's Algorithm

In 1994, inspired by Simon's algorithm, Peter Shor found a bounded probability polynomial-time quantum algorithm for factoring integers. Since the 1970s, researchers have searched for efficient algorithms for factoring integers. The most efficient classical algorithm known today, the number field sieve, is superpolynomial in the size of the input. The input to the algorithm is M, the number to be factored. The input M is given as a list M's digits, so the size of the input is taken to be $m = \lceil \log M \rceil$. The number field sieve requires $O(\exp(m^{1/3}))$ steps. People were confident enough that factoring could not be done efficiently that the security of many cryptographic systems, such as the widely used RSA algorithm, depends on the computational difficulty of this problem. Shor's result surprised the community at large, prompting widespread interest in quantum computing.

Shor's factoring algorithm provides a fast means for finding the period of a function. A standard classical reduction of the factoring problem to the problem of finding the period of a certain function has long been known. Shor's algorithm uses quantum parallelism to produce a superposition of all the values of this function in one step; it then uses the quantum Fourier transform to create efficiently a state in which most of the amplitude is in states close to multiples of the reciprocal of the period. With high probability, measuring the state yields information from which, by classical means, the period can be extracted. The period is then used to factor M.

Section 7.8.2 covered the crux of the quantum part of Shor's algorithm: the quantum Fourier transform. The remaining complications are classical, particularly the extraction of the period from the measured value.

Section 8.1 explains the classical reduction of factoring to the problem of finding the period of a function. Section 8.2 explains the details of Shor's algorithm, and section 8.3 walks through Shor's algorithm in a specific case. Section 8.4 analyzes the efficiency of Shor's algorithm. Section 8.5 describes a variant of Shor's algorithm in which a measurement performed in the course of the algorithm is omitted. Section 8.6 defines two problems that are solved by generalizations of Shor's factoring algorithm: the discrete logarithm problem and the Abelian hidden subgroup problem. Appendix B describes the generalizations of Shor's algorithm that solve these problems and discusses the difficulty of the general hidden subgroup problem.

8.1 Classical Reduction to Period-Finding

The *order* of an integer a modulo M is the smallest integer $r > 0$ such that $a^r = 1 \bmod M$; if no such integer exists, the order is said to be infinite. Two integers are *relatively prime* if they share no prime factors. As long as a and M are relatively prime, the order of a is finite. Consider the function $f(k) = a^k \bmod M$. Because $a^k = a^{k+r} \bmod M$ if and only if $a^r = 1 \bmod M$, for a relatively prime to M, the order r of a modulo M is the period of f. If $a^r = 1 \bmod M$ and r is even, we can write

$$(a^{r/2} + 1)(a^{r/2} - 1) = 0 \bmod M.$$

As long as neither $a^{r/2} + 1$ nor $a^{r/2} - 1$ is a multiple of M, both $a^{r/2} + 1$ and $a^{r/2} - 1$ have nontrivial common factors with M. Thus, if r is even, $a^{r/2} + 1$ and $a^{r/2} - 1$ are likely to have a nontrivial common factor with M. This property suggests a strategy for factoring M:

- Randomly choose an integer a and determine the period r of $f(k) = a^k \bmod M$.
- If r is even, use the Euclidean algorithm to compute efficiently the greatest common divisor of $a^{r/2} + 1$ and M.
- Repeat if necessary.

In this way, factoring M has been converted to a different hard problem, that of computing the period of the function $f(k) = a^k \bmod M$. Shor's quantum algorithm attacks the problem of efficiently finding the period of a function.

8.2 Shor's Factoring Algorithm

Before giving the details of Shor's factoring algorithm in sections 8.2.1 and 8.2.2, we give a high-level outline. Quantum computation is required only for parts 2 and 3; the other parts would most likely be carried out on a classical computational device.

1. Randomly choose an integer a such that $0 < a < M$. Use the Euclidean algorithm to determine whether a and M are relatively prime. If not, we have found a factor of M. Otherwise, apply the rest of the algorithm.

2. Use quantum parallelism to compute $f(x) = a^x \bmod M$ on the superposition of inputs, and apply a quantum Fourier transform to the result. Section 8.2.2 shows that it suffices to consider input values $x \in \{0, \ldots, 2^n - 1\}$, where n is such that $M^2 \leq 2^n < 2M^2$.

3. Measure. With high probability, a value v close to a multiple of $\frac{2^n}{r}$ will be obtained.

4. Use classical methods to obtain a conjectured period q from the value v.

5. When q is even, use the Euclidean algorithm to check efficiently whether $a^{q/2} + 1$ (or $a^{q/2} - 1$) has a nontrivial common factor with M.

6. Repeat all steps if necessary.

Sections 8.2.1 and 8.2.2 describe Shor's algorithm in more detail. Section 8.3 runs through an example with specific values of M and a.

8.2.1 The Quantum Core

After using quantum parallelism to create the superposition $\sum_x |x, f(x)\rangle$, part 2 of Shor's algorithm applies the quantum Fourier transform.

Since $f(x) = a^x \bmod M$ can be computed efficiently classically, the results of chapter 6 imply that the transformation

$$U_f : |x\rangle|0\rangle \rightarrow |x\rangle|f(x)\rangle$$

has an efficient implementation. (We discuss the efficiency of the entire algorithm in section 8.4.) We use quantum parallelism with U_f to obtain the superposition

$$\frac{1}{\sqrt{2^n}} \sum_{x=0}^{2^n-1} |x\rangle|f(x)\rangle. \tag{8.1}$$

The analysis simplifies slightly if we now measure the second register. Section 8.5 shows how the measurement can be omitted without affecting the efficiency or the result of the algorithm.

Measuring the second register randomly returns a value u for $f(x)$ and the state becomes

$$C \sum_x g(x)|x\rangle|u\rangle, \tag{8.2}$$

where

$$g(x) = \begin{cases} 1 & \text{if } f(x) = u \\ 0 & \text{otherwise,} \end{cases}$$

and C is the appropriate scale factor. The value of u is of no interest and, since the second register is no longer entangled with the first, we can ignore it. Because the function $f(x) = a^x \bmod M$ has the property that $f(x) = f(y)$ if and only if x and y differ by a multiple of the period, the values of x that remain in the sum, those with $g(x) \neq 0$, differ from each other by multiples of the period. Thus, the function g has the same period as the function f. If we could somehow obtain the value of two successive terms in the sum, we would have the period. Unfortunately, the laws of quantum physics permit only one measurement from which we can obtain only one random value of x. Repeating the process does not help because we would be unlikely to measure the same value u of $f(x)$, so the two values of x obtained from two runs would have no relation to each other.

Applying the quantum Fourier transform to the first register of this state produces

$$U_F \left(C \sum_x g(x)|x\rangle \right) = C' \sum_c G(c)|c\rangle, \tag{8.3}$$

where $G(c) = \sum_x g(x) \exp(\frac{2\pi i c x}{2^n})$. The analysis of section 7.8.2 tells us that when the period r of the function $g(x)$ is a power of two, $G(c) = 0$ except when c is a multiple of $2^n/r$. When the period r does not divide 2^n, the transform approximates the exact case, so most of the amplitude is attached to integers close to multiples of $\frac{2^n}{r}$. For this reason, measurement yields, with high probability, a value v close to a multiple of $\frac{2^n}{r}$. The quantum core of the algorithm has now been completed. The next section examines the classical use of v to obtain a good guess for the period.

8.2.2 Classical Extraction of the Period from the Measured Value

This section sketches a purely classical algorithm for extracting the period from the measured value v obtained from the quantum core of Shor's algorithm. When the period r happens to be a power of 2, the quantum Fourier transform gives exact multiples of $2^n/r$, which makes the period easy to extract. In this case, the measured value v is equal to $j\frac{2^n}{r}$ for some j. Most of the time j and r will be relatively prime, in which case reducing the fraction $\frac{v}{2^n}$ to its lowest terms will yield a fraction $\frac{j}{r}$ whose denominator is the period r. The rest of this section explains how to obtain a good guess for r when it is not a power of 2.

In general the quantum Fourier transform gives only approximate multiples of the scaled frequency, which complicates the extraction of the period from the measurement. When the period is not a power of 2, a good guess for the period can be obtained from the continued fraction expansion of $\frac{v}{2^n}$ described in box 8.1. Shor shows that with high probability v is within $\frac{1}{2}$ of some multiple of $\frac{2^n}{r}$, say $j\frac{2^n}{r}$. The reason why n was chosen to satisfy $M^2 \leq 2^n < 2M^2$ becomes apparent when we try to extract the period r from the measured value v. In the high-probability case that

$$\left| v - j\frac{2^n}{r} \right| < \frac{1}{2}$$

for some j, the left inequality $M^2 \leq 2^n$ implies that

$$\left| \frac{v}{2^n} - \frac{j}{r} \right| < \frac{1}{2 \cdot 2^n} \leq \frac{1}{2M^2}.$$

In general, the difference between two distinct fractions $\frac{p}{q}$ and $\frac{p'}{q'}$ with denominators less than M is bounded:

$$\left| \frac{p}{q} - \frac{p'}{q'} \right| = \left| \frac{pq' - p'q}{qq'} \right| > \frac{1}{M^2}.$$

Thus, there is at most one fraction $\frac{p}{q}$ with denominator $q < M$ such that $\left| \frac{v}{2^n} - \frac{p}{q} \right| < \frac{1}{M^2}$. In the high probability case that v is within $\frac{1}{2}$ of $j\frac{2^n}{r}$, this fraction will be $\frac{j}{r}$. The fraction $\frac{p}{q}$ can be computed using a continued fraction expansion (see box 8.1). We take the denominator q of the obtained fraction. as our guess for the period. This guess will be correct whenever j and r are relatively prime.

Box 8.1
Continued Fraction Expansion

The unique fraction with denominator less than M that is within $\frac{1}{M^2}$ of $\frac{v}{2^n}$ can be obtained efficiently from the continued fraction expansion of $\frac{v}{2^n}$ as follows. Let $[x]$ be the greatest integer less than x. Using the sequences

$$a_0 = \left[\frac{v}{2^n}\right]$$

$$\epsilon_0 = \frac{v}{2^n} - a_0$$

$$a_i = \left[\frac{1}{\epsilon_{i-1}}\right]$$

$$\epsilon_i = \frac{1}{\epsilon_{i-1}} - a_i$$

$$p_0 = a_0$$

$$p_1 = a_1 a_0 + 1$$

$$p_i = a_i p_{i-1} + p_{i-2}$$

$$q_0 = 1$$

$$q_1 = a_1$$

$$q_i = a_i q_{i-1} + q_{i-2},$$

compute the first fraction $\frac{p_i}{q_i}$ such that $q_i < M \leq q_{i+1}$.

8.3 Example Illustrating Shor's Algorithm

This section illustrates the operation of Shor's algorithm as it attempts to factor the integer $M = 21$. Since $M^2 = 441 \leq 2^9 < 882 = 2M^2$, take $n = 9$. Since $\lceil \log M \rceil = m = 5$, the second register requires five qubits. Thus, the state

$$\frac{1}{\sqrt{2^9}} \sum_{x=0}^{2^9-1} |x\rangle |f(x)\rangle \tag{8.4}$$

is a 14-qubit state, with nine qubits in the first register and five in the second.

Suppose the randomly chosen integer is $a = 11$ and that quantum measurement of the second register of the superposition of equation 8.1

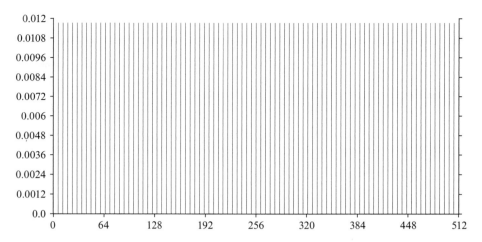

Figure 8.1
Probabilities for measuring x when measuring the state $C \sum_{x \in X} |x, 8\rangle$ obtained in equation 8.2, where $X = \{x | 11^x \bmod 21 = 8\}$.

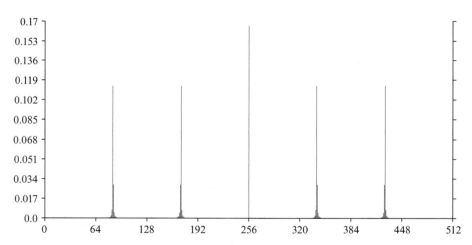

Figure 8.2
Probability distribution of the quantum state after Fourier transformation.

$$\frac{1}{\sqrt{2^9}} \sum_{x=0}^{2^9-1} |x\rangle |f(x)\rangle \tag{8.5}$$

produces $u = 8$. The state of the first register after this measurement is shown in figure 8.1, which clearly shows the periodicity of f.

Figure 8.2 shows the result of applying the quantum Fourier transform to this state; it is the graph of the fast Fourier transform of the function shown in Figure 8.1. In this particular example, the period of f does not divide 2^n, which is why the probability distribution has some spread around multiples of $2^n/r$ instead of having a single spike at each of these values.

Suppose that measurement of the state returns $v = 427$. Since v and 2^n are relative prime, we use the continued fraction expansion of box 8.1 to obtain a guess q for the period. The following table shows a trace of the continued fraction algorithm:

i	a_i	p_i	q_i	ϵ_i
0	0	0	1	0.8339844
1	1	1	1	0.1990632
2	5	5	6	0.02352941
3	42	211	253	0.5

The algorithm terminates with $6 = q_2 < M \le q_3$. Thus, $q = 6$ is our guess for the period of f.

Since 6 is even, $a^{6/2} - 1 = 11^3 - 1 = 1330$ and $a^{6/2} + 1 = 11^3 + 1 = 1332$ are likely to have a common factor with M. In this particular example, $\gcd(21, 1330) = 7$ and $\gcd(21, 1332) = 3$.

8.4 The Efficiency of Shor's Algorithm

This section considers the efficiency of Shor's algorithm, examining both the efficiency of each part in terms of the number of gates or classical steps needed to implement the part and the expected number of times the algorithm would need to be repeated.

The Euclidean algorithm on integers $x > y$ requires at most $O(\log x)$ steps, so both parts 1 and 5 require $O(\log M) = O(m)$ steps. The continued fraction algorithm used in part 4 is related to the Euclidean algorithm and also requires $O(m)$ steps. Part 3 is a measurement of m qubits or, as section 8.5 shows, can be omitted altogether. Part 2 consists of the computation of U_f and the computation of the quantum Fourier transform. Section 7.8.2 showed that the quantum Fourier transform on m qubits requires $O(m)$ steps. The algorithm for modular exponentiation given in section 6.4 requires $O(n^3)$ steps could be used to implement U_f. The transformation U_f can be implemented more efficiently using an algorithm for modular exponentiation, described by Shor, that is based on the most efficient classical method known, and runs in $O(n^2 \log n \log \log n)$ time and $O(n \log n \log \log n)$ space. These results show that the overall runtime of a single iteration of Shor's algorithm is dominated by the computation of U_f, and that the overall time complexity for a single iteration of the algorithm is $O(n^2 \log n \log \log n)$.

To show that Shor's algorithm is efficient, we also need to show that the parts do not need to be repeated too many times. Four things can go wrong:

- The period of $f(x) = a^x \bmod M$ could be odd.
- Part 4 could yield M as M's factor.
- The value v obtained in part 3 might not be close enough to a multiple of $\frac{2^n}{r}$.
- A multiple $j\frac{2^n}{r}$ of $\frac{2^n}{r}$ is obtained from v, but j and r could have a common factor, in which case the denominator q is actually a factor of the period, not the period itself.

The first two problems appear in the classical reduction, and standard classical arguments bound the probabilities as at most $1/2$. For the case in which the period r divides 2^n, problem 3 does not arise. Shor shows that, in the general case, v is within $1/2$ of a multiple of $\frac{2^n}{r}$ with high probability. As for problem 4, when r divides 2^n, it is not hard to see that every outcome $v = j\frac{2^n}{r}$ is equally likely: the state after taking the quantum Fourier transform is

$$C' \sum_{c=0}^{2^n-1} G(c)|c\rangle,$$

where

$$G(c) = \sum_{x \in X_u} \exp(2\pi i \frac{cx}{2^n}) = \sum_{y=0}^{2^n/r} \exp(2\pi i \frac{cry}{2^n})$$

where $X_u = \{x \,|\, f(x) = u\}$. As we mentioned in section 7.8.1, the final sum is 1 when c is a multiple of $2^n/r$, and 0 otherwise. Thus, in this case, any $j \in \{0, \ldots, r-1\}$ is equally likely. From j, we obtain the period r exactly when r and j are relatively prime, $\gcd(r, j) = 1$. The number of positive integers less than r that are relatively prime to r is given by the famous Euler ϕ function, which is known to satisfy $\phi(r) \geq \delta/\log\log r$ for some constant δ. Thus we need to repeat the parts only $O(\log\log r)$ times in order to achieve a high probability of success. The argument for the general case in which r does not divide 2^n is somewhat more involved but yields the same result.

8.5 Omitting the Internal Measurement

Part 3 of Shor's algorithm, the measurement of the second register of the state in equation 8.1 to obtain u, can be skipped entirely. This section first describes the intuition for why this measurement can be omitted and then gives a formal argument.

If the measurement is omitted, the state consists of a superposition of several periodic functions, one for each value of $f(x)$, all of which have the same period. By the linearity of quantum transformations, applying the quantum Fourier transformation leads to a superposition of the Fourier transforms of these functions. The different functions remain distinct parts of the superposition and do not interfere with each other because each one corresponds to a different value u of the second register. Measuring the first register gives a value from one of these Fourier transforms, which as before will be close to $j\frac{2^n}{r}$ for some j and so can be used to obtain the period in the same way as before. Seeing how this argument can be formalized illustrates some of the subtleties of working with quantum superpositions.

Let $X_u = \{x\,|\,f(x) = u\}$. The state of equation 8.1 can be written as

$$\frac{1}{\sqrt{2^n}} \sum_{x=0}^{2^n-1} |x\rangle |f(x)\rangle = \frac{1}{\sqrt{2^n}} \sum_{u\in R} \sum_{x\in X_u} |x\rangle |u\rangle$$

$$= \frac{1}{\sqrt{2^n}} \sum_{u\in R} \left(\sum_{x=0}^{2^n-1} g_u(x)|x\rangle \right) |u\rangle,$$

where R is the range of $f(x)$ and g_u is the family of functions indexed by u such that

$$g_u(x) = \begin{cases} 1 & \text{if } f(x) = u \\ 0 & \text{otherwise.} \end{cases}$$

The amplitudes in states with different u in the second register can never interfere (add or cancel) with each other. The result of applying the transform $U_F \otimes I$ to the preceding state can be written

$$U_F \otimes I \left(\frac{1}{\sqrt{2^n}} \sum_{u\in R} \left(\sum_{x=0}^{2^n-1} g_u(x)|x\rangle \right) |u\rangle \right) = \frac{1}{\sqrt{2^n}} \sum_{u\in R} \left(U_F \sum_x g_u(x)|x\rangle \right) |u\rangle$$

$$= C' \sum_{u\in R} \left(\sum_{c=0}^{2^n-1} G_u(c)|c\rangle \right) |u\rangle,$$

where $G_u(c)$ is the discrete Fourier transform of $g_u(x)$. This results is a superposition of the possible states of equation 8.3 over all possible u. Since the g_u all have the same period, measuring the first part of this state returns a c close to a multiple of $2^n/r$, just as happened when the second register was measured as part of the original algorithm.

8.6 Generalizations

Shor's original paper contained not only a quantum factoring algorithm, but also a related algorithm for the *discrete logarithm problem*. Further generalizations of Shor's quantum algorithms have been obtained for problems falling in the general class of hidden subgroup problems. The next two sections, sections 8.6.1 and 8.6.2, require knowledge of group theory. Readers unfamiliar with group theory should just skim these sections; the results they contain will not be used later in the book, apart from appendix B and the section of the final chapter that reviews more recent algorithmic results. The basics of group theory are reviewed in boxes.

8.6.1 The Discrete Logarithm Problem

The discrete logarithm problem is also of cryptographic importance; the security of Diffie-Hellman and El Gamal and elliptic curve public key encryption, for example, rest on the classical difficulty of this problem. In fact, all standard public key encryption systems and digital signature schemes are based on either factoring or the discrete logarithm problem. Electronic commerce and communication rely on public key encryption and digital signature schemes for their security and efficiency. It is currently unclear whether a public key encryption system believed to be secure against classical and quantum attacks can be established before quantum computers are built. If quantum computers win this race, the practical implications will be substantial. Once quantum computers become a reality, all currently accepted public key encryption systems will be completely insecure.

Let $\mathbf{Z}_p^*$ be the group of integers $\{1, \ldots, p-1\}$ under multiplication modulo p, and let b be a generator for this group (any b relatively prime to $p-1$ will do). The *discrete logarithm* of $y \in \mathbf{Z}_p^*$ with respect to base b is the element $x \in \mathbf{Z}_p^*$ such that $b^x = y \bmod p$.

Discrete Logarithm Problem   Given a prime p, a base $b \in \mathbf{Z}_p^*$, and an arbitrary element $y \in \mathbf{Z}_p^*$, find an $x \in \mathbf{Z}_p^*$ such that $b^x = y \bmod p$.

For large p, this problem is computationally difficult to solve. The discrete logarithm problem can be generalized to arbitrary finite cyclic groups G, though for some large G it is is not difficult to solve classically. The discrete logarithm is a special case of the Abelian hidden subgroup problem. Appendix B describes a general algorithm for the Abelian hidden subgroup problem that yields essentially Shor's original discrete logarithm algorithm in the special case. The next section discusses hidden subgroup problems.

8.6.2 Hidden Subgroup Problems

The hidden subgroup framework subsumes many of the problems and quantum algorithms we have discussed. Understanding this framework requires experience with group theory. The definition of a group is reviewed in box 8.2, which also contains examples. Box 8.3 defines some properties of groups and subgroups. Box 8.4 discusses Abelian groups.

The Hidden Subgroup Problem   Let G be a group. Suppose a subgroup $H < G$ is implicitly defined by a function f on G in that f is constant and distinct on every coset of H. Find a set of generators for H.

The aim is to find a polylogarithmic algorithm that computes a set of generators for H in $O((\log|G|)^k)$ steps for some k. The difficulty of the problem depends not only on G and F but also on what is meant by *given a group G*. Some useful properties may be expensive to compute from certain descriptions of a group and immediate from others. For example, computing the size of a group from certain types of descriptions, such as a defining set of generators and relations, is known to be computationally hard. Also, we can hope to find a solution in poly-log time only if f itself is computable in poly-log time.

Box 8.2
Groups

A *group* is a non-empty set G with an associative binary operation, denoted $\circ$, satisfying

- (*closure*) for any two elements g_1 and g_2 of G, the product $g_1 \circ g_2$ is also in G,
- an *identity* element $e \in G$ such that $e \circ g = g \circ e = g$, and
- every element $g \in G$ has an *inverse* $g^{-1} \in G$ such that $g \circ g^{-1} = g^{-1} \circ g = e$.

The associative binary operation $\circ$ is generally referred to as the group's *product*. Often the product is indicated simply by juxtaposition, with the $\circ$ omitted: $g_1 \circ g_2$ is written simply as $g_1 g_2$. For some groups, other notation is used for the binary operation.

Some examples of groups:

- The integers $\{0, 1, \ldots, n-1\}$ form a group under addition modulo n. This group is denoted $\mathbf{Z}_n$, with binary operator $+$.
- The set of k-bit strings, $\mathbf{Z}_2^k$, forms a group under bitwise addition modulo 2.
- For p prime, the set of integers $\{1, \ldots, n-1\}$ forms a group $\mathbf{Z}_p^*$ under multiplication modulo p.
- The set $\mathcal{U}(n)$ of all unitary operators on an n-dimensional vector space V forms a group.
- The Pauli group consisting of the eight elements $\pm I$, $\pm X$, $\pm Y$, and $\pm Z$ forms a group.
- The extended Pauli group consisting of the sixteen elements ωI, ωX, ωY, and ωZ, where $\omega \in \{1, -1, -\mathbf{i}, \mathbf{i}\}$, forms a group.

Box 8.3
Properties of Groups and Subgroups

The number of elements $|G|$ of a group is called its *order*. A group is said to be *finite* if its order is a finite number; otherwise it is an infinite group.

A subset H of G that is a group in its own right, under the restriction of G's product to H, is called a *subgroup* of G. The subgroup relation is written $H < G$. For example, for any integer m dividing n, the set of multiples of m forms a subgroup of $\mathbf{Z}_n$. Also, any subspace W of a vector space V is a subgroup of the group V under vector addition. The Pauli group is a subgroup of the unitary group $U(n)$.

The *order of an element* g is the size of the subgroup of G that it generates. The order of an element must divide the order of a group.

A set of *generators of a group* G is a subset of G such that all elements of G can be written as a finite product of the generators and their inverses (in any order and allowing repeats). A set of generators of a group is *independent* if no generator can be written as a product of the other generators. A group is *finitely generated* if a finite set of generators exists. If a group can be generated by a single element it is *cyclic*. The set of generators for a given group is not unique in general.

The *centralizer*, $Z(H)$, of a subgroup H of G is the set of elements of G that commute with all elements of H:

$$Z(H) = \{g \in G | gh = hg \text{ for all } h \in H\}.$$

For $H < G$, the centralizer $Z(H)$ of H is a subgroup of G.

Box 8.4
Abelian Groups

A group is *Abelian* if its group product ∘ is commutative: $g_1 \circ g_2 = g_2 \circ g_1$.

The group $\mathbf{Z}_n$ is Abelian, but the set of unitary operators $\mathcal{U}(n)$ is not Abelian.

The *product* $G \times H$ of two groups G and H, with products $\circ_G$ and $\circ_H$ respectively, is the set of pairs $\{(g, h) | g \in G, h \in H\}$ with the product $(g_1, h_1) \circ (g_2, h_2) = (g_1 \circ_G g_2, h_1 \circ_H h_2)$.

The structure of finite Abelian groups is well understood. Every finite Abelian group is isomorphic to a product of one or more cyclic groups $\mathbf{Z}_{n_i}$. For example, for the product n of two relatively prime integers p and q, the group $\mathbf{Z}_n$ is isomorphic to $\mathbf{Z}_p \times \mathbf{Z}_q$. Any finite Abelian group A has a unique decomposition (up to the ordering of the factors) into cyclic groups of prime power order. The decomposition depends only on its order $|A|$. Let $|A| = \Pi_i c_i$ be the prime factorization of $|A|$, where $c_i = p_i^{s_i}$ and the p_i are distinct primes. Then

$$A \cong \mathbf{Z}_{c_1} \times \mathbf{Z}_{c_2} \times \cdots \times \mathbf{Z}_{c_k}.$$

While the general hidden subgroup problem remains unsolved, a polylogarithmic bounded probability quantum algorithm for the general case of finite Abelian groups, specified in terms of their cyclic decomposition, exists. The cyclic decomposition for Abelian groups is described in box 8.4.

Finite Abelian Hidden Subgroup Problem   Let G be a finite Abelian group with cyclic decomposition $G = \mathbf{Z}_{n_0} \times \cdots \times \mathbf{Z}_{n_L}$. Suppose G contains a subgroup $H < G$ that is implicitly defined by a function f on G in that f is constant and distinct on every coset of H. Find a set of generators for H.

Example 8.6.1 *Period-finding as a hidden subgroup problem.* Period-finding can be rephrased as a hidden subgroup problem. Let f be a periodic function on $\mathbf{Z}_N$ with period r that divides N. The subgroup $H < \mathbf{Z}_N$ generated by r is the hidden subgroup. Once a generator h for H has been found, the period r can be found by taking the greatest common divisor of h and N: $r = \gcd(h, N)$.

In addition to period-finding, both Simon's problem and the discrete logarithm problem are instances of the finite Abelian hidden subgroup problem. Recognizing how Simon's problem can be viewed as a hidden subgroup problem is relatively easy. Understanding how the discrete logarithm problem is a special case of the hidden subgroup problem requires some ingenuity.

Example 8.6.2 *The discrete logarithm problem as a hidden subgroup problem.* The discrete log problem asks: Given the group $G = \mathbf{Z}_p^*$, where p is prime, a base $b \in G$, and an arbitrary element $y \in G$, find an $x \in G$ such that $b^x = y \bmod p$. Consider $f : G \times G \to G$ where $f(g, h) = b^{-g} y^h$. The set of elements satisfying $f(g, h) = 1$ is the hidden subgroup H of $G \times G$ consisting of tuples of the form (mx, m). From any generator of H, the element $(x, 1)$ can be computed. Thus, solving this hidden subgroup problem yields x, the solution to the discrete logarithm problem.

A crucial ingredient of Shor's algorithm is the quantum Fourier transform. The quantum algorithm for Simon's problem also uses a quantum Fourier transform; quantum Fourier transforms can be defined for all finite Abelian groups (and more generally all finite groups), and the quantum Fourier transform for the group $\mathbf{Z}_2^n$ is the Walsh-Hadamard transformation W. The solution to the hidden subgroup problem over an Abelian group G uses the quantum Fourier transform over the group G. The Fourier transformation over an general finite group G is defined in terms of the group representations of G. These ingredients are described in appendix B, which also describes the general solution to the finite Abelian hidden subgroup problem. It makes use of deeper group theory results than the rest of the book. No one knows how to solve the hidden subgroup problem over general non-Abelian groups. What progress has been made toward understanding the non-Abelian hidden subgroup problem is discussed in chapter 13.

8.7 References

Lenstra and Lenstra [193] describes the best currently known classical factoring algorithm, the number field sieve, including its $O(\exp(n^{1/3}))$ complexity. Some simpler but less efficient classical factoring algorithms are described in Knuth [182].

Shor's algorithm first appeared in 1994 [250]. Shor later published an expanded version [253] that contains a detailed analysis of the complexity and the probability of success.

The continued fraction expansion, and the approximations it gives, is described in detail in most standard number theory texts including Hardy and Wright [149]. Its efficiency and relation to the Euclidean algorithm is discussed in Knuth [182]. The Euler ϕ function and its properties are also discussed in standard number theory books such as Hardy and Wright [149].

Kitaev solved the general finite Abelian hidden subgroup problem [172]. Jozsa [165] and [112] provide accessible accounts of the quantum Fourier transform in the context of the hidden subgroup problem. The general hidden subgroup problem was introduced by Mosca and Ekert in [214].

Koblitz and Menezes, in their 2004 survey [183], give a detailed overview of proposed public key encryption schemes, including ones not based on factoring or the discrete logarithm problem, as well as the more standard public key schemes. Rieffel [242] discusses the practical implications of quantum computing for security. There are conferences in the field of *post-quantum cryptography*. The book *Post-Quantum Cryptography* [47] contains a compilation of papers on the implications of quantum computing for cryptography and overviews of some of the more promising directions. Perlner and Cooper [228] survey public key encryption and digital signature schemes that are not known to be vulnerable to quantum attacks and discuss design criteria that need to be met if such a system were to be deployed in the future.

8.8   Exercises

Exercise 8.1.   Give the exact value of the scale factor C in equation 8.2 in terms of properties of f and u.

Exercise 8.2.   Show that with high probability v, the value obtained from the quantum core of Shor's algorithm described in section 8.2.1 is within $\frac{1}{2}$ of some multiple of $\frac{2^n}{r}$,

Exercise 8.3.   Determine the efficiency of Shor's algorithm in the general case when r does not divide 2^n.

Exercise 8.4.   Show that the probability that the period of $f(x) = a^x \bmod M$ is odd is at most $1/2$.

Exercise 8.5.   Show that in the general case in which r does not divide 2^n, the parts of Shor's algorithm need to be repeated only $O(\log \log r)$ times in order to achieve a high probability of success.

Exercise 8.6.   Explain how Deutsch's problem of section 7.3.1 is an instance of the hidden subgroup problem.

Exercise 8.7.   Explain how Simon's problem is an instance of the hidden subgroup problem.

9 Grover's Algorithm and Generalizations

Grover's algorithm is the most famous algorithm in quantum computing after Shor's algorithm. Its status, however, differs from that of Shor's in a number of respects. Shor's algorithm solves a problem with clear practical consequences, but its application is focused on a narrow, if important, range of problems. In contrast, Grover's algorithm and its many generalizations can be applied to a broad range of problems, although, as section 9.6 explains, there is debate as to how far-reaching the practical implications of Grover's algorithm and its generalizations are.

Grover's algorithm solves a black box problem. It succeeds in finding a solution with $O(\sqrt{N})$ calls to the oracle, whereas the best possible classical approaches require $O(N)$ calls. Thus, unlike Shor's algorithm, Grover's algorithm is provably better than any possible classical algorithm. This query complexity improvement over the classical case translates to a speedup only under certain conditions; it depends on the efficiency with which the black box can be implemented, and on whether there is additional structure to the problem that can be exploited by classical and quantum algorithms. This issue will be discussed in section 9.6. Even when the query complexity result translates to a time complexity improvement, the speedup is much less than for Shor's algorithm.

The $O(\sqrt{N})$ query complexity of Grover's algorithm is known to be optimal; no quantum algorithm can do better. This restriction is as important as the algorithm itself. It places a severe restriction on the power of quantum computation. Although Grover's algorithm is usually presented as succeeding with high probability, unlike for Shor's algorithm, variations that succeed with certainty are known. Grover's algorithm is simpler and easier to grasp than Shor's, and has an elegant geometric interpretation.

Section 9.1 describes Grover's algorithm and determines its query complexity. Section 9.2 covers amplitude amplification, a generalization of Grover's algorithm. It also provides a simple geometric view of the algorithm. The optimality of Grover's algorithm is proved in section 9.3. Section 9.4 shows how to derandomize Grover's algorithm while preserving its efficiency. Section 9.5 generalizes Grover's algorithm to handle cases in which the number of solutions is not known. Section 9.6 discusses black box implementability, explains under what circumstances the query complexity results translate into a speedup, and evaluates the extent of practical potential applications for Grover's algorithm.

9.1   Grover's Algorithm

Grover's algorithm uses *amplitude amplification* to search an unstructured set of N elements. The problem is usually stated in terms of a Boolean function, or predicate, $P : \{0, \ldots, N-1\} \rightarrow \{0, 1\}$ that captures the property being searched for. The goal of the problem is to find a *solution*, an element x with $P(x) = 1$. As in Simon's problem and the Deutsch-Jozsa problem, the predicate P is viewed an oracle, or black box, and we will concern ourselves with the query complexity, the number of calls made to the oracle P. Given a black box that outputs $P(x)$ upon input of x, the best classical approaches must, in the single solution case, inspect an average of $N/2$ values; it requires an average of $N/2$ evaluations of the predicate $P(x)$. Given a quantum black box U_P that outputs

$$\sum_x c_x |x\rangle |P(x)\rangle$$

upon input of

$$\sum_x c_x |x\rangle |0\rangle,$$

Grover's algorithm finds a solution with only $O(\sqrt{N})$ calls to U_P in the single solution case. Grover's algorithm iteratively increases the amplitudes c_x of those values x with $P(x) = 1$, so that a final measurement will return a value x of interest with high probability. For practical applications of Grover's algorithm, the predicate P must be efficiently computable, but without enough structure to enable classical methods to gain on the quantum algorithm.

9.1.1   Outline

Grover's algorithm starts with an equal superposition $|\psi\rangle = \frac{1}{\sqrt{N}} \sum_x |x\rangle$ of all N values of the search space and repeatedly performs the same sequence of transformations:

1. Apply U_P to $|\psi\rangle$.

2. Flip the sign of all basis vectors that represent a solution.

3. Perform *inversion about the average*, a transformation that maps every amplitude $A - \delta$ to $A + \delta$, where A is the average of the amplitudes.

For the case of a single solution, figure 9.1 illustrates how these steps increase the amplitude of the basis vector of a solution. We now look at this process in detail.

9.1.2   Setup

Without loss of generality, let $N = 2^n$ for some integer n, and let X be the state space generated by $\{|0\rangle, \ldots, |N-1\rangle\}$. Let U_P be a quantum black box that acts as

$$U_P : |x, a\rangle \rightarrow |x, P(x) \oplus a\rangle,$$

for all $x \in X$ and single-qubit states $|a\rangle$.

a)

b)

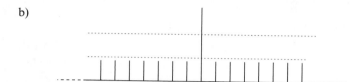

Figure 9.1
The iteration step of Grover's algorithm is achieved by (a) changing the sign of the good elements and (b) inverting about the average. The case of a single solution is illustrated.

Let $G = \{x|P(x)\}$ and $B = \{x|\neg P(x)\}$ denote the good and bad values respectively, and let the number of good states be a small fraction of the total number of states:

$$|G| << N.$$

Let

$$|\psi_G\rangle = \frac{1}{\sqrt{|G|}} \sum_{x \in G} |x\rangle$$

be an even superposition of all the good states, and

$$|\psi_B\rangle = \frac{1}{\sqrt{|B|}} \sum_{x \in B} |x\rangle$$

be an even superposition of the bad ones. Then $|\psi\rangle = W|0\rangle$, an equal superposition of all N values, can be written as a superpositions of $|\psi_G\rangle$ and $|\psi_B\rangle$

$$|\psi\rangle = \frac{1}{\sqrt{2^n}} \sum_{x=0}^{2^n-1} |x\rangle = g_0|\psi_G\rangle + b_0|\psi_B\rangle$$

where $g_0 = \sqrt{|G|/N}$ and $b_0 = \sqrt{|B|/N}$.

The core of Grover's algorithm is the repeated application of a unitary transformation

$$Q : g_i|\psi_G\rangle + b_i|\psi_B\rangle \rightarrow g_{i+1}|\psi_G\rangle + b_{i+1}|\psi_B\rangle$$

that increases the amplitude g_i of good states (and decreases b_i) until a maximal value is reached. After applying the amplitude amplifying transformation Q an appropriate number of times j, almost all amplitude will have shifted to good states, so that $|b_j| \ll |g_j|$. At this point, measurement will return an $x \in G$ with high probability. The exact number of times Q needs to be applied is on the order of $\sqrt{N}$ and depends on both N and $|G|$. Section 9.1.4 presents a detailed analysis.

9.1.3 The Iteration Step

The transformation Q is achieved by changing the sign of the good elements and then inverting about the average. This section describes the implementation of these two steps in detail. Both steps take real amplitudes to real amplitudes, so we will refer only to real amplitudes throughout the argument.

Changing the Sign of the Good Elements   To change the sign in a superposition $\sum c_x|x\rangle$ of exactly those $|x\rangle$ such that $x \in G$, apply S_G^π. A sign change is simply a phase shift by $e^{i\pi} = -1$. Section 7.4.2 showed that

$$U_P(|\psi\rangle \otimes H|1\rangle) = (S_G^\pi|\psi\rangle) \otimes H|1\rangle.$$

Changing the sign of the good elements is accomplished by

$$U_P : (g_i|\psi_G\rangle + b_i|\psi_B\rangle) \otimes H|1\rangle \rightarrow (-g_i|\psi_G\rangle + b_i|\psi_B\rangle) \otimes H|1\rangle.$$

The number of gates needed to change the sign on the good elements does not depend on N, but rather on how many gates it takes to compute U_P.

Inversion About the Average   Inversion about the average sends $a|x\rangle$ to $(2A - a)|x\rangle$ where A is the average of the amplitudes of all basis vectors in the superposition. (See figure 9.1.) It is easy to see that the transformation

$$\sum_{i=0}^{N-1} a_i|x_i\rangle \rightarrow \sum_{i=0}^{N-1} (2A - a_i)|x_i\rangle$$

is performed by the unitary transformation

$$D = \begin{pmatrix} \frac{2}{N} - 1 & \frac{2}{N} & \cdots & \frac{2}{N} \\ \frac{2}{N} & \frac{2}{N} - 1 & \cdots & \frac{2}{N} \\ \cdots & \cdots & \cdots & \cdots \\ \frac{2}{N} & \frac{2}{N} & \cdots & \frac{2}{N} - 1 \end{pmatrix}.$$

This paragraph shows how to implement this transformation with $O(n) = O(\log_2(N))$ quantum gates. Following Grover, define $D = -W S_0^\pi W$, where W is the Walsh-Hadamard transform and

$$S_0^\pi = \begin{pmatrix} -1 & 0 & \cdots & 0 \\ 0 & 1 & 0 & \cdots \\ 0 & \cdots & \cdots & 0 \\ 0 & \cdots & 0 & 1 \end{pmatrix}$$

is the phase shift by π of the basis vector $|0\rangle$ described in section 7.4.2. To see that $D = -W S_0^\pi W$, let

$$R = \begin{pmatrix} 2 & 0 & \ldots & 0 \\ 0 & 0 & 0 & \ldots \\ 0 & \ldots & \ldots & 0 \\ 0 & \ldots & 0 & 0 \end{pmatrix}.$$

Since $S_0^\pi = I - R$,

$$-W S_0^\pi W = W(R - I)W = WRW - I.$$

Since $R_{ij} = 0$ for $i \neq 0$ or $j \neq 0$,

$$(WRW)_{ij} = W_{i0} R_{00} W_{0j} = \frac{2}{N}$$

and $-W S_0^\pi W = WRW - I = D$.

Putting inversion about the average together with changing the sign of the good elements yields the iteration transformation

$$Q = -W S_0^\pi W S_G^\pi.$$

9.1.4 How Many Iterations?

This section examines the result of multiple application of the iteration step Q, which combines changing the sign and inverting about the average, in order to determine the optimal number of times to apply Q. It shows that Q is a fixed rotation and that the amplitude g_i of good states varies periodically with the number of iterations. To find a solution with high probability, the number of iterations i must be chosen carefully. To determine the correct number of iterations to use, we describe the result of applying Q in terms of recurrence relations on g_i and b_i.

The iteration step $Q = D S_G^\pi$ transforms $g_i |\psi_G\rangle + b_i |\psi_B\rangle$ to $g_{i+1} |\psi_G\rangle + b_{i+1} |\psi_B\rangle$. First,

$$S_G^\pi : g_i |\psi_G\rangle + b_i |\psi_B\rangle \rightarrow -g_i |\psi_G\rangle + b_i |\psi_B\rangle.$$

To compute the average amplitude, A_i, the term $-g_i |\psi_G\rangle$ contributes $|G|$ amplitudes

$$\frac{-g_i}{\sqrt{|G|}}$$

and $b_i |\psi_B\rangle$ contributes $|B|$ amplitudes

$$\frac{b_i}{\sqrt{|B|}}.$$

Thus, altogether

$$A_i = \frac{\sqrt{|B|} b_i - \sqrt{|G|} g_i}{N}.$$

Inversion about the average transforms

$$D: -g_i|\psi_G\rangle + b_i|\psi_B\rangle \rightarrow \sum_{x \in G}\left(2A_i + \frac{g_i}{\sqrt{|G|}}\right)|x\rangle + \sum_{x \in B}\left(2A_i - \frac{b_i}{\sqrt{|B|}}\right)|x\rangle$$

$$= (2A_i\sqrt{|G|} + g_i)|\psi_G\rangle + (2A_i\sqrt{|B|} - b_i)|\psi_B\rangle$$

$$= g_{i+1}|\psi_G\rangle + b_{i+1}|\psi_B\rangle$$

where

$$g_{i+1} = 2A_i\sqrt{|G|} + g_i,$$

$$b_{i+1} = 2A_i\sqrt{|B|} - b_i.$$

Let t be the probability that a random value in $\{0, \ldots, N-1\}$ satisfies P. Then $t = |G|/N$ and $1 - t = |B|/N$. Then

$$A_i\sqrt{|G|} = \frac{\sqrt{|B||G|}b_i - |G|g_i}{N} = \sqrt{t(1-t)}b_i - tg_i,$$

$$A_i\sqrt{|B|} = \frac{|B|b_i - \sqrt{|B||G|}g_i}{N} = (1-t)b_i - \sqrt{t(1-t)}g_i.$$

The recurrence relation can be written in terms of t:

$$g_{i+1} = (1-2t)g_i + 2\sqrt{t(1-t)}b_i,$$

$$b_{i+1} = (1-2t)b_i - 2\sqrt{t(1-t)}g_i$$

where $g_0 = \sqrt{t}$ and $b_0 = \sqrt{1-t}$. It is easy to verify that

$$g_i = \sin((2i+1)\theta)$$

$$b_i = \cos((2i+1)\theta)$$

is a solution to these equations with $\sin\theta = \sqrt{t} = \sqrt{|G|/N}$.

We are now ready to compute the optimum number of iterations of Q. To maximize the probability of measuring a good state, and thus finding an element with the desired property P, we wish to choose i such that $\sin((2i+1)\theta) \approx 1$ or $(2i+1)\theta \approx \pi/2$. For $|G| \ll N$ the angle θ becomes very small and $\sqrt{|G|/N} = \sin\theta \approx \theta$. Thus, g_i will be maximal for $i \approx \frac{\pi}{4}\sqrt{N/|G|}$.

Additional iteration will reduce the success probability of the algorithm. This situation is in contrast to many classical algorithms in which the greater the number of iterations the better the results. Using the equations for g_i and b_i, for $t = 1/4$, the optimum number of iterations is 1, and for $t = 1/2$, no amount of iteration will improve the situation.

Since every step of the iteration process has been written as a linear combination of $|\psi_G\rangle$ and $|\psi_B\rangle$ with real coefficients, Grover's algorithm can be viewed as acting in the real two-dimensional subspace spanned by $|\psi_G\rangle$ and $|\psi_B\rangle$. The algorithm simply shifts amplitude from $|\psi_B\rangle$ to $|\psi_G\rangle$. This picture leads to an elegant geometric interpretation of Grover's algorithm discussed in section 9.2.1. First, we describe a generalization of Grover's algorithm, amplitude amplification, to which this geometric picture also applies.

9.2 Amplitude Amplification

The first step of Grover's algorithm applies the iteration operator $Q = -W S_0^\pi W S_G^\pi$ to the initial state $W|0\rangle$. We can look at W as a trivial algorithm that maps $|0\rangle$ to all possible values and thus to a solution with probability $|G|/N$. Suppose we have an algorithm U such that $U|0\rangle$ gives an initial solution with a higher probability. This section shows that the analysis of 9.1.4 generalizes directly to any algorithm U such that $U|0\rangle$ has some amplitude in the good states. Amplitude amplification generalizes Grover's algorithm by replacing the iteration operator $Q = -W S_0^\pi W S_G^\pi$ with

$$Q = -U S_0^\pi U^{-1} S_G^\pi.$$

The rest of this section generalizes the argument of section 9.1.4 to obtain the same recurrence relations for this more general case.

Let $\mathcal{G}$ and $\mathcal{B}$ be the subspaces spanned by $\{|x\rangle \,|\, x \in G\}$ and $\{|x\rangle \,|\, x \notin G\}$ respectively, and let $P_\mathcal{G}$ and $P_\mathcal{B}$ be the associated projection operators. Let $|\psi\rangle = U|0\rangle$ be written as

$$|\psi\rangle = g_0|\psi_G\rangle + b_0|\psi_B\rangle$$

where $|\psi_G\rangle$ and $|\psi_B\rangle$ are the normalized projections of $|\psi\rangle$ onto the good and bad subspaces,

$$|\psi_G\rangle = \frac{1}{g_0} P_\mathcal{G}|\psi\rangle,$$

and

$$|\psi_B\rangle = \frac{1}{b_0} P_\mathcal{B}|\psi\rangle,$$

with

$$g_0 = |P_\mathcal{G}|\psi\rangle|,$$

and

$$b_0 = |P_\mathcal{B}|\psi\rangle|.$$

For $U = W$, $|\psi_G\rangle$, $|\psi_B\rangle$, g_0, and b_0 are as in section 9.1.4. Here g_0 and b_0 are not determined by the number of solutions, but rather by the properties of U relative to the good states. The states

$|\psi_G\rangle$ and $|\psi_B\rangle$ need not be equal superpositions of the good and bad states respectively, but g_0 and b_0 are still real. Again, we let $t = g_0^2$ with $1 - t = b_0^2$, where t should be thought of as the probability that measurement of the superposition $U|0\rangle$ yields a state that satisfies predicate P. The operator U can be viewed as a reversible algorithm that maps $|0\rangle$ to a set of solutions in G with a probability $t = |g_0|^2$.

To understand the effect of $Q = -U S_0^\pi U^{-1} S_G^\pi$, recall from section 7.4.2 that $S_0^\pi |\varphi\rangle$ can be written as $|\varphi\rangle - 2\langle 0|\varphi\rangle|0\rangle$. For an arbitrary state $|\psi\rangle$,

$$
\begin{aligned}
U S_0^\pi U^{-1} |\psi\rangle &= U \left(U^{-1}|\psi\rangle - 2\langle 0|U^{-1}|\psi\rangle|0\rangle \right) \\
&= |\psi\rangle - 2\langle 0|U^{-1}|\psi\rangle U|0\rangle \\
&= |\psi\rangle - 2\overline{\langle \psi|U|0\rangle} U|0\rangle.
\end{aligned}
$$

Since $S_G^\pi |\psi_G\rangle = -|\psi_G\rangle$ and $S_B^\pi |\psi_B\rangle = |\psi_B\rangle$,

$$
\begin{aligned}
Q|\psi_G\rangle &= -U S_0^\pi U^{-1} S_G^\pi |\psi_G\rangle \\
&= U S_0^\pi U^{-1} |\psi_G\rangle \\
&= |\psi_G\rangle - 2\overline{g_0} U|0\rangle \\
&= |\psi_G\rangle - 2\overline{g_0} g_0 |\psi_G\rangle - 2\overline{g_0} b_0 |\psi_B\rangle \\
&= (1 - 2t)|\psi_G\rangle - 2\sqrt{t(1-t)}|\psi_B\rangle
\end{aligned}
$$

and

$$
\begin{aligned}
Q|\psi_B\rangle &= -|\psi_B\rangle + 2\overline{b_0} U|0\rangle \\
&= -|\psi_B\rangle + 2\overline{b_0} g_0 |\psi_G\rangle + 2\overline{b_0} b_0 |\psi_B\rangle \\
&= -|\psi_B\rangle + 2(1-t)\frac{g_0}{b_0}|\psi_G\rangle + 2(1-t)|\psi_B\rangle \\
&= (1 - 2t)|\psi_B\rangle + 2\sqrt{t(1-t)}|\psi_G\rangle.
\end{aligned}
$$

An arbitrary real superposition of $|\psi_G\rangle$ and $|\psi_B\rangle$ is transformed by Q as follows:

$$
\begin{aligned}
&Q(g_i|\psi_G\rangle + b_i|\psi_B\rangle) \\
&= (g_i(1 - 2t) + 2b_i\sqrt{t(1-t)})|\psi_G\rangle + (b_i(1 - 2t) - 2g_i\sqrt{t(1-t)})|\psi_B\rangle,
\end{aligned}
$$

which leads to the same recurrence relation as in the previous section,

$$
g_{i+1} = (1 - 2t)g_i + 2\sqrt{t(1-t)}b_i
$$
$$
b_{i+1} = (1 - 2t)b_i - 2\sqrt{t(1-t)}g_i,
$$

with the solution

$$g_i = \sin((2i + 1)\theta)$$

$$b_i = \cos((2i + 1)\theta)$$

for $\sin\theta = \sqrt{t} = g_0$.

Thus, for small g_0, the amplitude g_i will be maximal after $i \approx \frac{\pi}{4}\frac{1}{g_0}$ iterations. If the algorithm U succeeds with probability t, then simple classical repetition of U requires an average of $1/t$ iterations to find a solution. Amplitude amplification speeds up this process so that it takes only $O(\sqrt{1/t})$ tries to find a solution. If U has no amplitude in the good states, g_0 will be zero and amplitude amplification will have no effect. Furthermore, just as no amount of iteration in Grover's algorithm improves the probability if $t = 1/2$, if g_0 is large, amplitude amplification cannot improve the situation. For this reason, amplitude amplification applied to an algorithm U that is the result of amplitude amplification does not improve the results.

9.2.1 The Geometry of Amplitude Amplification

The reasoning behind amplitude amplification, including the optimal number of iterations of Q to perform, can be reduced to a simple argument in two-dimensional Euclidean geometry. Let $|\psi_G\rangle$, $|\psi_B\rangle$, and $Q = -US_0^\pi U^{-1} S_G^\pi$ be as defined before. This section shows that the entire discussion of amplitude amplification, and Grover's algorithm in particular, reduces to a simple geometric argument about rotations in the two-dimensional real subspace generated by $\{|\psi_G\rangle, |\psi_B\rangle\}$.

By the definition of $|\psi_G\rangle$ and $|\psi_B\rangle$, the initial state $U|0\rangle = g_0|\psi_G\rangle + b_0|\psi_B\rangle$ has real amplitudes g_0 and b_0, so is in the two-dimensional real plane spanned by $\{|\psi_G\rangle, |\psi_B\rangle\}$. The smaller the success probability t, the closer $U|0\rangle$ is to $|\psi_B\rangle$. Let β be the angle between $U|0\rangle$ and $|\psi_G\rangle$ illustrated in figure 9.2. The angle β depends only on the probability $t = g_0^2$ that the initial state $U|0\rangle$, if measured, gives a solution: $\cos(\beta) = \langle\psi_G|U|0\rangle = g_0$. The rest of this section explains how each iteration of Grover's algorithm rotates the state by a fixed angle in the direction of the desired

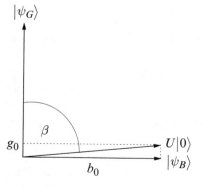

Figure 9.2
The initial state $U|0\rangle$ in the basis $\{|\psi_G\rangle, |\psi_B\rangle\}$.

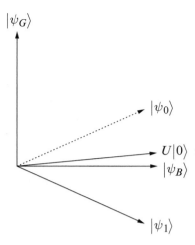

Figure 9.3
The transformation S_G^π reflects $|\psi_0\rangle$ about $|\psi_B\rangle$, result in the state $|\psi_1\rangle$

state. To maximize the amplitude in the good states, we iterate until the state is close to $|\psi_G\rangle$. From the simple geometry of the situation, we can determine both the optimal number of iterations and the probability that the run succeeds.

Amplitude amplification, and Grover's algorithm as the special case when $U = W$, consists of repeated applications of $Q = -U S_0^\pi U^{-1} S_G^\pi$. To understand this transformation geometrically, recall from section 7.4.2 that the transformation S_G^π can be viewed as a reflection about the hyperplane perpendicular to $|\psi_G\rangle$. In the plane spanned by $\{|\psi_G\rangle, |\psi_B\rangle\}$, this hyperplane reduces to the one-dimensional space spanned by $|\psi_B\rangle$. Figure 9.3 illustrates how S_G^π maps an arbitrary state $|\psi_0\rangle$ in the $\{|\psi_G\rangle, |\psi_B\rangle\}$ subspace to $|\psi_1\rangle = S_G^\pi |\psi_0\rangle$. Similarly, the transformation S_0^π is a reflection about the hyperplane orthogonal to $|0\rangle$. Since $U S_0^\pi U^{-1}$ differs from S_0^π by a change of basis, it is a reflection about the hyperplane orthogonal to $U|0\rangle$. The effect of this transformation on $|\psi_1\rangle$ is shown in figure 9.4. The final negative sign reverses the direction of the state vector, shown in figure 9.5. (Strictly speaking, this negative sign is unnecessary, since it does nothing to the quantum state: it is a global phase change, so it is physically irrelevant. However, since we are drawing our pictures in the plane, not in projective space, the negative sign makes it easier to see what is going on.) Recall from Euclidean geometry that the concatenation of two reflections is a rotation of twice the angle between the axes of the two reflections. The two axes of reflection in this case are perpendicular to $U|0\rangle$ and $|\psi_G\rangle$ respectively, so the angle between the axes of reflection is $-\beta$ where $\cos \beta = g_0$ as before. The two reflections perform a rotation by -2β, and the final negation amounts to a rotation by π. Thus, each step Q performs a rotation by $\pi - 2\beta$.

Let $\theta = \frac{\pi}{2} - \beta$, the angle between $U|0\rangle$ and $|\psi_B\rangle$, so $\sin \theta = g_0$ as it did in the analyses of the previous sections. Each iteration of Q rotates the state by 2θ, so the angle after i steps is $(2i + 1)\theta$.

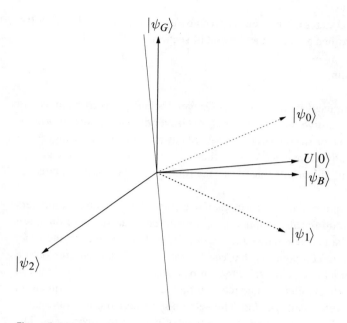

Figure 9.4
The transformation $U S_0^\pi U^{-1}$ reflects $|\psi_1\rangle$ about a line perpendicular to $U|0\rangle$, resulting in the state $|\psi_2\rangle$.

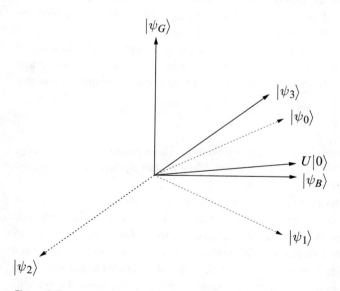

Figure 9.5
The final negative sign maps $|\psi_2\rangle$ to $|\psi_3\rangle$ for a total rotation of $\pi - 2\beta$.

As before, the amplitude in the good states after i steps is given by $g_i = \sin((2i + 1)\theta)$. We solve for the optimal number of iterations just as we did at the end of section 9.1.4.

9.3 Optimality of Grover's Algorithm

As important as Grover's algorithm itself is the proof that Grover's algorithm is as good as any possible quantum algorithm for exhaustive search. Even before Grover discovered his algorithm, researchers had proved a lower bound on the query complexity of any possible quantum algorithm for exhaustive search: no quantum algorithm can use fewer than $\Omega(\sqrt{N})$ calls to the predicate U_P. Thus, Grover's algorithm is optimal. This result places a severe limit on what quantum computers can ever hope to do.

The exponential size of the quantum state space gives naive hope that quantum computers could provide an exponential speedup for all computations; popular press accounts of quantum computers still widely make this claim. A less naive guess would be that quantum computers can provide exponential speedup for any computation that can be parallelized and requires only a single answer output. But the optimality of Grover's algorithm shows that even that hope is too optimistic; exhaustive search is easily parallelized and requires only a single answer, but quantum computers can provide only a relatively small speedup. This section sketches a proof of optimality in the case of a single solution x. The proof bounds the number of calls to the *oracle U_P*. The argument generalizes to the case of multiple solutions.

Section 7.4.2 shows how S_x^π can be computed from U_P. We use S_x^π as the interface to the oracle. We do not lose any generality in doing so; the process of computing S_x^π from U_P is reversible, so any algorithm using S_x^π could be rewritten in terms of U_P and vice versa.

Since the oracle U_P provides us with the only way to access any information about the element x we are searching for, an arbitrary quantum search algorithm can be viewed as an algorithm that alternates between unitary transformations independent of x and calls to S_x^π; any quantum search algorithm can be written as

$$|\psi_k^x\rangle = U_k S_x^\pi U_{k-1} S_x^\pi \ldots U_1 S_x^\pi U_0 |0\rangle,$$

where the U_i are unitary transformations that do not depend on x. The argument does not change if we allow the use of additional qubits; we simply use $I \otimes S_x^\pi$ instead of S_x^π and, since N is now larger, the algorithm will be less efficient.

It is important to recognize that the algorithm must work no matter which x is the solution. For any particular x, there are transformations that find x very quickly. We want an algorithm that finds x quickly no matter what x is. Any search algorithm worth the name must return x with reasonable probability for all possible values of x. We consider only quantum search algorithms that return x with at least probability $p = 1/2$. It is easy for the reader to check that any value $0 < p < 1$ results in an $O(\sqrt{N})$ bound, just with a different constant. More formally, we will show that if the state $|\psi_k^x\rangle$, obtained after k steps of the form $U_i S_x^\pi$, satisfies

$$|\langle x|\psi_k^x\rangle|^2 \geq \frac{1}{2}$$

for all x, then k must be $\Omega(\sqrt{N})$.

This paragraph describes the rough strategy and intuition behind the proof. The requirement that the algorithm work for any x means that if the oracle interface is S_x^π, then the result of applying the algorithm $U_k S_x^\pi U_{k-1} S_x^\pi \ldots U_1 S_x^\pi U_0 |0\rangle$ must be a state $|\psi_k^x\rangle$ sufficiently close to $|x\rangle$ that x will be obtained upon measurement with high probability. Since two elements of the standard basis $|x\rangle$ and $|y\rangle$ cannot be closer than a certain constant, the final states of the algorithm for different S_x^π and S_y^π must be sufficiently far apart. Since the U_i are all the same, any difference in the result of running the algorithm arises from calls to S_x^π. The algorithms all start with the same state $U_0|0\rangle$, so if we can bound from above the amount each step increases the distance between $|\psi_i^x\rangle$ and $|\psi_i^y\rangle$, then we can obtain a bound on k, the number of calls to the oracle interface S_x^π. In other words, we want to bound from above the amount this distance can increase by applying $U_i S_x^\pi$ to $|\psi_{i-1}^x\rangle$ and $U_i S_y^\pi$ to $|\psi_{i-1}^y\rangle$. To obtain this bound, we compare both $|\psi_i^x\rangle$ and $|\psi_i^y\rangle$ with $|\psi_i\rangle$, the state obtained by applying U_0 up through U_i without any intervening calls to S_x^π. We first give the details of how to use inequalities based on these ideas to prove that $\Omega(\sqrt{N})$ calls to the oracle are required, and then give detailed proofs of each of the inequalities.

9.3.1 Reduction to Three Inequalities

The proof considers the relation between three classes of quantum states: the desired result $|x\rangle$, the state of the computation $|\psi_k^x\rangle$ after k steps, and the state $|\psi_k\rangle = U_k U_{k-1} \ldots U_1 U_0 |0\rangle$ obtained by performing the sequence of transformations U_i without consulting the oracle. The analysis simplifies if we sometimes consider, instead of $|x\rangle$, a phase-adjusted version of $|x\rangle$, namely $|x_k'\rangle = e^{i\theta_k^x}|x\rangle$, where $e^{i\theta_k^x} = \langle x|\psi_k^x\rangle/|\langle x|\psi_k^x\rangle|$. The phase adjustment is chosen so that $\langle x_k'|\psi_k^x\rangle$ is positive real for all k. Since $|x_k'\rangle$ differs from $|x\rangle$ only in a phase, whenever $|\langle x|\psi_k^x\rangle|^2 \geq \frac{1}{2}$, we have a similar inequality for $|x_k'\rangle$, namely

$$|\langle x_k'|\psi_k^x\rangle|^2 \geq \frac{1}{2},$$

in which case $\langle x_k'|\psi_k^x\rangle \geq \frac{1}{\sqrt{2}}$.

We consider the distances between certain pairs of these states:

$$d_{kx} = ||\psi_k^x\rangle - |\psi_k\rangle|$$
$$a_{kx} = ||\psi_k^x\rangle - |x_k'\rangle|$$
$$c_{kx} = ||x_k'\rangle - |\psi_k\rangle|.$$

The proof establishes bounds involving the sum, or average, of these distances squared:

$$D_k = \frac{1}{N}\sum_x d_{kx}^2, \qquad A_k = \frac{1}{N}\sum_x a_{kx}^2, \qquad C_k = \frac{1}{N}\sum_x c_{kx}^2.$$

The reason for considering the sum, or equivalently the average, is that any generally useful search algorithm must efficiently find x for all possible x. The proof relies on three inequalities involving D_k, A_k, and C_k, which we will prove in section 9.3.2. Before proving the inequalities, we describe them and show how they imply a lower bound on the number of calls to the oracle.

The first inequality bounds from above A_k, the average squared distance between the state $|\psi_k^x\rangle$ obtained after k steps and and the phase adjusted solution state $|x_k'\rangle$; section 9.3.2 shows that in order to obtain a success probability of $|\langle x|\psi_k^x\rangle|^2 \geq \frac{1}{2}$, the following inequality must hold:

$$A_k \leq 2 - \sqrt{2}.$$

The second inequality bounds C_k, the sum of the squared distances between the vector $|\psi_k\rangle$ and all basis vectors $|j\rangle$, from below as long as $N \geq 4$:

$$C_k \geq 1.$$

The third inequality bounds the growth of D_k, the average squared distance between $|\psi_k^x\rangle$ and $|\psi_k\rangle$ as k increases:

$$D_k \leq \frac{4k^2}{N}.$$

The three quantities d_{kx}, a_{kx}, and c_{kx} are related as follows:

$$d_{kx} = \big| |\psi_k^x\rangle - |\psi_k\rangle \big| = \big| |\psi_k^x\rangle - e^{i\theta_x^k}|x\rangle + e^{i\theta_x^k}|x\rangle - |\psi_k\rangle \big| \geq a_{kx} - c_{kx}.$$

To relate the quantities D_{kx}, A_{kx}, and C_{kx}, we use the Cauchy-Schwarz inequality (see box 9.1) to obtain

Box 9.1
The Cauchy-Schwarz Inequality

We use the Cauchy-Schwarz inequality in two forms, the general form

$$\sum_i u_i v_i \leq \sqrt{\left(\sum_i u_i^2\right)\left(\sum_i v_i^2\right)}, \tag{9.1}$$

and a specialization for $v_i = 1$ in an N dimensional space

$$\sum_i u_i \leq \sqrt{N}\sqrt{\sum_i u_i^2}. \tag{9.2}$$

$$D_k = \frac{1}{N} \sum_x d_{kx}^2$$

$$\geq \frac{1}{N} \left(\sum_x a_{kx}^2 - 2 \sum_x a_{kx} c_{kx} + \sum_x c_{kx}^2 \right)$$

$$\geq \frac{1}{N} \sum_x a_{kx}^2 - \frac{2}{N} \sqrt{\left(\sum_x a_{kx}^2 \right) \left(\sum_x c_{kx}^2 \right)} + \frac{1}{N} \sum_x c_{kx}^2$$

$$\geq A_k - 2\sqrt{A_k C_k} + C_k.$$

Making use of this inequality and the three earlier ones, we bound $\frac{4k^2}{N}$ from below by a constant:

$$\frac{4k^2}{N} \geq D_k$$

$$\geq A_k - 2\sqrt{A_k C_k} + C_k$$

$$= \left(\sqrt{C_k} - \sqrt{A_k} \right)^2$$

$$\geq \left(1 - \sqrt{2 - \sqrt{2}} \right)^2,$$

since $1 \geq 2 - \sqrt{2} \geq A_k$. Thus, for $N \geq 4$ (needed for the second inequality), and taking $q = 1 - \sqrt{2 - \sqrt{2}}$, at least $k \geq \frac{q}{2}\sqrt{N}$ iterations are required for a success probability of $|\langle x | \psi_k^x \rangle|^2 \geq \frac{1}{2}$ for all x.

We now turn to the proofs of the three inequalities.

9.3.2 Proofs of the Three Inequalities

The inequality for A_k By assumption, $|\langle \psi_k^x | x \rangle|^2 \geq \frac{1}{2}$. By the choice of phase $e^{i\theta_k^x}$ relating $|x\rangle$ and $|x_k'\rangle$,

$$\langle \psi_k^x | x_k' \rangle \geq \frac{1}{\sqrt{2}},$$

so

$$a_{kx}^2 = \| |\psi_k^x\rangle - |x_k'\rangle \|^2$$

$$= \| |\psi_k^x\rangle \|^2 - 2\langle x_k' | \psi_k^x \rangle + \| |x_k'\rangle \|^2$$

$$\leq 2 - \sqrt{2},$$

from which it follows that

$$A_k = \frac{1}{N} \sum_x a_{kx}^2 \leq 2 - \sqrt{2}.$$

Bound on sum squared distance to all basis vectors   The terms c_{kx}^2 can be bounded as follows:

$$c_{kx}^2 = ||x_k'\rangle - |\psi_k\rangle|^2$$

$$= |e^{i\theta_k^x}|x\rangle - |\psi_k\rangle|^2$$

$$= ||\psi_k\rangle|^2 - \overline{e^{i\theta_k^x}\langle\psi_k|x\rangle} - e^{i\theta_k^x}\langle\psi_k|x\rangle + ||x\rangle|^2$$

$$= 2 - 2Re(e^{i\theta_k^x}\langle\psi_k|x\rangle)$$

$$\geq 2 - 2|\langle x|\psi_k\rangle|.$$

We can now bound the average of these terms:

$$C_k = \frac{1}{N}\sum_x c_{kx}^2$$

$$\geq 2 - \frac{2}{N}\sum_x |\langle x|\psi_k\rangle|$$

$$\geq 2 - \frac{2}{\sqrt{N}}\sqrt{\sum_x |\langle x|\psi_k\rangle|^2} \tag{9.3}$$

$$= 2 - \frac{2}{\sqrt{N}}, \tag{9.4}$$

where inequality 9.3 follows from the Cauchy-Schwarz inequality (box 9.1), and equation 9.4 holds because $|\psi_k\rangle$ is a unit vector and $\{|x\rangle\}$ forms a basis. Thus, the second inequality $C_k \geq 1$ holds as long as $N \geq 4$.

As an aside, since this argument made no assumption about $|\psi_k\rangle$, the bound on the sum of the distances to all basis vectors holds for any quantum state:

$$\frac{1}{N}\sum_x ||x\rangle - |\psi\rangle|^2 \geq 2 - \frac{2}{\sqrt{N}}$$

for any $|\psi\rangle$.

The inequality for D_k   First, we bound how much the distance between $|\psi_k^x\rangle$ and $|\psi_k\rangle$ can increase each step. Consider the following relation between d_{kx} and $d_{k+1,x}$:

$$d_{k+1,x} = ||\psi_{k+1}^x\rangle - |\psi_{k+1}\rangle|$$

$$= |U_{k+1}S_x^\pi|\psi_k^x\rangle - U_{k+1}|\psi_k\rangle|$$

$$= |S_x^\pi|\psi_k^x\rangle - |\psi_k\rangle|$$

$$= |S_x^\pi(|\psi_k^x\rangle - |\psi_k\rangle) + (S_x^\pi - I)|\psi_k\rangle|$$

$$\leq |S_x^{\pi}(|\psi_k^x\rangle - |\psi_k\rangle)| + |(S_x^{\pi} - I)|\psi_k\rangle|$$

$$= d_{kx} + 2|\langle x|\psi_k\rangle|.$$

This inequality shows that with each step the distance between $|\psi_k^x\rangle$ and $|\psi_k\rangle$ can increase by at most $2|\langle x|\psi_k\rangle|$. Using this bound, we prove by induction that

$$D_k = \frac{1}{N}\sum_x d_{kx}^2 \leq \frac{4k^2}{N}.$$

Base case   For $k = 0$, for all x, we have $|\psi_0^x\rangle = U_0|0\rangle = |\psi_0\rangle$, so $d_{0x} = 0$ and therefore $D_0 = 0$.

Induction step

$$D_{k+1} = \frac{1}{N}\sum_x d_{k+1,x}^2$$

$$\leq \frac{1}{N}\sum_x (d_{kx} + 2|\langle x|\psi_k\rangle|)^2$$

$$= \frac{1}{N}\sum_x d_{kx}^2 + \frac{4}{N}\sum_x |\langle x|\psi_k\rangle|^2 + \frac{4}{N}\sum_x d_{kx}|\langle x|\psi_k\rangle|$$

$$= D_k + \frac{4}{N} + \frac{4}{N}\sum_x d_{kx}|\langle x|\psi_k\rangle|.$$

The Cauchy-Schwarz inequality gives

$$\frac{1}{N}\sum_x d_{kx}\,|\langle x|\psi_k\rangle| \leq \frac{1}{N}\sqrt{\left(\sum_x d_{kx}^2\right)\left(\sum_x |\langle x|\psi_k\rangle|^2\right)} = \sqrt{\frac{D_k}{N}}.$$

Using the induction assumption $D_k \leq \frac{4k^2}{N}$, we have

$$D_{k+1} \leq D_k + \frac{4}{N} + 4\sqrt{\frac{D_k}{N}} \leq \frac{4(k+1)^2}{N}.$$

9.4   Derandomization of Grover's Algorithm and Amplitude Amplification

Unlike Shor's algorithm, Grover's algorithm is not inherently probabilistic. With a little cleverness, Grover's algorithm can be modified in such a way that it is guaranteed to find a solution while still preserving the quadratic speedup. More generally, amplitude amplification can be derandomized. Brassard, Høyer, and Tapp suggest two approaches. In the first, each iteration rotates by an angle that is slightly smaller than the one used in section 9.2.1, while the second changes only the last step to a smaller rotation. This section describes each approach in turn.

9.4.1 Approach 1: Modifying Each Step

Suppose the angle θ in Grover's algorithm or amplitude amplification happened to be such that $\frac{\pi}{4\theta} - \frac{1}{2}$ were an integer. In this case, after $i = \frac{\pi}{4\theta} - \frac{1}{2}$ iterations, the amplitude g_i would be 1 and the algorithm would output a solution with certainty. Recall from section 9.2 that θ satisfies $\sin \theta = \sqrt{t} = g_0$. To derandomize amplitude amplification for algorithm U with success probability g_0, we modify U to obtain an algorithm U' with success probability $g_0' < g_0$ such that, for θ' satisfying $\sin \theta' = g_0'$, the quantity $\frac{\pi}{4\theta'} - \frac{1}{2}$ is an integer.

Intuitively, it seems as though it should not be hard to modify an algorithm U so that it is less successful, but we must make sure that we can compute such a U' efficiently from U. The trick is to allow the use of an additional qubit b. Given an algorithm U with success probability g_0 acting on an n-qubit register $|s\rangle$, define U' to be the transformation $U \otimes B$ on an $(n+1)$-qubit register $|s\rangle|b\rangle$, where B is the single-qubit transformation

$$B = \sqrt{1 - \frac{g_0'}{g_0}}|0\rangle + \sqrt{\frac{g_0'}{g_0}}|1\rangle.$$

Let G' be the set of basis states $|x\rangle \otimes |b\rangle$ such that $|x\rangle \in G$ and $|b\rangle = |1\rangle$. The reader may check that the initial success probability $|P_{G'}U'|0\rangle|$ is indeed g_0'. Amplitude amplification, now on an $(n+1)$-qubit state, with U' for U, $S_{G'}^\pi$ for S_G^π, and iteration operator $Q' = -U'S_0^\pi (U')^{-1} S_{G'}^\pi$, succeeds with certainty after $i = \frac{\pi}{4\theta'} - \frac{1}{2}$ steps.

This modified algorithm obtains a solution with certainty, using $O(\sqrt{\frac{1}{t}})$ calls to the oracle, at the at the cost of a single additional qubit.

9.4.2 Approach 2: Modifying Only the Last Step

This approach is more complicated to describe, but results in a solution in $O\left(\sqrt{\frac{1}{t}}\right)$ time with certainty without the need for an additional qubit. The idea is to modify S_G^π and S_0^π in the last step so that exactly the desired final state is obtained. To this end, we begin by analyzing general properties of transformations of the form

$$Q(\phi, \tau) = -U S_0^\phi U^{-1} S_G^\tau,$$

where ϕ and τ are both arbitrary angles and

$$S_X^\phi |x\rangle = \begin{cases} e^{i\phi}|x\rangle & \text{if } |x\rangle \in X \\ |x\rangle & \text{if } |x\rangle \notin X. \end{cases}$$

Section 7.4.2 showed how to implement S_X^ϕ efficiently.

First, we show that for any quantum state $|v\rangle$,

$$U S_0^\phi U^{-1}|v\rangle = |v\rangle - \left(1 - e^{i\phi}\right) \overline{\langle v|U|0\rangle} U|0\rangle.$$

Write

$$|v\rangle = \sum_{i=1}^{N-1} \overline{\langle v|U|i\rangle} U|i\rangle + \overline{\langle v|U|0\rangle} U|0\rangle.$$

Then

$$U S_0^\phi U^{-1} |v\rangle = U S_0^\phi \left(\sum_{i=1}^{N-1} \overline{\langle v|U|i\rangle} |i\rangle + \overline{\langle v|U|0\rangle} |0\rangle \right)$$

$$= U \left(\sum_{i=1}^{N-1} \overline{\langle v|U|i\rangle} |i\rangle + \overline{\langle v|U e^{i\phi}|0\rangle} |0\rangle \right)$$

$$= \sum_{i=1}^{N-1} \overline{\langle v|U|i\rangle} U|i\rangle + e^{i\phi} \overline{\langle v|U|0\rangle} U|0\rangle$$

$$= |v\rangle - \left(1 - e^{i\phi}\right) \overline{\langle v|U|0\rangle} U|0\rangle.$$

Using this result, we now can see the effect of $Q(\phi, \tau) = U S_0^\phi U^{-1} S_G^\tau$ on any superposition $|v\rangle = g|v_G\rangle + b|v_B\rangle$ in the subspace spanned by $|v_G\rangle$ and $|v_B\rangle$. We have

$$Q(\phi, \tau)|v\rangle = g(-e^{i\tau}|v_G\rangle + e^{i\tau}(1 - e^{i\phi})\overline{\langle v_G|U|0\rangle} U|0\rangle)$$

$$+ b(-|v_B\rangle + (1 - e^{i\phi})\overline{\langle v_B|U|0\rangle} U|0\rangle).$$

After $s = \lfloor \frac{\pi}{4\theta} - \frac{1}{2} \rfloor$ iterations of amplitude amplification we have the state $|\psi_s\rangle = \sin((2s+1)\theta)|\psi_G\rangle + \cos((2s+1)\theta)|\psi_B\rangle$, where $\sin\theta = \sqrt{t} = g_0$. Applying $Q(\phi, \tau)$ to the states $|\psi_G\rangle$ and $|\psi_B\rangle$, we obtain

$$Q(\phi, \tau)|\psi_G\rangle = e^{i\tau}\left((1 - e^{i\phi})g_0^2 - 1\right)|\psi_G\rangle + e^{i\tau}(1 - e^{i\phi})g_0 b_0 |\psi_B\rangle),$$

$$Q(\phi, \tau)|\psi_B\rangle = (1 - e^{i\phi})b_0 g_0 |\psi_G\rangle + \left((1 - e^{i\phi})b_0^2 - 1\right)|\psi_B\rangle).$$

So

$$Q(\phi, \tau)|\psi\rangle = g(\phi, \tau)|\psi_G\rangle + b(\phi, \tau)|\psi_B\rangle,$$

where

$$g(\phi, \tau) = \sin((2s+1)\theta) e^{i\tau}\left((1 - e^{i\phi})g_0^2 - 1\right) + \cos((2s+1)\theta)(1 - e^{i\phi})b_0 g_0$$

$$b(\phi, \tau) = \sin((2s+1)\theta) e^{i\tau}(1 - e^{i\phi})g_0 b_0 + \cos((2s+1)\theta)\left((1 - e^{i\phi})b_0^2 - 1\right).$$

Our aim now is to show that there exist ϕ and τ such that if $Q(\phi, \tau) = U S_0^\phi U^{-1} S_G^\tau$ is applied as a final step, a solution is obtained with certainty.

To show that ϕ and τ can be chosen so that $Q(\phi, \tau)|\psi\rangle$ has all of its amplitude in the good states, we want $b(\phi, \tau) = 0$ or

$$\left(\sin\left((2s+1)\theta\right)e^{\mathbf{i}\tau}(1-e^{\mathbf{i}\phi})g_0b_0\right)+\cos\left((2s+1)\theta\right)\left((1-e^{\mathbf{i}\phi})b_0^2-1\right)=0,$$

or

$$e^{\mathbf{i}\tau}(1-e^{\mathbf{i}\phi})g_0\sqrt{1-g_0^2}\sin\left((2s+1)\theta\right)=\left(1-(1-e^{\mathbf{i}\phi})(1-g_0^2)\right)\cos\left((2s+1)\theta\right),$$

since $b_0=\sqrt{1-g_0^2}$. Since the right-hand side equals

$$\left(g_0^2(1-e^{\mathbf{i}\phi})+e^{\mathbf{i}\phi}\right)\cos\left((2s+1)\theta\right),$$

we want ϕ and τ to satisfy

$$\cot\left((2s+1)\theta\right)=\frac{e^{\mathbf{i}\tau}(1-e^{\mathbf{i}\phi})g_0\sqrt{1-g_0^2}}{g_0^2(1-e^{\mathbf{i}\phi})+e^{\mathbf{i}\phi}}. \tag{9.5}$$

Once ϕ is chosen, we choose τ to make the right-hand side real. To find ϕ, compute the magnitude squared of the right-hand side of equation 9.5

$$\frac{g_0^2b_0^2(2-2\cos\phi)}{g_0^4(2-2\cos\phi)-g_0^2(2-2\cos\phi)+1}.$$

The maximum value of the magnitude squared, obtained when $\cos\phi=-1$, is

$$\frac{4g_0^2b_0^2}{4g_0^4-4g_0^2+1}=\frac{4g_0^2b_0^2}{(2g_0^2-1)^2}.$$

So the maximum magnitude is

$$\frac{2g_0b_0}{2g_0^2-1}=\frac{2g_0b_0}{g_0^2-b_0^2}=\tan(2\theta),$$

where $\sin\theta=\sqrt{t}=g_0$ as before. Thus, ϕ and τ can be chosen to make the right-hand side of equation 9.5 any real number between $[0,\tan(2\theta)]$. By the geometric interpretation of section 9.2.1, after $s=\lfloor\frac{\pi}{4\theta}-\frac{1}{2}\rfloor$ iterations, the state has been rotated to within 2θ of the desired state. Thus, we have shown that ϕ and τ can be chosen so that applying s iterations of Q, followed by one application of $Q(\phi,\tau)$, yields a solution with certainty.

9.5 Unknown Number of Solutions

Grover's algorithm requires that we know the relative number of solutions $t=|G|/N$ in order to determine how many times we should apply the transformation Q. More generally, amplitude amplification requires as input the success probability $t=|g_0|^2$ of $U|0\rangle$. This section sketches two approaches to handling cases in which we do not know t. The first approach repeats Grover's algorithm multiple times, choosing a random number of iterations of Q in each run. While inelegant, this approach does succeed in finding a solution with high probability. The second approach, called *quantum counting*, uses the quantum Fourier transform to estimate t. Both approaches require $O(\sqrt{N})$ calls to U_P.

9.5.1 Varying the Number of Iterations

Consider Grover's algorithm applied to a problem with tN solutions in a space of cardinality N. When t is unknown, a simple strategy is to repeatedly execute Grover's algorithm with a number of iteration steps picked randomly between 0 and $\frac{\pi}{4}\sqrt{N}$. For large values of t, this simple approach is clearly not optimal. Nevertheless, as we show, this simple strategy succeeds with at most $O(\sqrt{N})$ calls to U_P regardless of the value of t.

The results of section 9.1.4 imply that the average probability of success for a run with i iterations of Q, where i is randomly chosen between 0 and r, is given by

$$Pr(i < r) = \frac{1}{r}\sum_{i=0}^{r-1}\sin^2((2i+1)\theta),$$

where $\sin\theta = \sqrt{t}$ as before. A plot of the average success probability for different values of r is shown in figure 9.6. The graph will be identical for all values of t as long as $t \ll 1$. For comparison, the graph of the success probability after exactly r iteration steps of Grover's algorithm is also given.

It is easy to see from the graph of this function that there is a constant c such that $Pr(i < r) > c$ for all $r \geq \frac{\pi}{4}\sqrt{\frac{1}{t}}$. For $\frac{1}{t} \leq N$, guaranteeing at least one solution, if we choose $r = \frac{\pi}{4}\sqrt{N}$, then $Pr(i < \pi/4\sqrt{N}) \geq c$. Thus, a single run of the algorithm, where the number of iterations of Q

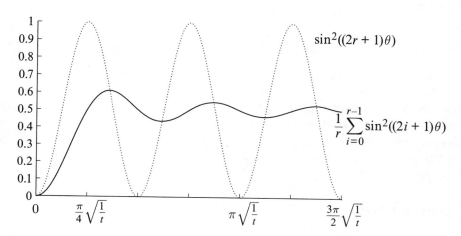

Figure 9.6
The average success probability $Pr(i < r)$ over runs with a random number of iterations chosen between 0 and r plotted as a function of r where $\sin\theta = \sqrt{t}$ as usual. For reference, the dotted curve gives the success probability for a run with exactly r iterations.

is chosen randomly between 0 and $\pi/4\sqrt{N}$, finds the solution with probability at least c. The expected number of calls to the oracle during such a run is therefore $O(\sqrt{N})$. For any probability $c' > c$, there is a constant K such that if Grover's algorithm is run K times, with the number of iterations for each run chosen as above, then a solution will be found with probability c'. Thus, for any c', the total number of times Q is applied, and therefore the total number of calls to the oracle, is $O(\sqrt{N})$.

9.5.2 Quantum Counting

Instead of repeating Grover's algorithm with randomly varying numbers of iterations of Q, quantum counting takes a more quantum approach: create a superposition of results for different numbers of applications of Q and then use the quantum Fourier transform on that superposition to obtain a good estimate for the relative number of solutions t. The same strategy can be used for the amplitude amplification algorithm to estimate the success probability t of $U|0\rangle$. This approach also has query complexity $O(\sqrt{N})$.

The algorithm itself is easy to describe, though determining the size of the superposition needed is more involved. Let U and Q be as defined in the amplitude amplification algorithm of section 9.2. Define a transformation **RepeatQ**, with input $|k\rangle$ and $|\psi\rangle$, that performs k iterations of Q on $|\psi\rangle$:

$$\textbf{RepeatQ} : |k\rangle \otimes |\psi\rangle \rightarrow |k\rangle \otimes Q^k|\psi\rangle.$$

This transformation is more powerful than the classical ability to repeat Q because **RepeatQ** can be applied to a superposition. We apply **RepeatQ** to a superposition of all $k < M = 2^m$ tensored with the state $U|0\rangle$ to obtain

$$\frac{1}{\sqrt{M}} \sum_{k=0}^{M-1} |k\rangle \otimes U|0\rangle \rightarrow \frac{1}{\sqrt{M}} \sum_{k=0}^{M-1} |k\rangle \otimes (g_k|\psi_G\rangle + b_k|\psi_B\rangle),$$

where we ignore for the moment how M was chosen.

A measurement of the right register in the standard basis produces a state $|x\rangle$ that is either a good state (orthogonal to $|\psi_B\rangle$) or a bad state (orthogonal to $|\psi_G\rangle$). Thus, the state of the left register collapses to either $|\psi\rangle = C \sum_{k=0}^{M-1} b_k|k\rangle$ or $|\psi\rangle' = C' \sum_{k=0}^{M-1} g_k|k\rangle$. Let us suppose the former state $|\psi\rangle$ is obtained; the reasoning for the latter case is analogous. From section 9.2, $b_k = \cos((2k+1)\theta)$, so

$$|\psi\rangle = C \sum_{k=0}^{M-1} \cos((2k+1)\theta)|k\rangle.$$

Apply the quantum Fourier transform to this state to obtain

$$\mathcal{F} : C \sum_{k=0}^{M-1} b_k|k\rangle \rightarrow \sum_{j=0}^{M-1} B_j|j\rangle.$$

Section 7.8.1 explained that, for a cosine function of period $\frac{\pi}{\theta}$, most of the amplitude is in those B_j that are close the single value $\frac{M\theta}{\pi}$. If we measure the state now, from the measured value $|j\rangle$ we obtain, with high probability, a good approximation of θ by taking $\theta = \frac{\pi j}{M}$. Thus, with high probability, the value $t = \sqrt{\sin \theta}$ is a good approximation for the ratio of solutions in the Grover's algorithm case, or the success probability of $U|0\rangle$ in the case of amplitude amplification.

There is, of course, one issue remaining: we do not know a priori a proper value for M. This problem can be addressed by repeating the algorithm for increasing M until a meaningful value for j is read. Since $\theta = \frac{j}{M}\pi$, for a given θ we will likely read an integer value $j \sim \frac{\theta M}{\pi}$ and j will be measured as 0 with high probability when M is chosen too small for the given problem.

9.6 Practical Implications of Grover's Algorithm and Amplitude Amplification

The introduction to this chapter mentioned that there is debate as to the extent of the practical impact of Grover's algorithm and its generalization, amplitude amplification. Although the quadratic reduction in the query complexity provided by Grover's algorithm and amplitude amplification over classical algorithms may seem minor compared to the superpolynomial speed up of Shor's algorithm, a quadratic speedup can be of practical importance. For example, even though the fast Fourier transform is only a quadratic speedup over the straightforward way of implementing the Fourier transform, it is viewed as a significant improvement. That the speedup provided by Grover's algorithm is no greater is the least of our concerns in terms of the practical impact of these algorithms.

A major concern is the efficiency with which U_P can be computed for a given practical problem. Unless U_P is efficiently computable, the $O(\sqrt{N})$ speedup of the search is swamped by the amount of time it takes to compute U_P. If U_P takes $O(N)$ time to compute, which is true for a generic P, then a run of Grover's algorithm takes $O(N)$ time, even though it uses U_P only $O(\sqrt{N})$ times. Furthermore, there is no savings for multiple searches over the same space; the measurement at the end of the algorithm destroys the superposition, so U_P must be computed afresh for each search.

Another concern is that most searches done in practice are over spaces with a lot of structure, which in many cases enables fast classical algorithms that amplitude amplification cannot improve upon. For example, classical algorithms can find an element of an alphabetical list of N elements in $O(\log_2 N)$ time. Furthermore, that algorithm is not of a form amenable to speedup by amplitude amplification. There are relatively few practical search problems for which the search space has no structure, so Grover's algorithm on its own has few practical applications. Its generalization, amplitude amplification, is more widely applicable in that it can be used to speed up certain, but by no means most, classes of heuristics.

Grover's algorithm applies to *exhaustive search* of a search space. Grover's algorithm is commonly called a *database* search algorithm, but that appellation is misleading. Grover's search algorithm gives a speedup only over classical algorithms for unstructured search. Databases, which are generally highly structured, can be searched rapidly classically. Most databases, be they of employee records or experimental results, are structured and yet at the same time hard to compute from first principles (the relevant U_P is expensive to compute); for example, an alphabetical list of names is structured, yet computing it most likely cannot be done any faster than by separately adding each entry, an $O(N)$ time operation. For this reason, it is an unfortunate historical accident that Grover's algorithm was ever called a *database* search algorithm. Contrary to popular claims, Grover's algorithm will not aid standard database or Internet searches, since it takes longer to place the elements in a quantum superposition, which gets destroyed in each run of the search, than it takes to perform the classical search in the first place: re-creating the superposition is often linear in N which negates the $O(\sqrt{N})$, benefit of the search algorithm. In fact, Childs et al. showed that for ordered data, quantum computation can give no more than a constant factor improvement over optimal classical algorithms.

When the candidate solutions to a problem can be enumerated easily and there is an efficient test for whether a given value x represents a solution or not, U_P can be computed efficiently, thus avoiding that concern. The amplitude amplification technique used in Grover's algorithm has been extended to provide small speedups for a number of problems, including approximating the mean of a sequence and other statistics finding collisions in r-to-1 functions, string matching, and path integration. NP-complete problems also fall into the class of problems for which the relevant U_P for such problems are efficiently computable. Unfortunately, amplitude amplification gives only a quadratic speedup, so problems that require an exponential number of queries classically remain exponential for Grover's algorithm. In particular, Grover's search does not provide a means to solve NP-complete problems efficiently. Moreover, NP-complete problems have structure that is exploited in classical heuristic algorithms, and only some of these can be improved upon using amplitude amplification.

9.7 References

Grover's search algorithm was first presented in [143]. Grover extended his algorithm to achieve quadratic speedup for other non-search problems, such as computing the mean and median of a function [144]. Using similar techniques, Grover has also shown that certain search problems that classically run in $O(\log N)$ can be solved in $O(1)$ on a quantum computer [143]. Amplitude amplification can be used as a subroutine in other quantum computations in light of a result of Biron et al. [52] that shows how amplitude amplification works with essentially any initial amplitude distribution while still maintaining $O(\sqrt{N})$ complexity.

Jozsa [166] provides a complementary description of the geometric interpretation of Grover's algorithm and amplitude amplification.

Bennett, Bernstein, Brassard, and Vazirani [41] give the earliest proof of optimality of Grover's algorithm. Boyer et al. [58] provide a detailed analysis of the performance of Grover's algorithm and give a solution to the recurrence relation of section 9.1.4. A tighter version of the optimality of Grover's algorithm is given by Zalka [290].

Boyer et al. [58] present the strategy for choosing a random number of iterations when t is unknown. They also present a more efficient algorithm for large t based on the same principle. The idea of quantum counting is due to Brassard et al. [62]. Their paper contains a detailed analysis that includes a strategy for iterating to find M. The two modifications of Grover's algorithm that show that it is not inherently probabilistic are in this same paper.

Childs et al. [81] showed that quantum computation can give no more than a constant factor improvement over optimal classical algorithms for searches of ordered data. Both Viamontes et al. [277] and Zalka [291] discuss issues related to practical use of Grover's search algorithm and its generalizations.

Extensions to Grover include approximating the mean of a sequence and other statistics [144, 216], finding collisions in r-to-1 functions [61], string matching [234], and path integration [271]. Grover's algorithm has been generalized to support nonbinary labelings [58], arbitrary initial conditions [52], and nested searches [77].

9.8 Exercises

Exercise 9.1. Verify that

$$g_i = \sin((2i + 1)\theta)$$

$$b_i = \cos((2i + 1)\theta),$$

with $\sin \theta = \sqrt{t} = \sqrt{|G|/N}$, is a solution to the recurrence relations of section 9.1.4.

Exercise 9.2. Show that applying Grover's algorithm in the case $t = |G|/N = 1/2$ results in no improvement.

Exercise 9.3. What happens if we try to apply Grover's algorithm to the case $t = |G|/N = 3/4$?

Exercise 9.4.

a. How many iterations should Grover's algorithm use in order to find one item among sixteen?

b. If we apply one fewer than the optimal number of iterations and then measure, how does the success probability compare to the optimal case?

c. If we apply one more than the optimal number of iterations and then measure, how does the success probability compare to the optimal case?

Exercise 9.5. Suppose $P : \{0, \ldots, N - 1\} \to \{0, 1\}$ is zero except at $x = t$, and suppose we are given not only a quantum oracle U_P, but also the information that the solution t differs from a known string s in exactly k bits. Exhibit an algorithm that finds the solution with $O(\sqrt{2^k})$ calls to U_P.

Exercise 9.6. Suppose $P : \{0, \ldots, N - 1\} \to \{0, 1\}$ is zero except at $x = t$, and suppose we are given not only a quantum oracle U_P, but also the information that all suffixes except 010 and 100 have been ruled out. In other words, the solution t must end with either 010 and 100. Exhibit an algorithm that finds the solution with fewer calls to U_P than Grover's algorithm.

Exercise 9.7. Suppose $P : \{0, \ldots, N - 1\} \rightarrow \{0, 1\}$ is zero except at $x = t$, and suppose we are given not only a quantum oracle U_P, but also the information that the solution t differs from a known string s in at most k bits. Exhibit an algorithm that is more efficient, in terms of the number of calls needed, than $O(\sqrt{2^n})$?

Exercise 9.8. Suppose $P : \{0, \ldots, N - 1\} \rightarrow \{0, 1\}$ is zero except at $x = t$, and suppose we are given a quantum oracle U_P and told that with probability 0.9 the first $n/2$ bits of the solution t are zero. How can we take advantage of this information to obtain an algorithm that is more efficient, in terms of the number of calls needed, than $O(\sqrt{2^n})$?

Exercise 9.9. Given a quantum black box for a function $f : \{0, \ldots, N - 1\} \rightarrow \{0, \ldots, N - 1\}$, design a quantum algorithm that finds the minimum with $O(\sqrt{N} \log N)$ queries, where $N = 2^n$.

Exercise 9.10. Suppose there is an error in the initial state, so that instead of starting with $|00 \ldots 0\rangle$ we run Grover's algorithm, starting with the state

$$\frac{1}{\sqrt{1 + \epsilon^2}}(|00 \ldots 0\rangle + \epsilon|11 \ldots 1\rangle).$$

How does this error affect the results of Grover's algorithm?

Exercise 9.11. Why does applying amplitude amplification to the output of a first application of amplitude amplification not result in an additional square root reduction in the query complexity?

Exercise 9.12. Prove the optimality of Grover's algorithm in the multiple solution case.

Exercise 9.13. For the quantum counting procedure of section 9.5.2, show how the estimate of t is obtained in the case that a bad state is measured.

III ENTANGLED SUBSYSTEMS AND ROBUST QUANTUM COMPUTATION

10 Quantum Subsystems and Properties of Entangled States

The power of quantum computation is most often attributed to *entanglement*. Indeed, Jozsa and Linden [167] have shown that any quantum algorithm that achieves exponential speedup over classical algorithms must make use of entanglement between a number of qubits that increases with the size of the input to the algorithm. Nevertheless, as section 13.9 explains, the exact role of entanglement in quantum computation, and in quantum information processing more generally, remains unclear. While entanglement is generally viewed as an important resource for quantum computation, entanglement, particularly multipartite entanglement, is still only poorly understood.

This chapter surveys some of what is known about entanglement, particularly multipartite entanglement, entanglement between three or more subsystems. It also illustrates some of the complexities that make developing a deeper understanding of entanglement challenging. For example, there are many distinctly different types of multipartite entanglement; for entanglement between four (or more) subsystems, the different *types* of entanglement are uncountably infinite! Much work remains to be done to understand which types of entanglement are useful, and for what.

As chapters 3 and 4 emphasize, the notion of entanglement is well defined only with respect to a particular tensor decomposition of the system into subsystems. A deeper understanding of entanglement comes from studying these quantum subsystems. Of particular interest will be entanglement between the n subsystems consisting of the single qubits making up a n qubit system and entanglement between registers of a quantum computer during a computation. Density operators, introduced in section 10.1, are used to model quantum systems, and more particularly how a state appears when only one subsystem is accessible. Density operators are also useful for describing measurements yet to be performed, or for which the outcome is not yet known. The mathematics of density operators enables a more detailed examination of entanglement, including issues related to quantifying how much entanglement a state contains and distinguishing different types of entanglement.

Section 10.2 uses the density operator formalization to quantify bipartite entanglement and to examine properties of multipartite entanglement. How density operators model measurements is the concern of section 10.3, and section 10.4 discusses transformations of quantum systems. Properties of quantum subsystems give insight not only into entanglement, but also into robustness issues in quantum computation. From a practical standpoint, any quantum system such as a quantum computer is always really a quantum subsystem: any experimental setup is never completely isolated from the rest of the universe, so any experiment is rightly viewed as only one part of a larger quantum system. The final section of this chapter, section 10.4.4, shows how decoherence, errors caused by interaction with the environment, can be modeled. This model forms the foundation for the discussion of quantum error correction and fault-tolerant quantum computation in chapters 11 and 12.

10.1 Quantum Subsystems and Mixed States

It commonly arises that we have access to, or interest in, only one part of a larger system. This section develops concepts and notation to support the study of quantum subsystems and entanglement between subsystems.

Some of the issues in modeling quantum subspaces are illustrated by an EPR pair distributed between two parties. Imagine that Alice has the first qubit of an EPR pair $\frac{1}{\sqrt{2}}(|00\rangle + |11\rangle)$, and Bob has the second. How would Alice describe her qubit? It is not in a single-qubit quantum state, a quantum state of the form $a|0\rangle + b|1\rangle$. Were Alice to measure her qubit in the standard basis, she would have a 50 percent chance of seeing $|0\rangle$ or $|1\rangle$. So it might appear that her qubit's state must be an even superposition of $|0\rangle$ and $|1\rangle$. But if she measured it in the basis $\{|+\rangle, |-\rangle\}$, she would have a 50 percent chance of seeing $|+\rangle$ or $|-\rangle$. In fact, in any basis whatsoever, it appears to be an even superposition of the two basis states. But no single-qubit state has this property. For example, the state $\frac{1}{\sqrt{2}}(|0\rangle + |1\rangle)$ is an even superposition in the standard basis but is an uneven superposition in most bases and is deterministic in the basis $\{|+\rangle, |-\rangle\}$. So what can Alice say about her qubit?

To answer that question, it is worth looking carefully at what is meant by a state of a system. A state captures all information that could conceivably be learned about the system. Since information can only be gained by measurement, and measurement changes the quantum state, imagine an infinite supply of identically prepared quantum systems. The quantum state encapsulates all information that could be gained from any number of measurements on this infinite supply of identical quantum systems.

Another way of saying that most states of a multiple-qubit quantum system cannot be described in terms of the states of each of its single-qubit subsystems separately is that a single qubit of a multiple-qubit system is generally not in a well-defined quantum state. Alice's qubit of the entangled pair is such a case. An n-qubit quantum state captures all of the information that could conceivably be learned from measurements on an infinite supply of identically prepared quantum systems. For an infinite supply of m-qubit subsystems of identically prepared n-qubit systems, it is interesting to ask what can be learned from measurements of the m-qubit subsystem alone. The structure that encapsulates that information is called the *mixed state* of the m-qubit subsystem, and it will be modeled by the mathematics of *density operators*. So far we have considered only systems that are universes unto themselves; the states of such systems, all the states we have studied so far, are called *pure states*.

The meaning of *state* in *mixed state* should be interpreted with care. That a subsystem always has a well-defined mixed state should not be interpreted to mean that when a state of a system is entangled with respect to a decomposition into subsystems, the states of the subsystems are well defined after all; mixed states are not quantum states in the conventional sense, the sense we have used up to now. Knowing the mixed states of all the subsystems making up a system does not enable us to know the state of the entire system; many different states of a system give the same set of mixed states on the subsystems. Knowing the mixed states of all the subsystems gives full knowledge of the state of the whole system precisely when the state of the entire system is unentangled with respect to that subspace decomposition. In exactly this case, the mixed states of the subsystems can be viewed as pure states. The relationship of a mixed state for a subsystem to the pure state of the whole system is analogous to the relationship of a marginal distribution to a joint distribution. This analogy can be made precise; see appendix A.

The following section develops the mathematics of density operators for modeling mixed states. It concludes with a description of Alice's qubit.

10.1.1 Density Operators

For an m-qubit subsystem A of a larger n-qubit system $X = A \otimes B$, the mixed state for subsystem A must capture all possible results of measurement by operators of the form $O \otimes I$, where O is a measurement operator on just the m qubits of A and I is the identity on the $n - m$ qubits of B. Let $|x\rangle$ be a state of the entire n-qubit system. The next few paragraphs culminate in the description of an operator on the 2^m-dimensional complex vector space A, called the *density operator* $\rho_x^A : A \to A$, that captures all of the information that can be gained about $|x\rangle$ from measurements on the m-qubit subsystem A alone. For this reason, density operators are used to model mixed states.

Let $M = 2^m$ and $L = 2^{n-m}$. Given bases $\{|\alpha_0\rangle, \ldots, |\alpha_{M-1}\rangle\}$ and $\{|\beta_0\rangle, \ldots, |\beta_{L-1}\rangle\}$ for A and B, respectively, $\{|\alpha_i\rangle \otimes |\beta_j\rangle\}$ is a basis for $X = A \otimes B$. A state $|x\rangle$ of X can be written

$$|x\rangle = \sum_{i=0}^{M-1} \sum_{j=0}^{L-1} x_{ij} |\alpha_i\rangle |\beta_j\rangle.$$

Measurements on system A alone are modeled by observables O^A with associated projectors $\{P_i^A\}$, $0 \le i < 2^m$. On the whole space X, such measurements have the form $O^A \otimes I^B$ with projectors $P_i^A \otimes I^B$. For any particular projector P^A, $\langle x|P^A \otimes I|x\rangle$ gives the probability that measurement of $|x\rangle$ by $O^A \otimes I^B$ results in a state in the subspace associated to P^A. Writing this probability in terms of the bases $\{|\alpha_0\rangle, \ldots, |\alpha_{M-1}\rangle\}$ and $\{|\beta_0\rangle, \ldots, |\beta_{L-1}\rangle\}$ yields

$$\langle x|P^A \otimes I|x\rangle = \left(\sum_{ij} \overline{x_{ij}} \langle \alpha_i| \otimes \langle \beta_j| \right) (P^A \otimes I) \left(\sum_{kl} x_{kl} |\alpha_k\rangle \otimes |\beta_l\rangle \right)$$

$$= \sum_{ijkl} \overline{x_{ij}} x_{kl} \langle \alpha_i|P^A|\alpha_k\rangle \langle \beta_j|\beta_l\rangle,$$

where indices i and k are summed over $[0 \ldots M - 1]$, and j and l are summed over $[0 \ldots L - 1]$. Since $\langle \beta_j | \beta_l \rangle = \delta_{lj}$, each term is zero except those for which $j = l$, so the probability that the measurement outcome is the one associated with P^A can be written more concisely as

$$\langle x | P^A \otimes I | x \rangle = \sum_{ijk} \overline{x_{ij}} x_{kj} \langle \alpha_i | P^A | \alpha_k \rangle. \tag{10.1}$$

This formula, together with facts about the partial trace found in box 10.3, yields the density operator that encapsulates all information that can be gained by measurements of the form $O^A \otimes I^B$. Since $\{|\alpha_u\rangle\}$ is a basis for A,

$$\sum_{u=0}^{M-1} |\alpha_u\rangle\langle\alpha_u| = I$$

is the identity operator on A. We write

$$\langle x | P^A \otimes I | x \rangle = \sum_{ijk} \overline{x_{ij}} x_{kj} \langle \alpha_i | P^A | \alpha_k \rangle$$

$$= \sum_{ik} \sum_{j} \overline{x_{ij}} x_{kj} \langle \alpha_i | P^A \left(\sum_{u} |\alpha_u\rangle\langle\alpha_u| \right) |\alpha_k\rangle$$

$$= \sum_{u} \sum_{ik} \sum_{j} \overline{x_{ij}} x_{kj} \langle \alpha_u | \alpha_k \rangle \langle \alpha_i | P^A | \alpha_u \rangle$$

$$= \sum_{u} \langle \alpha_u | \left(\sum_{ik} \sum_{j} \overline{x_{ij}} x_{kj} |\alpha_k\rangle\langle\alpha_i| P^A \right) |\alpha_u\rangle$$

$$= \mathbf{tr}(\rho_x^A P^A),$$

where we define

$$\rho_x^A = \sum_{ik} \sum_{j} \overline{x_{ij}} x_{kj} |\alpha_k\rangle\langle\alpha_i| \tag{10.2}$$

and call ρ_x^A the *density operator* for $|x\rangle$ on subsystem A. By box 10.3,

$$\rho_x^A = \mathbf{tr}_B(|x\rangle\langle x|). \tag{10.3}$$

Since O^A is a general observable on A, and P^A a general projector associated with O^A, this calculation shows that all information from measurements on subsystem A alone can be gained from the density operator ρ_x^A. Thus the density operator ρ_x^A models the mixed state corresponding to the part of $|x\rangle$ in A.

This definition of a density operator of $|x\rangle$ is physically reasonable only if it does not depend on the choice of basis $\{|\alpha_i\rangle\}$, since physically no basis is preferred. The next two paragraphs show that density operators are well defined in the sense that calculating ρ_x^A in different bases gives

Box 10.1
The Trace of an Operator

To define the *trace* of an operator O acting on a vector space V, we first define the trace of a matrix for O and then show that the trace is basis independent and therefore a property of an operator, not of the specific matrix representation used. The *trace* of a matrix M for $O : V \to V$ is the sum of its diagonal elements:

$$\mathbf{tr}(M) = \sum_i \langle v_i | M | v_i \rangle$$

where $\{|v_i\rangle\}$ is the basis for V with respect to which the matrix M is written. The following identities are easily verified:

$$\mathbf{tr}(M_1 + M_2) = \mathbf{tr}(M_1) + \mathbf{tr}(M_2),$$
$$\mathbf{tr}(\alpha M) = \alpha \mathbf{tr}(M),$$
$$\mathbf{tr}(M_1 M_2) = \mathbf{tr}(M_2 M_1).$$

The last equality implies $\mathbf{tr}(C^{-1} M C) = \mathbf{tr}(M)$ for any invertible matrix C, which means that the trace is invariant under basis change. Thus the notion of a trace of an operator is independent of basis, so we can simply talk about $\mathbf{tr}(O)$ without specifying a basis.

Useful fact: For any $|\psi_1\rangle$ and $|\psi_2\rangle$ in a space V, and any operator O acting on V,

$$\langle \psi_1 | O | \psi_2 \rangle = \mathbf{tr}(|\psi_2\rangle\langle\psi_1| O). \tag{10.4}$$

The proof of this fact illustrates a common way of reasoning about traces. For any basis $\{|\alpha_i\rangle\}$ for V,

$$\mathbf{tr}(|\psi_2\rangle\langle\psi_1| O) = \sum_i \langle \alpha_i | \psi_2 \rangle \langle \psi_1 | O | \alpha_i \rangle$$

$$= \sum_i \langle \psi_1 | O | \alpha_i \rangle \langle \alpha_i | \psi_2 \rangle$$

$$= \langle \psi_1 | O \left(\sum_i |\alpha_i\rangle\langle\alpha_i| \right) |\psi_2\rangle.$$

Since $\sum_i |\alpha_i\rangle\langle\alpha_i|$ is the identity matrix, the result follows.

the same operator. We first prove the result for density operators of pure states and then use that result to prove the general case.

Suppose the subsystem under consideration is the whole system, $A = X$. The system is in a pure state $|x\rangle$, written as $|x\rangle = \sum_i x_i |\psi_i\rangle$ in the basis $\{|\psi_i\rangle\}$ for X. The density operator (equation 10.2) becomes

$$\rho_x^X = \rho_x^A = \sum_{ik} \overline{x_i} x_k |\psi_k\rangle\langle\psi_i| = |x\rangle\langle x|.$$

Box 10.2
Restricting Operators to Subsystems

Corresponding to any operator O_{AB} on $A \otimes B$, there is a family of operators on subsystem A that is parametrized by pairs of elements from B. Any pair of states $|b_1\rangle$ and $|b_2\rangle$ in B defines an operator on A denoted by $\langle b_1|O_{AB}|b_2\rangle$. We first define the operator $\langle b_1|O_{AB}|b_2\rangle$ in terms of a basis $\{|\alpha_i\rangle\}$ for A, and then show that it is independent of basis, so any basis for A defines the same operator. Operator $\langle b_1|O_{AB}|b_2\rangle$ acts as follows:

$$\langle b_1|O_{AB}|b_2\rangle : A \;\; \rightarrow \;\; A$$

$$|x\rangle \mapsto \sum_i \langle\alpha_i|\langle b_1|O_{AB}|x\rangle|b_2\rangle \, |\alpha_i\rangle. \qquad (10.5)$$

This notation takes some getting used to. It may help the reader to begin by writing the operator $\langle b_1|O_{AB}|b_2\rangle$ as $\langle\_, b_1|O_{AB}|\_, b_2\rangle$.

To prove basis independence, let $\{|a'_j\rangle\}$ be another basis for A with $|a'_j\rangle = \sum_i a_{ij}|\alpha_i\rangle$. Then

$$\langle b_1|O_{AB}|b_2\rangle|a\rangle = \sum_j \langle a'_j|\langle b_1|O_{AB}|b_2\rangle|a\rangle \, |a'_j\rangle$$

$$= \sum_j \left(\sum_i \overline{a_{ij}}\langle\alpha_i|\right) \langle b_1|O_{AB}|a\rangle|b_2\rangle \left(\sum_k a_{kj}|\alpha_k\rangle\right)$$

$$= \sum_i \sum_k \sum_j \overline{a_{ij}}a_{kj}\langle\alpha_i|\langle b_1|O_{AB}|a\rangle|b_2\rangle \, |\alpha_k\rangle$$

$$= \sum_i \langle\alpha_i|\langle b_1|O_{AB}|a\rangle|b_2\rangle \, |\alpha_i\rangle$$

where the last line follows because $\{|\alpha_i\rangle\}$ is a basis so $\sum_j \overline{a_{ij}}a_{kj} = \delta_{ik}$. These restricted operators are useful for defining the partial trace (box 10.3), the canonical restriction of O_{AB} to subsystem A, and the operator sum decomposition discussed in section 10.4.

Thus, the density operator of a pure state $\rho_x^X = |x\rangle\langle x|$ is independent of the basis for X. As with any operator, a matrix representation for the operator does depend on the basis. In basis $\{|\psi_i\rangle\}$, the ijth entry of the matrix for ρ_x^X is $\overline{x_j}x_i$. The diagonal elements $\overline{x_i}x_i$ of the matrix have special meaning for measurements in the basis $\{|\psi_i\rangle\}$: the probability that $|x\rangle$ will be measured as being in the basis state $|\psi_i\rangle$ with projector $P_i = |\psi_i\rangle\langle\psi_i|$ is

$$\langle x|P_i|x\rangle = \langle x|\psi_i\rangle\langle\psi_i|x\rangle = \overline{x_i}x_i.$$

In the general case ($A \neq X$), let $X = A \otimes B$, and let $\{|\alpha_i\rangle\}$ and $\{|\beta_j\rangle\}$ be bases for A and B respectively. The matrix for the density operator ρ_x^X of the state $|x\rangle = \sum x_{ij}|\alpha_i\rangle|\beta_j\rangle$ of X in the basis $\{|\alpha_i\beta_j\rangle\}$ has entries $\overline{x_{ij}}x_{kl}$:

$$\rho_x^X = \sum_{i,k=0}^{M-1} \sum_{j,l}^{L-1} \overline{x_{ij}} x_{kl} |\alpha_k\rangle |\beta_l\rangle \langle \alpha_i | \langle \beta_j |.$$

To obtain the density matrix ρ_x^A, we use equation 10.3, which says that ρ_x^A is simply the partial trace over B of ρ_x^X (see box 10.3):

Box 10.3
The Partial Trace

For any operator O_{AB} on $A \otimes B$, the *partial trace* of O_{AB} with respect to subsystem B is an operator $\mathbf{tr}_B O_{AB}$ on subsystem A defined by

$$\mathbf{tr}_B O_{AB} = \sum_i \langle \beta_i | O_{AB} | \beta_i \rangle,$$

where $\{|\beta_i\rangle\}$ is a basis for B. The operators $\langle \beta_i | O_{AB} | \beta_i \rangle$ were defined in box 10.2. The partial trace $\mathbf{tr}_B O_{AB}$ is basis independent by an argument similar to the one given for $\langle \beta_1 | O_{AB} | \beta_2 \rangle$ in box 10.2. In terms of bases $\{|\alpha_i\rangle\}$ and $\{|\beta_j\rangle\}$ for A and B respectively, the matrix for $\mathbf{tr}_B O_{AB}$ has entries

$$(\mathbf{tr}_B O_{AB})_{ij} = \sum_{k=0}^{M-1} \langle \alpha_i | \langle \beta_k | O_{AB} | \alpha_j \rangle | \beta_k \rangle,$$

so the matrix for $\mathbf{tr}_B O_{AB}$ is

$$\mathbf{tr}_B O_{AB} = \sum_{i,j=0}^{N-1} \left(\sum_{k=0}^{M-1} \langle \alpha_i | \langle \beta_k | O_{AB} | \alpha_j \rangle | \beta_k \rangle \right) |\alpha_i\rangle \langle \alpha_j |$$

where N and M are the dimensions of A and B respectively. In the special case in which $O_{AB} = |x\rangle \langle x|$, let $x_{ij} \overline{x_{kl}}$ be the entries of O_{AB} in the basis $|\alpha_i\rangle |\beta_j\rangle$, so

$$O_{AB} = \sum_{ij} x_{ij} |\alpha_i\rangle |\beta_j\rangle \sum_{kl} \overline{x_{kl}} \langle \alpha_k | \langle \beta_l |$$

$$= \sum_{ijkl} x_{ij} \overline{x_{kl}} |\alpha_i\rangle |\beta_j\rangle \langle \alpha_k | \langle \beta_l |.$$

Then

$$\mathbf{tr}_B(O_{AB}) = \mathbf{tr}_B(|x\rangle \langle x|) = \sum_{i,k=0}^{N-1} \sum_{j=0}^{M-1} x_{ij} \overline{x_{kj}} |\alpha_i\rangle \langle \alpha_k |.$$

In the special case in which an operator is the tensor product of operators on the separate subsystems, the partial trace has the simple form $\mathbf{tr}_B(O_A \otimes O_B) = O_A \, \mathbf{tr}(O_B)$.

$$\rho_x^A = \mathbf{tr}_B(\rho_x^X)$$

$$= \mathbf{tr}_B \left(\sum_{i,k=0}^{M-1} \sum_{j,l}^{L-1} \overline{x_{ij}} x_{kl} |\alpha_k\rangle |\beta_l\rangle \langle \alpha_i | \langle \beta_j | \right)$$

$$= \sum_{u,v=0}^{M-1} \left(\sum_{w}^{L-1} \langle \alpha_u | \langle \beta_w | \left(\sum_{i,k=0}^{M-1} \sum_{j,l}^{L-1} \overline{x_{ij}} x_{kl} |\alpha_k\rangle |\beta_l\rangle \langle \alpha_i | \langle \beta_j | \right) |\alpha_v\rangle |\beta_w\rangle \right) |\alpha_u\rangle \langle \alpha_v |$$

$$= \sum_{u,v=0}^{M-1} \sum_{w}^{L-1} \overline{x_{vw}} x_{uw} |\alpha_u\rangle \langle \alpha_v |.$$

Since the partial trace is basis independent, so is the density operator.

Example 10.1.1 Let us return to Alice, who controls the first qubit of the EPR pair $|\psi\rangle = \frac{1}{\sqrt{2}}(|00\rangle + |11\rangle)$ while Bob controls the second. The density matrix for the pure state $|\psi\rangle \in A \otimes B$ is

$$\rho_\psi = |\psi\rangle \langle \psi |$$

$$= \frac{1}{2}(|00\rangle \langle 00| + |00\rangle \langle 11| + |11\rangle \langle 00| + |11\rangle \langle 11|)$$

$$= \frac{1}{2} \begin{pmatrix} 1 & 0 & 0 & 1 \\ 0 & 0 & 0 & 0 \\ 0 & 0 & 0 & 0 \\ 1 & 0 & 0 & 1 \end{pmatrix}.$$

The mixed state of Alice's qubit, which encapsulates all information that could be obtained from any sequence of measurements on Alice's qubit alone on a sequence of identical states $|\psi\rangle$), is modeled by the density matrix ρ_ψ^A obtained from ρ_ψ by tracing over Bob's qubit, $\rho_\psi^A = \mathbf{tr}_B \rho_\psi$. The four entries a_{00}, a_{01}, a_{10}, and a_{11} for a matrix representing ρ_ψ^A in the standard basis can be computed separately:

$$a_{00} = \sum_{j=0}^{1} \langle 0| \langle j | \; |\psi\rangle \langle \psi | \; |0\rangle |j\rangle = \left(\frac{1}{2} + 0 \right) = \frac{1}{2},$$

$$a_{01} = \sum_{j=0}^{1} \langle 0| \langle j | \; |\psi\rangle \langle \psi | \; |1\rangle |j\rangle = (0+0) = 0,$$

$$a_{10} = \sum_{j=0}^{1} \langle 1 | \langle j | \ |\psi\rangle \langle \psi| \ |0\rangle | j \rangle = (0+0) = 0,$$

$$a_{11} = \sum_{j=0}^{1} \langle 1 | \langle j | \ |\psi\rangle \langle \psi| \ |1\rangle | j \rangle = \left(0 + \frac{1}{2}\right) = \frac{1}{2}.$$

So

$$\rho_\psi^A = \frac{1}{2} \begin{pmatrix} 1 & 0 \\ 0 & 1 \end{pmatrix}.$$

By symmetry, the density operator for Bob's qubit is

$$\rho_\psi^B = \frac{1}{2} \begin{pmatrix} 1 & 0 \\ 0 & 1 \end{pmatrix}.$$

In general, it is not possible to recover the state of the entire system from the set of density operators for all of the subsystems; information has been lost. For example, for a two-qubit system, if the density matrices for each of the two qubits are $\rho_\psi^A = \frac{1}{2} \begin{pmatrix} 1 & 0 \\ 0 & 1 \end{pmatrix}$ and $\rho_\psi^B = \frac{1}{2} \begin{pmatrix} 1 & 0 \\ 0 & 1 \end{pmatrix}$, the state of the two-qubit system as a whole could be $\frac{1}{\sqrt{2}}(|00\rangle + |11\rangle)$, as in example 10.1.1, or it could be $\frac{1}{\sqrt{2}}(|00\rangle - |11\rangle)$ or $\frac{1}{\sqrt{2}}(|01\rangle + |10\rangle)$ among other possibilities.

10.1.2 Properties of Density Operators
Any density operator ρ_x^A satisfies

1. ρ_x^A is Hermitian (self-adjoint),
2. $\mathbf{tr}(\rho_x^A) = 1$, and
3. ρ_x^A is positive.

Property (1) follows immediately from the definition (equation 10.2). Since $|x\rangle$ is a unit vector, $\mathbf{tr}(\rho_x^A) = \sum \overline{x_{ij}} x_{ij} = 1$. An operator $O : V \to V$ being positive means that $\langle v|O|v \rangle$ is real with $\langle v|O|v \rangle \geq 0$ for all $|v\rangle$ in V. To show that $\rho_x^A : A \to A$ is positive, let $|v\rangle \in A$. Then

$$\langle v|\rho_x^A|v\rangle = \sum_{ik}\sum_{j} \langle v|(\overline{x_{ij}}x_{kj}|\alpha_k\rangle\langle\alpha_i|)|v\rangle$$

$$= \sum_{ik}\sum_{j} \overline{x_{ij}}\langle\alpha_i|v\rangle x_{kj}\langle v|\alpha_k\rangle$$

$$= \sum_{j} \left(\overline{\sum_{i} x_{ij}\langle v|\alpha_i\rangle}\right)\left(\sum_{k} x_{kj}\langle v|\alpha_k\rangle\right)$$

$$= \sum_j \left| \sum_i x_{ij} \langle v | \alpha_i \rangle \right|^2$$

$$\geq 0.$$

Positivity implies that all of the eigenvalues of ρ_x^A are real and non-negative: if λ is an eigenvalue of ρ_x^A with eigenvector $|v_\lambda\rangle$, then $\lambda = \langle v_\lambda | \rho_x^A | v_\lambda \rangle$ is real and non-negative. It follows from these properties that in any (orthonormal) eigenbasis $\{|v_0\rangle, \ldots, |v_{M-1}\rangle\}$ for ρ_x^A, the matrix for ρ_x^A is a diagonal matrix with non-negative real entries λ_i that sum to 1. Thus $\rho_x^A = \sum_i \lambda_i |v_i\rangle\langle v_i|$. In this way the mixed state with density operator ρ_x^A may be viewed as a mixture of pure states $|v_i\rangle\langle v_i|$ or, more precisely, as a probability distribution over these states.

It turns out that any operator satisfying (1), (2), and (3) is a density operator; in some expositions, that is how density operators are first defined. To establish this equivalence, we need to show that, for any operator $\rho : A \rightarrow A$ satisfying these conditions, there is a pure state $|\psi\rangle$ of a larger system $A \otimes B$ such that $\mathbf{tr}_B(|\psi\rangle\langle\psi|) = \rho$. The state $|\psi\rangle$ is called a *purification* of ρ. Let ρ be any operator acting on a subsystem A of dimension $M = 2^m$ that satisfies (1), (2), and (3). These properties mean that in its eigenbasis $\{|\psi_0\rangle, |\psi_1\rangle, \ldots, |\psi_{M-1}\rangle\}$, ρ is diagonal with non-negative real eigenvalues λ_i that sum to 1. Thus, for any ρ,

$$\rho = \lambda_0 |\psi_0\rangle\langle\psi_0| + \cdots + \lambda_{M-1} |\psi_{M-1}\rangle\langle\psi_{M-1}|,$$

for some $\{|\psi_0\rangle, |\psi_1\rangle, \ldots, |\psi_{M-1}\rangle\}$. Let B be a quantum system with associated vector space of dimension $2^n > M$, and let $\{|0\rangle, \ldots, |M-1\rangle\}$ be the first M elements of a (orthonormal) basis for B. Then the pure state $|x\rangle \in A \otimes B$

$$|x\rangle = \sqrt{\lambda_0} |\psi_0\rangle |0\rangle + \sqrt{\lambda_1} |\psi_1\rangle |1\rangle + \cdots + \sqrt{\lambda_{M-1}} |\psi_{M-1}\rangle |M-1\rangle$$

satisfies $\rho_x^A = \rho$.

For a pure state $|x\rangle$, the density operator $\rho_x^X = |x\rangle\langle x|$ has a particularly simple form in terms of a basis that contains $|x\rangle$ as its ith element: it is a matrix with all entries 0 except for a single 1 on the diagonal in the ith spot. It follows that the density operator of a pure state is a projector: $\rho_x^X \rho_x^X = \rho_x^X$. Conversely, any density operator that is a projector corresponds to a pure state: projection operators have only 0 and 1 as eigenvalues, and to obtain trace 1 the density operator must have only a single 1-eigenvector, which is the corresponding pure state.

Another nice property of density operators of pure states is that the non-uniqueness of the representation of states due to the global phase disappears. Let $|x\rangle = e^{i\theta}|y\rangle$. The density operator corresponding to $|x\rangle$ is $\rho_x = |x\rangle\langle x|$, which is also equal to $|y\rangle\langle y|$, since $\rho_x = |x\rangle\langle x| = e^{i\theta}|y\rangle\langle y|e^{-i\theta} = |y\rangle\langle y|$. Thus, any two vectors that differ by a global phase have the same density operator.

It is important not to confuse mixed states with superpositions. The mixed state that is an even probabilistic combination of $|0\rangle$ and $|1\rangle$ is *not* the same as the pure state superposition $|+\rangle = \frac{1}{\sqrt{2}}(|0\rangle + |1\rangle)$. Their density operators are different: in the standard basis, the density matrix for the former is

$$\rho_{ME} = \frac{1}{2} \begin{pmatrix} 1 & 0 \\ 0 & 1 \end{pmatrix},$$

whereas the density matrix for the latter is

$$\rho_+ = \frac{1}{2} \begin{pmatrix} 1 & 1 \\ 1 & 1 \end{pmatrix}.$$

The latter gives deterministic results when measured in an appropriate basis, whereas the former gives probabilistic results in all bases.

Mixed states are not viewed as *true* quantum states, but rather as a way of describing a subsystem whose state is not well defined, being only a mixed state, or a probabilistic mixture of well defined pure states. Therefore, *state* or *quantum state* will mean a pure state unless it is prefaced with the word *mixed*. Furthermore, when it is clear which subsystem is being talked about, we drop the superscript and just say ρ_x.

10.1.3 The Geometry of Single-Qubit Mixed States

The Bloch sphere (section 2.5.2) can be extended in an elegant way to include single-qubit mixed states. Mixed states are convex combinations of pure states, linear combinations of pure states with non-negative coefficients that sum to 1, so it is not surprising that single-qubit mixed states can be viewed as lying in the interior of the Bloch sphere. The precise connection with the geometry uses the fact that density operators are Hermitian (self-adjoint) operators with trace 1. Any self-adjoint 2×2-matrix is of the form

$$\begin{pmatrix} a & c - \mathbf{i}d \\ c + \mathbf{i}d & b \end{pmatrix},$$

where a, b, c, and d are real parameters. Requiring that the matrix have trace 1 means there are only 3 real parameters. Such matrices can be written as

$$\frac{1}{2} \begin{pmatrix} 1+z & x - \mathbf{i}y \\ x + \mathbf{i}y & 1 - z \end{pmatrix},$$

where x, y, and z are real parameters. Thus, any density matrix for a single-qubit system can be written as

$$\frac{1}{2}(I + x\sigma_x + y\sigma_y + z\sigma_z),$$

where

$$\sigma_x = X = \begin{pmatrix} 0 & 1 \\ 1 & 0 \end{pmatrix}, \sigma_y = -\mathbf{i}Y = \begin{pmatrix} 0 & -\mathbf{i} \\ \mathbf{i} & 0 \end{pmatrix}, \text{ and } \sigma_z = Z = \begin{pmatrix} 1 & 0 \\ 0 & -1 \end{pmatrix} \text{ are the Pauli spin}$$

matrices. (The Pauli spin matrices are related to the Pauli group elements X, Y, and Z of section 5.2.1 by $\sigma_x = X$, $\sigma_y = -\mathbf{i}Y$, and $\sigma_z = Z$.)

The determinant of a single-qubit density operator $\rho = \frac{1}{2}(I + x\sigma_x + y\sigma_y + z\sigma_z)$ has geometric meaning; it is easily computed to be

$$\det(\rho) = \frac{1}{4}(1 - r^2)$$

where $r = \sqrt{|x|^2 + |y|^2 + |z|^2}$ is the radial distance from the origin in x, y, z coordinates. Since the determinant of ρ is the product of its eigenvalues, which for a density operator must be non-negative, $\det(\rho) \geq 0$. So $0 \leq r \leq 1$. Thus, with x, y, and z acting as coordinates, the density matrices of single-qubit mixed states $\rho = \frac{1}{2}(I + x\sigma_x + y\sigma_y + z\sigma_z)$ all lie within a sphere of radius 1. The density matrices for states on the boundary of the sphere have $\det(\rho) = 0$; one of their eigenvalues must be 0. Since density operators have trace 1, the other eigenvalue must be 1. Thus, density operators on the boundary of the sphere are projectors, which means they are pure states. We have recovered the boundary of the Bloch sphere discussed in section 2.5 as the boundary of a ball for which the Pauli spin matrices provide the coordinates. This entire ball is called the *Bloch sphere* (though it ought to be called the *Bloch ball*).

The following table gives the density matrices, in the standard basis, and the Bloch sphere coordinates for some familiar states and mixed states.

(x, y, z) coordinate	state vector	density matrix
$(1, 0, 0)$	$|+\rangle$	$\frac{1}{2}(I + \sigma_x) = \frac{1}{2}\begin{pmatrix} 1 & 1 \\ 1 & 1 \end{pmatrix}$
$(0, 1, 0)$	$|i\rangle$	$\frac{1}{2}(I + \sigma_y) = \frac{1}{2}\begin{pmatrix} 1 & -i \\ i & 1 \end{pmatrix}$
$(0, 0, 1)$	$|0\rangle$	$\frac{1}{2}(I + \sigma_z) = \frac{1}{2}\begin{pmatrix} 2 & 0 \\ 0 & 0 \end{pmatrix}$
$(0, 0, 0)$		$\rho_0 = \frac{1}{2}I = \frac{1}{2}\begin{pmatrix} 1 & 0 \\ 0 & 1 \end{pmatrix}$

The set of all density operators for mixed states of an n-qubit system with $n \geq 2$ also forms a convex set, but its geometry is significantly more complicated than the simple Bloch sphere picture. As one example, in the single-qubit case the boundary of the Bloch sphere contains exactly the pure states, where as for $n \geq 2$, the boundary of the set of all mixed states contains both pure and mixed states. The reader may easily check that this statement must be true by computing the dimension of the space of n-qubit mixed states to be $2^{2n} - 1$ and comparing it to the dimension of the space of pure states, which is only $2^{n+1} - 2$.

10.1.4 Von Neumann Entropy

The density matrix of one qubit of an EPR pair,

$$\rho_{ME} = \frac{1}{2}\begin{pmatrix} 1 & 0 \\ 0 & 1 \end{pmatrix},$$

corresponds to the point $(0, 0, 0)$ in the center of the sphere, farthest from the boundary. In a technical sense, this state is the *least pure* single-qubit mixed state possible: it is the maximally uncertain state in that no matter in what basis it is measured, it gives the two possible answers with equal probability. In contrast, for any pure state, there is a basis in which measurement gives a deterministic result. For no state, mixed or pure, do measurements in two different bases give deterministic results, so pure states are as certain as possible.

This notion of uncertainty can be quantified for general n-qubit states by an extension of the classical information theoretic notion of entropy. The von Neumann entropy of a mixed state with density operator ρ is defined to be

$$S(\rho) = -\mathbf{tr}(\rho \log_2 \rho) = -\sum_i \lambda_i \log_2 \lambda_i,$$

where λ_i are the eigenvalues of ρ (with repeats). As is done for classical entropy, take $0 \log(0) = 0$.

The von Neumann entropy is zero for pure states; since the density operator ρ_x for a pure state $|x\rangle$ is a projector, it has a single 1-eigenvalue with $n - 1$ 0-eigenvalues, so $S(\rho_x) = 0$. Observe that the maximally uncertain single qubit mixed state ρ_{ME} has von Neumann entropy $S(\rho) = 1$. More generally, a maximally uncertain n-qubit state has a density operator that is diagonal with entries all 2^{-n}; a maximally uncertain n-qubit state ρ has von Neumann entropy $S(\rho) = n$.

For a single-qubit state with density operator ρ, the von Neumann entropy $S(\rho)$ is related to the distance between the point in the Bloch sphere corresponding to ρ and the center of the Bloch sphere. Let λ_1 and λ_2 be the eigenvalues of ρ. Since density operators have trace 1, $\lambda_2 = 1 - \lambda_1$. The von Neumann entropy of ρ can be deduced from its determinant: $\det(\rho) = \lambda_1 \lambda_2$, so $\det(\rho) = \lambda_1(1 - \lambda_1)$, so $\lambda_1^2 - \lambda_1 + \det(\rho) = 0$, which has solutions

$$\lambda_1 = \frac{1 + \sqrt{1 - 4 \det \rho}}{2}$$

and

$$\lambda_2 = \frac{1 - \sqrt{1 - 4 \det \rho}}{2}.$$

Using $\det(\rho) = \frac{1}{4}(1 - r^2)$ from section 10.1.3, we see that

$$\lambda_1 = \frac{1 + r}{2}$$

and

$$\lambda_2 = \frac{1 - r}{2}.$$

So, for single-qubit mixed states, the entropy is simply a function of the radial distance r:

$$S(\rho) = -\left(\left(\frac{1+r}{2}\right) \log_2 \left(\frac{1+r}{2}\right) + \left(\frac{1-r}{2}\right) \log_2 \left(\frac{1-r}{2}\right)\right). \tag{10.6}$$

10.2 Classifying Entangled States

The concepts and notation for quantum subsystems support a deeper study of entanglement. Ever since the beginning of the field, entanglement has been recognized as a fundamental resource for quantum information processing and a key to what distinguishes it from classical processing. Nevertheless, it is still only poorly understood. Only in the simplest case, that of pure states of a bipartite system $A \otimes B$, is entanglement well understood. What is known about multipartite entanglement is that it is complicated; there are many distinct types of multipartite entanglement whose utility and relation are only beginning to be understood. Even for bipartite mixed states there are distinct measures of entanglement. Each gives insight into the entanglement resources needed for various quantum information processing tasks; no single measure of entanglement will do.

Recall that entanglement is not an absolute property of a quantum state, but depends on the tensor decomposition of the system into subsystems. A (pure) state $|\psi\rangle$ of a quantum system with associated vector space V, is *separable* with respect to the tensor decomposition $V = V_1 \otimes \cdots \otimes V_n$ if it can be written as

$$|\psi\rangle = |\psi_1\rangle \otimes \cdots \otimes |\psi_n\rangle,$$

where $|\psi_i\rangle$ is contained in V_i. Otherwise, $|\psi\rangle$ is said to be *entangled* with respect to this decomposition. For n-qubit systems, we will generally speak of entanglement with respect to the decomposition into the n single-qubit systems. Thus, when we say that a state is entangled without further qualification, we mean that it is entangled with respect to this decomposition into individual qubits.

For bipartite pure states, it is possible to quantify the amount of entanglement a state contains. Any reasonable measure of entanglement should satisfy certain properties. For example, any measure of entanglement should take its minimal value, usually zero, on unentangled states. Furthermore, performing any sequence of operations, including measurements, on the subsystems individually should not increase the value of an entanglement measure. Even allowing the result of a measurement on one subsystem to influence which operations are performed on another subsystem should not increase the value. Imagine different people in control of each subsystem, with only classical communication channels between them. The restricted set of operations they can perform is often abbreviated LOCC, for *local operations with classical communication*. The LOCC requirement for any reasonable measure of entanglement means that nothing these people can do can increase the value of the entanglement measure.

10.2.1 Bipartite Quantum Systems

To find a good measure of entanglement for pure states of a bipartite system $X = A \otimes B$, let us look at the simplest of bipartite systems: two-qubit systems. The state $|\psi\rangle = \frac{1}{\sqrt{2}}(|00\rangle + |11\rangle)$ is maximally entangled in the sense that, when looked at separately, the state of each qubit is as uncertain as possible. Tracing over each qubit gives the mixed state $\rho_{ME} = \frac{1}{2}I$. This state has maximal von Neumann entropy among all two-qubit states. Similarly, unentangled states are the least entangled states possible in the sense that, when looked at separately, the state of each qubit is as certain as possible. Tracing over each qubit gives a pure state, a state with zero von Neumann entropy. These examples suggest that the von Neumann entropy of the partial trace with respect to one of the subsystems might make a good measure of entanglement in bipartite systems.

In order for this approach to make sense, the von Neumann entropy of the partial trace should be the same whether we look at subsystem A or subsystem B. The proof that the two quantities are the same relies on the *Schmidt decomposition*. The Schmidt decomposition also leads directly to a coarse measure of entanglement. For any pure state $|\psi\rangle$ of a bipartite system $A \otimes B$, there exist orthonormal sets of states $\{|\psi_i^A\rangle\}$ and $\{|\psi_i^B\rangle\}$ such that

$$|\psi\rangle = \sum_{i=1}^{K} \lambda_i |\psi_i^A\rangle \otimes |\psi_i^B\rangle$$

for some positive real λ_i such that $\sum_{i=1} \lambda_i^2 = 1$. Exercises 10.8 and 10.9 step through a proof that Schmidt decomposition exists for every state $|\psi\rangle$. The λ_i are called the *Schmidt coefficients*, and K, the number of λ_i, is called the *Schmidt rank* or *Schmidt number* of $|\psi\rangle$. For unentangled states, the Schmidt rank is 1.

We now use the Schmidt decomposition to check that $\mathbf{tr}_A \rho = \mathbf{tr}_B \rho$. Let $|\psi\rangle$ be a state in a bipartite system $X = A \otimes B$, where A and B are general multiple-qubit systems. Let $\rho = |\psi\rangle\langle\psi|$. Let

$$|\psi\rangle = \sum_{i=0}^{K-1} \lambda_i |\psi_i^A\rangle \otimes |\psi_i^B\rangle$$

be a Schmidt decomposition for $|\psi\rangle$. Then

$$\rho = |\psi\rangle\langle\psi| = \sum_{i=0}^{K-1} \sum_{j=0}^{K-1} \lambda_i \lambda_j |\psi_i^A\rangle\langle\psi_j^A| \otimes |\psi_i^B\rangle\langle\psi_j^B|,$$

so

$$\mathbf{tr}_B \rho = \sum_{i=0}^{K-1} \lambda_i^2 |\psi_i^A\rangle\langle\psi_i^A|$$

and

$$\mathbf{tr}_A \rho = \sum_{i=0}^{K-1} \lambda_i^2 |\psi_i^B\rangle\langle\psi_i^B|.$$

Since $\{|\psi_i^A\rangle\}$ is an orthonormal set, it follows that

$$S(\mathbf{tr}_A \rho) = -\sum_{i=0}^{K-1} \lambda_i^2 \log_2 \lambda_i^2.$$

Similarly,

$$S(\mathbf{tr}_B \rho) = -\sum_{i=0}^{K-1} \lambda_i^2 \log_2 \lambda_i^2.$$

Thus

$$S(\mathbf{tr}_A \rho) = S(\mathbf{tr}_B \rho).$$

The amount of entanglement between the two parts of a pure state $|\psi\rangle$ of a bipartite system $X = A \otimes B$ with density operator $\rho = |\psi\rangle\langle\psi|$ is defined to be

$$S(\mathbf{tr}_A \rho),$$

or equivalently $S(\mathbf{tr}_B \rho)$.

We compute this quantity for a variety of bipartite states. To begin with, it is zero on unentangled states.

Example 10.2.1   For $|x\rangle = \frac{1}{\sqrt{2}}(|00\rangle + |11\rangle)$, recall from example 10.1.1 that $\rho_x^1 = \mathbf{tr}_2|x\rangle\langle x| = \rho_{ME} = \frac{1}{2}I$. Thus, by the formula for the von Neumann entropy for single-qubit mixed states, equation 10.6, the amount of entanglement is $S(\rho_{ME}) = 1$. If we work out the density matrices for the first qubit of states $\frac{1}{\sqrt{2}}(|01\rangle + |10\rangle)$ and $\frac{1}{\sqrt{2}}(|00\rangle - i|11\rangle)$ we will find that these too are equal to ρ_{ME}. Such states are among the maximally entangled two-qubit states.

Example 10.2.2   Let $|x\rangle = \frac{7}{10}|00\rangle + \frac{1}{10}|01\rangle + \frac{1}{10}|10\rangle + \frac{7}{10}|11\rangle$ with density operator $\rho_x = |x\rangle\langle x|$. To obtain the density operator $\rho_x^1 = \mathbf{tr}_2|x\rangle\langle x|$, trace over the second qubit. The four terms that make up matrix ρ_x^1 in the standard basis are:

$$\sum_{j=0}^{1}\langle 0|\langle j||x\rangle\langle x||0\rangle|j\rangle \; |0\rangle\langle 0| = \left(\left(\frac{7}{10}\right)^2 + \left(\frac{1}{10}\right)^2\right)|0\rangle\langle 0| = \frac{1}{2}|0\rangle\langle 0|,$$

$$\sum_{j=0}^{1}\langle 0|\langle j||x\rangle\langle x||1\rangle|j\rangle \; |0\rangle\langle 1| = \left(\frac{7}{10}\frac{1}{10} + \frac{1}{10}\frac{7}{10}\right)|0\rangle\langle 1| = \frac{7}{50}|0\rangle\langle 1|,$$

$$\sum_{j=0}^{1}\langle 1|\langle j||x\rangle\langle x||0\rangle|j\rangle \; |1\rangle\langle 0| = \left(\frac{1}{10}\frac{7}{10} + \frac{7}{10}\frac{1}{10}\right)|1\rangle\langle 0| = \frac{7}{50}|1\rangle\langle 0|,$$

$$\sum_{j=0}^{1}\langle 1|\langle j||x\rangle\langle x||1\rangle|j\rangle \; |1\rangle\langle 1| = \left(\left(\frac{1}{10}\right)^2 + \left(\frac{7}{10}\right)^2\right)|1\rangle\langle 1| = \frac{1}{2}|1\rangle\langle 1|.$$

So

$$\rho_x^1 = \frac{1}{2}|0\rangle\langle 0| + \frac{7}{50}|1\rangle\langle 0| + \frac{7}{50}|0\rangle\langle 1| + \frac{1}{2}|1\rangle\langle 1|$$

$$= \frac{1}{100}\begin{pmatrix} 50 & 14 \\ 14 & 50 \end{pmatrix} = \frac{1}{2}(I + \frac{14}{50}X)$$

corresponding to the point $(14/50, 0, 0)$ in the Bloch sphere. To compute $S(\rho_x^1)$, we note that $\{|+\rangle, |-\rangle\}$ is the eigenbasis of ρ_x^1 with eigenvalues $\frac{16}{25}$ and $\frac{9}{25}$, so

$$S(\rho_x^1) = -\frac{16}{25}\log_2\frac{16}{25} - \frac{9}{25}\log_2\frac{9}{25} = 0.942\ldots.$$

More directly, we could have used equation 10.6 of section 10.1.4 to compute the eigenvalues from the distance $r = \frac{14}{50}$ of ρ_x^1 from the center of the sphere.

Example 10.2.3 Let $|y\rangle = \frac{i}{10}|00\rangle + \frac{\sqrt{99}}{10}|11\rangle$ with density operator $\rho_y = |y\rangle\langle y|$. Tracing over the second qubit, we obtain

$$\rho^1(y) = \mathbf{tr}_2(\rho_y) = \frac{1}{100}\begin{pmatrix} 1 & 0 \\ 0 & 99 \end{pmatrix} = \frac{1}{2}(I - \frac{49}{50}Z)$$

corresponding to the point $(0, 0, 49/50)$ in the Bloch sphere. Using the relation between $r = \frac{49}{50}$ and the eigenvalues given by equation 10.6 of section 10.1.3, we obtain

$$S(\rho_y^1) = -\frac{1}{100}\log_2\frac{1}{100} - \frac{99}{100}\log_2\frac{99}{100} = 0.0807\ldots.$$

To underscore how strongly the notion of entanglement depends on the subsystem decomposition under consideration, we give an example of a state with widely different von Neumann entropies with respect to two different system decompositions.

Example 10.2.4 The amount of entanglement in the four-qubit state

$$|\psi\rangle = \frac{1}{2}(|00\rangle + |11\rangle + |22\rangle + |33\rangle) = \frac{1}{2}(|0000\rangle + |0101\rangle + |1010\rangle + |1111\rangle)$$

differs greatly for two different bipartite system decompositions.

First, consider the decomposition into the first and third qubits and the second and fourth qubits. Trace over the first subsystem. The state $\rho_\psi^{24} = \mathbf{tr}_{1,3}(|\psi\rangle\langle\psi|)$ has von Neumann entropy 0 since, by example 3.2.3, $|\psi\rangle$ can be written as the tensor product of pure states in each of these subsystems. Thus, with respect to this decomposition, the state $|\psi\rangle$ is unentangled.

Now consider the decomposition into the first and second qubits and the third and fourth qubits. Tracing over the second system yields

$$\rho_\psi^{12} = \mathbf{tr}_{34}(|\psi\rangle\langle\psi|) = \sum_{i,j=0}^{3}\sum_{k=0}^{3}\langle j|\langle k||\psi\rangle\langle\psi||i\rangle|k\rangle|j\rangle\langle i|.$$

The coefficient of $|j\rangle\langle i|$ is $\frac{1}{4}\delta_{ij}$, so ρ_ψ^{12} is the 4×4 diagonal matrix with diagonal entries all $1/4$, so $S(\mathbf{tr}_{1,2}(|\psi\rangle\langle\psi|)) = 2$. The maximum value possible for the von Neumann entropy of states of a two-qubit system is 2. Thus, with respect to this decomposition, the state $|\psi\rangle$ is maximally entangled.

While the von Neumann entropy of the partial trace with respect to one of the subsystems is the most common measure of entanglement for bipartite pure states, the Schmidt rank K is also a useful measure of entanglement. Both are nonincreasing under local operations and classical communication (LOCC). The Schmidt rank is a much coarser measure of entanglement than the von Neumann entropy of the partial trace. For two-qubit systems, the Schmidt rank merely distinguishes between unentangled states, with Schmidt rank 1, and entangled states, with Schmidt rank 2. For bipartite systems $A \otimes B$, where A and B are multiple-qubit systems, the Schmidt rank is more interesting than in the single-qubit case, but it is still a coarser measure than the von Neumann entropy of the partial trace.

10.2.2 Classifying Bipartite Pure States up to LOCC Equivalence

A state $|\psi\rangle \in X$ can be converted to a state $|\phi\rangle \in X$ by local operations and classical communication (LOCC) with respect to a tensor decomposition $X = X_1 \otimes \cdots \otimes X_n$ if there exists a sequence of unitary operators and measurements on separate X_i that when applied to $|\psi\rangle$ are guaranteed to result in the state $|\phi\rangle$. Which transformations are applied is allowed to depend on the outcomes of previous measurements, but is otherwise deterministic. Two states $|\psi\rangle$ and $|\phi\rangle$ are said to be *LOCC equivalent* with respect to the decomposition $X = X_1 \otimes \cdots \otimes X_n$ if $|\psi\rangle$ can be converted to $|\phi\rangle$ via LOCC and vice versa. An unentangled state cannot be converted to an entangled one using only LOCC.

Example 10.2.5 The Bell states $\frac{1}{\sqrt{2}}(|00\rangle + |11\rangle)$ and $\frac{1}{\sqrt{2}}(|01\rangle + |10\rangle)$ are LOCC equivalent; simply apply X to the second qubit.

Example 10.2.6 The Bell state $\frac{1}{\sqrt{2}}(|00\rangle + |11\rangle)$ can be converted to $|00\rangle$ via LOCC, but not vice versa: measure the first qubit in the standard basis to obtain either $|00\rangle$ or $|11\rangle$, and if the result was $|1\rangle$ apply X to each of the qubits.

Nielsen provides an elegant classification of pure states of bipartite systems up to LOCC equivalence, in terms of *majorization* of the sets of eigenvalues of the density operators of the subsystems. Let $a = (a_1, \ldots, a_m)$ and $b = (b_1, \ldots, b_m)$ be two vectors in $\mathbf{R}^m$. Let $a^\downarrow$ be the reordered version of a such that $a_i \geq a_{i+1}$ for all i. We say that b *majorizes* a, written $b \succeq a$, if for each k, $1 \leq k \leq m$, $\sum_{j=1}^{k} a_j^{\downarrow} \leq \sum_{j=1}^{k} b_j^{\downarrow}$ with the additional requirement that $\sum_{j=1}^{k} a_j^{\downarrow} = \sum_{j=1}^{k} b_j^{\downarrow}$ when $k = m$. For a pure state $|\psi\rangle$ of a bipartite system $A \otimes B$, let $\lambda^{\psi} = (\lambda_1^{\psi}, \ldots, \lambda_m^{\psi})$ be the eigenvalues of $\mathbf{tr}_B |\psi\rangle\langle\psi|$. Nielsen has proved that the state $|\psi\rangle$ can be transformed to $|\phi\rangle$ by LOCC if and only if λ^{ψ} is majorized by λ^{ϕ}. Thus, $|\psi\rangle$ and $|\phi\rangle$ are LOCC equivalent if and only if $\lambda^{\psi} \succeq \lambda^{\phi}$ and $\lambda^{\phi} \succeq \lambda^{\psi}$.

In the case of a bipartite system consisting of two qubits, the majorization condition reduces to a simple one. Let $|\psi\rangle$ and $|\phi\rangle$ be two states of a two-qubit system with $\lambda^{\psi} = (\lambda, 1 - \lambda)$ and $\lambda^{\phi} = (\mu, 1 - \mu)$, where $\lambda \geq 1/2$ and $\mu \geq 1/2$. Then $\lambda^{\psi} \succeq \lambda^{\phi}$ if and only if $\lambda \geq \mu$. It follows that $\lambda^{\psi} \succeq \lambda^{\phi}$ if and only if $S(\mathbf{tr}_2|\psi\rangle\langle\psi|) \leq S(\mathbf{tr}_2|\phi\rangle\langle\phi|)$. Thus, $|\phi\rangle$ can be converted to $|\psi\rangle$ via LOCC if and only if $|\phi\rangle$ is more entangled than $|\psi\rangle$. Similarly $|\phi\rangle$ and $|\psi\rangle$ are LOCC equivalent if and only if the von Neumann entropies of the density operators for the partial trace over one of the subsystems are equal. Observe that there are infinitely many LOCC equivalence classes, and that these classes are parametrized by a continuous variable, $1/2 \leq \lambda \leq 1$.

For bipartite systems with subsystems larger than single-qubit systems, the classification is more complicated in that there are incomparable states. For example, if A and B are both two-qubit systems, the states

$$|\psi\rangle = \frac{3}{4}|0\rangle|0\rangle + \frac{2}{4}|1\rangle|1\rangle + \frac{\sqrt{2}}{4}|2\rangle|2\rangle + \frac{1}{4}|3\rangle|3\rangle$$

and

$$|\phi\rangle = \frac{2\sqrt{2}}{4}|0\rangle|0\rangle + \frac{\sqrt{6}}{4}|1\rangle|1\rangle + \frac{1}{4}|2\rangle|2\rangle + \frac{1}{4}|3\rangle|3\rangle$$

are incomparable because

$$\lambda_1^{\psi} = \frac{9}{16} > \frac{1}{2} = \lambda_1^{\phi},$$

but

$$\lambda_1^{\psi} + \lambda_2^{\psi} = \frac{13}{16} < \frac{14}{16} = \lambda_1^{\phi} + \lambda_2^{\phi}.$$

Nevertheless, in any bipartite system, no matter how large, the vector for any unentangled state majorizes all others. Furthermore, in any bipartite system there are *maximally entangled* states $|\psi\rangle$ for which λ^{ψ} is majorized by λ^{ϕ} for all states $|\phi\rangle$. Let X be a bipartite system $X = A \otimes B$ where A and B have dimensions n and m respectively, with $n \geq m$. Let $|\psi\rangle$ be a state of the form

$$|\psi\rangle = \frac{1}{\sqrt{m}} \sum_{i=1}^{m} |\phi_i^A\rangle \otimes |\phi_i^B\rangle$$

where the $\{|\phi_i^A\rangle\}$ and $\{|\phi_i^B\rangle\}$ are orthonormal sets, and since m is the dimension of B, the set $\{|\phi_i^B\rangle\}$ is a basis for B. The vector λ^{ψ} is majorized by λ^{ϕ} for all states $|\phi\rangle \in A \otimes B$. Furthermore, as one would expect, these maximally entangled states have maximal Schmidt rank, and the von Neumann entropy after tracing over either subsystem is the maximum possible value. These states fulfill our current expectations for maximally entangled states in every way. We shall see, however, that for multipartite states it is nevertheless highly unclear what maximally entangled should mean.

10.2.3 Quantifying Entanglement in Bipartite Mixed States

Before discussing entanglement in multipartite quantum systems, we take a brief look at the meaning of entanglement for mixed states. A mixed state ρ of a quantum system $V_1 \otimes \cdots \otimes V_n$ is *separable* with respect to this tensor decomposition if it can be written as a probabilistic mixture of unentangled states: ρ is separable if it can be written as

$$\rho = \sum_{j=1}^{m} p_j |\phi_j^{(1)}\rangle\langle\phi_j^{(1)}| \otimes \cdots \otimes |\phi_j^{(n)}\rangle\langle\phi_j^{(n)}|,$$

where $|\phi_j^{(i)}\rangle \in V_i$ and $p_i \geq 0$ with $\sum_i p_i = 1$. For a given i, the various $|\phi_j^{(i)}\rangle$ need not be orthogonal. If a mixed state ρ cannot be written as above, it is said to be *entangled*.

This definition may appear more complicated than expected; why not say a mixed state ρ is entangled if it cannot be written as $\rho_1 \otimes \cdots \otimes \rho_n$? The more involved definition distinguishes entanglement from mere classical correlation. For example, the mixed state $\rho_{cc} = \frac{1}{2}|00\rangle\langle00| + \frac{1}{2}|11\rangle\langle11|$ is classically correlated (it cannot be written as $\rho_1 \otimes \cdots \otimes \rho_n$), but is not entangled. The state $\rho_{\Phi+} = \frac{1}{2}(|00\rangle + |11\rangle)(\langle00| + \langle11|)$ is entangled. Appendix A discusses quantum entanglement versus classical correlations in more detail.

If a mixed state ρ can be written as a probabilistic mixture of entangled states, it is not necessarily entangled; it still may be separable. For example, consider

$$\rho = \frac{1}{2}|\Phi^+\rangle\langle\Phi^+| + \frac{1}{2}|\Phi^-\rangle\langle\Phi^-|,$$

where $|\Phi^+\rangle$ and $|\Phi^-\rangle$ are the Bell states $|\Phi^+\rangle = 1/\sqrt{2}(|00\rangle + |11\rangle)$ and $|\Phi^-\rangle = 1/\sqrt{2}(|00\rangle - |11\rangle)$. We defined the mixed state ρ as a probabilistic mixture of maximally entangled states, but it is easy to check that it can also be written as

$$\rho = \frac{1}{2}|00\rangle\langle00| + \frac{1}{2}|11\rangle\langle11|,$$

a probabilistic mixture of product states, so ρ is actually separable.

There are a number of useful measures of entanglement for mixed bipartite states, all of which coincide with the standard measure of entanglement on pure states, the von Neumann entropy of the density operator for one of the subsystems. We give a rough description of a few of these measures. The amount of *distillable entanglement* contained in a mixed state ρ is the asymptotic ratio m/n of the maximum number m of maximally entangled states ρ_{ME} that can be obtained from n copies of ρ by LOCC. Conversely, the *entanglement cost* is the asymptotic ratio m/n of the minimum number n of copies of a maximally entangled state ρ_{ME} needed to produce m copies of ρ using only LOCC. The *relative entropy of entanglement* can be thought of as measuring how close ρ is to a separable state ρ_S, and is defined to be

$$\inf_{\rho_S \in S} \mathbf{tr}[\rho(\log \rho - \log \rho_S)],$$

where the infimum is over all separable states ρ_S. It is known that not only is the distillable entanglement never greater than the cost of entanglement, but also for most mixed states the distillable entanglement is strictly less than the cost of entanglement. In particular, there exist *bound entangled states* from which no entanglement can be distilled, but whose entanglement cost is non-zero. The study of entanglement in mixed bipartite states is a rich area of research, with many known results not described here, but also with many remaining open questions. Even the relationship between the measures we just described is not fully understood.

10.2.4 Multipartite Entanglement

Researchers are continuing to develop new measures of entanglement and explore properties of states entangled with respect to tensor decompositions into more than two subsystems. For quantum computation, we are particularly interested in properties of n-qubit states for large n, and measures of entanglement for these states with respect to the decomposition into the individual qubit systems. In spite of broad recognition that understanding multipartite entanglement is crucial for understanding the power and limitations of quantum computation, much remains unknown. Entangled states provide a fundamental resource for other types of quantum information processing, such teleportation and dense coding, as well as quantum computation. Which types of entangled states are most useful for which types of quantum information processing tasks, is an active area of research.

Even for pure states of the simplest multipartite systems, three-qubit systems, quantifying entanglement is complicated. We just saw that for two-qubit systems there are infinitely many LOCC equivalence classes. However, relaxing the LOCC condition simplifies the picture somewhat. A state $|\psi\rangle$ can be converted to $|\phi\rangle$ by *stochastic local operations and classical communication (SLOCC)* if there is a sequence of local operations with classical communications that with non-zero probability turns $|\phi\rangle$ into $|\psi\rangle$. States $|\psi\rangle$ and $|\phi\rangle$ are SLOCC equivalent if $|\psi\rangle$ can be converted to $|\phi\rangle$ by SLOCC and vice versa. Under SLOCC equivalence, the two-qubit case reduces to two classes: entangled states and unentangled states.

For a three-qubit system $X = A \otimes B \otimes C$, the SLOCC classification of states with respect to the decomposition into the three systems, has six distinct SLOCC classes:

- unentangled states,
- A-BC decomposable states,
- B-AC decomposable states,
- C-AB decomposable states,
- SLOCC equivalent to $|GHZ_3\rangle = \frac{1}{\sqrt{2}}(|000\rangle + |111\rangle)$, and
- SLOCC equivalent to $|W_3\rangle = \frac{1}{\sqrt{3}}(|001\rangle + |010\rangle + |100\rangle)$.

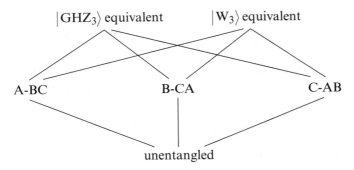

Figure 10.1
Partial order of SLOCC classes for a three-qubit system, where states in upper classes can be converted to states in lower classes using SLOCC, but the two uppermost classes cannot be converted to each other.

A state $|\psi\rangle$ being in the A-BC decomposable class means that it can be written as $|\psi\rangle = |\psi_A\rangle \otimes |\psi_{BC}\rangle$ for $|\psi_A\rangle \in A$ and $|\psi_{BC}\rangle \in B \otimes C$, but it cannot be fully decomposed into $|\psi\rangle = |\psi_A\rangle \otimes |\psi_B\rangle \otimes |\psi_C\rangle$, where $|\psi_A\rangle \in A$, $|\psi_B\rangle \in B$ and $|\psi_C\rangle \in C$.

A partial order on these six classes is shown in figure 10.1: a state $|\psi\rangle$ is contained in a class above that of state $|\phi\rangle$ if there exists an SLOCC sequence taking $|\psi\rangle$ to $|\phi\rangle$ but not vice versa. There are two inequivalent classes of states at the top of the hierarchy; $|GHZ_3\rangle$ cannot be converted to $|W_3\rangle$ or the other way around. It is not clear whether $|GHZ_3\rangle$ or $|W_3\rangle$ should be considered more entangled; each appears to be highly entangled in some ways and less so in others. To illustrate the distinct types of entanglement these states embody, we look at these states in terms of the *persistency* and *connectedness* of their entanglement.

Persistency of entanglement   The *persistency* of entanglement of $|\psi\rangle \in V \otimes \cdots \otimes V$ is the minimum number of qubits, P_e, that need to be measured to guarantee that the resulting state is unentangled.

Maximal connectedness   A state $|\psi\rangle \in V \otimes \cdots \otimes V$ is *maximally connected* if for any two qubits there exists a sequence of single-qubit measurements on the other qubits that when performed guarantee that the two qubits end up in a maximally entangled state.

Let $|GHZ_n\rangle$ be the n-qubit state

$$|GHZ_n\rangle = \frac{1}{\sqrt{2}}(|00\ldots0\rangle + |11\ldots1\rangle),$$

and $|W_n\rangle$ be the n-qubit state

$$|W_n\rangle = \frac{1}{\sqrt{n}}(|0\ldots001\rangle + |0\ldots010\rangle + |0\ldots100\rangle + \cdots + |1\ldots000\rangle).$$

Because only one qubit needs to be measured to reduce $|GHZ_n\rangle$ to an unentangled state, the persistency of entanglement of $|GHZ_n\rangle$ is only 1, so in this sense it is not very entangled. On the other hand, $|GHZ_n\rangle$ is maximally connected. It is relatively easy to check that the states $|W_n\rangle$ are not maximally connected. Yet, they do have high persistency: $P_e(|W_n\rangle) = n - 1$. Thus, whether $|GHZ_n\rangle$ or $|W_n\rangle$ should be considered more entangled depends on what properties of entanglement one is interested in.

For $n \geq 4$, the situation becomes far more complicated. For $n \geq 4$ there are infinitely many SLOCC equivalence *classes*, and these classes are parametrized by continuous parameters. As n increases, it becomes less and less clear which states should be considered maximally entangled.

Cluster States   A class of n-qubit entangled states, cluster states, combine properties of both $|GHZ_n\rangle$ and $|W_n\rangle$ states. The $|GHZ_n\rangle$ states are maximally connected but have persistency of only 1. The persistency of the $|W_n\rangle$ states increases with n, but they are not maximally connected. Cluster states are maximally connected and have persistency increasing with n. Cluster states form a universal entanglement resource for quantum computation that is the basis for *cluster state*, or *one-way*, *quantum computing*, an alternative model of quantum computing discussed in chapter 13.

Let G be any finite graph whose vertices are qubits. The neighborhood of any vertex $v \in G$, $nbhd(v)$ is the set of vertices w connected to v by an edge of the graph. An operator O stabilizes a state $|\psi\rangle$ if $O|\psi\rangle = |\psi\rangle$. The *graph states* $|G\rangle$ corresponding to a graph G is the state stabilized by the set of operators, one for each vertex of G,

$$X^v \otimes \bigotimes_{i \in nbhd(v)} Z^i, \tag{10.7}$$

where $X = |1\rangle\langle 0| + |0\rangle\langle 1|$ and $Z = |0\rangle\langle 0| - |1\rangle\langle 1|$ are the familiar Pauli operators, and the superscript on these operators indicates to which qubit the operator is applied. If the graph G is a d-dimensional rectangular lattice, then $|G\rangle$ is called a *cluster state* (see figure 10.2). There is some discrepancy in terminology in the literature; sometimes *cluster states* is taken to be synonymous with *graph states*. Graph states, including cluster states, can be constructed as follows. For each vertex, begin with a qubit in state $|+\rangle$. Then for each edge in the graph apply the controlled phase operator $C_P = |00\rangle\langle 00| + |01\rangle\langle 01| + |10\rangle\langle 10| - |11\rangle\langle 11|$. Since the controlled phase operator is symmetric on the qubits, and the applications of the controlled phase all commute with each other, it does not matter in which order the operators are applied. Here we consider only the states stabilized by the operators $X^v \otimes \bigotimes_{i \in nbhd(v)} Z^i$, but some expositions consider all states which are joint eigenstates of these operators.

Example 10.2.7   *Construction of the cluster state for a 1×2 lattice.* Apply C_P to $|+\rangle|+\rangle$ to obtain the cluster state

$$|\phi_2\rangle = \frac{1}{2}(|00\rangle + |01\rangle + |10\rangle - |11\rangle) = \frac{1}{\sqrt{2}}(|+\rangle|0\rangle + |-\rangle|1\rangle) = \frac{1}{\sqrt{2}}(|0\rangle|+\rangle + |1\rangle|-\rangle).$$

This state is LOCC equivalent to a Bell state.

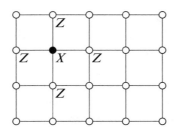

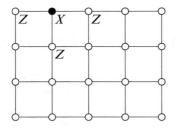   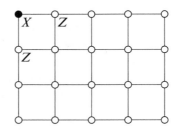

Figure 10.2
A 4×5 rectangular lattice. Sample operators that define a cluster state are shown, one for an internal node, one for a node on the boundary but not at a corner, and one for a corner node. A cluster state is a state that is a simultaneous eigenstate of all such operators, one for each node of the lattice.

Example 10.2.8 *Cluster state for a* 1×3 *lattice.* The operator $(C_P \otimes I)(I \otimes C_P)$ applied to $|+\rangle|+\rangle|+\rangle$ results in the cluster state

$$|\phi_3\rangle = (C_P \otimes I)(\frac{1}{\sqrt{2}}(|+\rangle(|0\rangle|+\rangle + |1\rangle|-\rangle)))$$

$$= \frac{1}{2}(|0\rangle|0\rangle|+\rangle + |0\rangle|1\rangle|-\rangle + |1\rangle|0\rangle|+\rangle - |1\rangle|1\rangle|-\rangle)$$

$$= \frac{1}{2}(|+\rangle|0\rangle|+\rangle + |-\rangle|1\rangle|-\rangle).$$

This state is LOCC equivalent to $|GHZ_3\rangle$.

Example 10.2.9 *Cluster state for a* 1×4 *lattice.*

$$|\phi_4\rangle = \frac{1}{2}(|0\rangle|+\rangle|0\rangle|+\rangle + |1\rangle|-\rangle|0\rangle|+\rangle + |0\rangle|-\rangle|1\rangle|-\rangle + |1\rangle|+\rangle|1\rangle|-\rangle)$$

$$= \frac{1}{2}(|+\rangle|0\rangle|+\rangle|0\rangle + |-\rangle|0\rangle|-\rangle|0\rangle + |+\rangle|0\rangle|-\rangle|1\rangle + |-\rangle|1\rangle|+\rangle|1\rangle)$$

The reader should check that each of these states is stabilized by all of the operators of equation 10.7.

Briegel and Raussendorf give a straightforward proof that all cluster states are maximally connected. A more involved argument shows that cluster states $|\phi_n\rangle$ have persistency $\lfloor n/2 \rfloor$. Thus, while the persistency of $|\phi_n\rangle$ is not as great as the persistency of $|W_n\rangle$, the persistency of cluster states does increase linearly with the number of qubits, and unlike $|W_n\rangle$, cluster states are maximally connected. Thus cluster states combine entanglement strengths of both the $|GHZ_n\rangle$ states and the $|W_n\rangle$ states. In section 13.4.1, we briefly return to cluster states to describe the use of their entanglement as a quantum computational resource. The following table summarizes the situation:

	max connected	P_e
$|GHZ_n\rangle$	YES	1
$|\phi_n\rangle$	YES	$\lfloor n/2 \rfloor$
$|W_n\rangle$	NO	$n-1$

10.3 Density Operator Formalism for Measurement

An analysis of a quantum algorithm or protocol that involves measurement must take into account all possible outcomes of any measurement. Up to now we have had only an awkward way of describing the result of a future measurement: listing the possible outcomes and their respective probabilities. Density operators provide a compact and elegant way to model the probabilistic outcomes of a measurement yet to be performed or of a measurement for which the outcome is unknown.

Density operators provide a means of compactly expressing a probability distribution over quantum states or the statistical properties of an ensemble of quantum states. If the reader has not yet read appendix A on the relations between probability theory and quantum mechanics, now would be a good time to do so. The following game motivates this use of density operators in this context. Keep in mind the definition of a state given in section 10.1: a state encapsulates "all information about the system that can be gained from any number of measurements on a supply of identical quantum systems." Suppose you are told you will be sent a sequence of qubits and that either all members of the sequence are the first qubit of a Bell state $\frac{1}{\sqrt{2}}(|00\rangle + |11\rangle)$ or that a random sequence of $|0\rangle$ and $|1\rangle$ are sent, with $|0\rangle$ and $|1\rangle$ having equal probability. Your job is to determine which type of sequence you are receiving. What is your strategy?

It is impossible to do better than guessing randomly; without access to more information, there is no way to distinguish the two sequences. If you were given access to the second qubit of each Bell pair in the first case and a second copy of each qubit in the second, a winning strategy is possible. But without access to the second qubit, the two sequences are indistinguishable. To see why, recall from section 10.1.1 that the density operator for one qubit of a Bell pair is $\frac{1}{2}I$, and that the density operators for $|0\rangle$ and $|1\rangle$ are $|0\rangle\langle 0| = \begin{pmatrix} 1 & 0 \\ 0 & 0 \end{pmatrix}$ and $|1\rangle\langle 1| = \begin{pmatrix} 0 & 0 \\ 0 & 1 \end{pmatrix}$ respectively.

From appendix A, the density operator ρ for a 50–50 probability distribution over the states $|0\rangle$ and $|1\rangle$ is

$$\rho = \frac{1}{2}\begin{pmatrix} 1 & 0 \\ 0 & 0 \end{pmatrix} + \frac{1}{2}\begin{pmatrix} 0 & 0 \\ 0 & 1 \end{pmatrix}$$

$$= \frac{1}{2}|0\rangle\langle 0| + \frac{1}{2}|1\rangle\langle 1|$$

$$= \frac{1}{2}I.$$

So the density operator of one qubit of a Bell pair is the same as the mixed state of a 50-50 probability distribution over the states $|0\rangle$ and $|1\rangle$.

More generally, a probability distribution over quantum states where $|\psi_i\rangle$ has probability p_i is represented by the density operator

$$\rho = \sum_{i=0}^{K-1} p_i |\psi_i\rangle\langle\psi_i|.$$

This representation works even if the states $|\psi_i\rangle$ are not mutually orthogonal. Probability distributions over quantum states have appeared frequently in this book to describe the possible outcomes of a measurement. Density operators provide a concise representation, one that can be manipulated directly to see the effects of subsequent unitary transformations and measurements. Given an orthogonal set $\{|x_i\rangle\}$ of the possible outcomes of a measurement of a specific state $|x\rangle$, with p_i being the probability of each outcome, the density operator representing this probability distribution over quantum states is

$$\rho = \sum p_i |x_i\rangle\langle x_i|.$$

It is easy to check that ρ is Hermitian, trace 1, and positive, so ρ is a density operator. The density operator $\rho = \sum p_i |x_i\rangle\langle x_i|$ summarizes the possible results of a measurement as a probabilistic mixture of the density operators for the possible resulting pure states weighted by the probability of the outcomes.

10.3.1 Measurement of Density Operators

This section discusses the meaning of and notation for measurement of density operators. The measurement of mixed states directly generalizes that of pure states. First, we write the familiar measurement of pure states in terms of density operators. Let $|x\rangle$ be an element of a $N = 2^n$ dimensional vector space X with corresponding density operator $\rho_x = |x\rangle\langle x|$. Measuring $|x\rangle$ with an operator O that has K associated projectors P_j, yields with probability $p_j = \langle x|P_j|x\rangle$ the state

$$\frac{P_j|x\rangle}{|P_j|x\rangle|} = \frac{1}{\sqrt{p_j}} P_j|x\rangle.$$

The density operator for each of these states is

$$\rho_x^j = \frac{1}{p_j} P_j|x\rangle\langle x|P_j^\dagger = \frac{1}{p_j} P_j \rho_x P_j^\dagger,$$

so the density operator ρ_x^O summarizing the possible outcomes of the measurement is

$$\rho_x^O = \sum_j p_j \rho_x^j = \sum_j P_j \rho_x P_j^\dagger.$$

When ρ_x is written in an eigenbasis for the measuring operator O, the result ρ_x^O of measurement with O is particularly easy to see. Let $\{|\alpha_i\rangle\}$ be an eigenbasis for O that contains the vectors $\frac{P_j|x\rangle}{|P_j|x\rangle|}$ as the first K elements of an N-element basis. In this basis,

$$|x\rangle = \sum_{j=0}^{K-1} \frac{P_j|x\rangle}{|P_j|x\rangle|} = \sum_{i=0}^{N-1} x_i|\alpha_i\rangle,$$

where $x_i = \sqrt{p_j}$ for $i < K$, and $x_i = 0$ for $i \geq K$. So

$$\rho_x = |x\rangle\langle x| = \left(\sum_{i=0}^{N-1} x_i|\alpha_i\rangle\right)\left(\sum_{j=0}^{N-1} x_j|\alpha_j\rangle\right)^\dagger = \sum_j \sum_i x_i \overline{x_j}|\alpha_i\rangle\langle\alpha_j|;$$

the ij^{th} entry of the matrix for ρ_x in basis $\{|\alpha_k\rangle\}$ is $\bar{x}_i x_j$. The density operator ρ_x^O is

$$\rho_x^O = \sum_j x_j \overline{x_j}|\alpha_j\rangle\langle\alpha_j| = \sum_j P_j|x\rangle\langle x|P_j^\dagger,$$

so ρ_x^O is obtained from ρ_x by removing all the cross terms; the matrix for ρ_x^O in the basis $\{|\alpha_i\rangle\}$ is the matrix ρ_x with the off-diagonal entries replaced with zeros.

Measurement of mixed states is easily derived from that of pure states. Let ρ be a density operator. Using results from section 10.1.1, ρ can be viewed as a probabilistic mixture $\rho = \sum_i q_i|\psi_i\rangle\langle\psi_i|$ of of pure states $|\psi_i\rangle$. Measuring the mixed state ρ can be viewed as measuring $|\psi_i\rangle\langle\psi_i|$ with probability q_i, so the measurement outcomes are encapsulated by the density operator ρ', a probabilistic mixture of the density operators representing the possible outcomes of measuring each $|\psi_i\rangle\langle\psi_i|$:

$$\rho_i' = \sum_j P_j|\psi_i\rangle\langle\psi_i|P_j^\dagger.$$

Thus, the density operator ρ' for the possible outcomes of measuring the mixed state ρ is

$$\rho' = \sum_i q_i \sum_j P_j|\psi_i\rangle\langle\psi_i|P_j^\dagger = \sum_j P_j\left(\sum_i q_i|\psi_i\rangle\langle\psi_i|\right)P_j^\dagger = \sum_j P_j \rho P_j^\dagger.$$

The term $P_j \rho P_j^\dagger$ is not a density operator in general; it is positive and Hermitian, but its trace may be less than one. Because the trace of a positive, Hermitian operator is zero only if the operator is the zero operator, ρ' may be viewed as a probabilistic mixture of density operators $\rho_j = \dfrac{P_j \rho P_j^\dagger}{\mathbf{tr}(P_j \rho P_j^\dagger)}$ with weighting $p_j = \mathbf{tr}(P_j \rho P_j^\dagger)$:

$$\rho' = \sum_j p_j \rho_j = \sum_j p_j \frac{P_j \rho P_j^\dagger}{\mathbf{tr}(P_j \rho P_j^\dagger)}$$

where we ignore the zero terms. For a pure state $|\psi\rangle$ with density operator $\rho = |\psi\rangle\langle\psi|$,

$$\rho' = \sum_j p_j \frac{P_j \rho P_j^\dagger}{\langle\psi|P_j|\psi\rangle}$$

because

$$\mathbf{tr}(P_j |\psi\rangle\langle\psi| P_j^\dagger) = \langle\psi|P_j^\dagger P_j|\psi\rangle = \langle\psi|P_j|\psi\rangle$$

by the trace trick of box 10.1 and the properties of projection operators.

Both measurements with known outcome and measurements that have yet to be performed or for which the outcome is not known can be concisely represented by density operators. Suppose we measure $|x\rangle$ with operator O and obtain outcome $|\psi\rangle = \frac{P_j|x\rangle}{|P_j|x\rangle|}$ with density operator $\rho = |\psi\rangle\langle\psi|$. There are two different representations for the result of this measurement, ρ_ψ and ρ_x^O. Which should we use? If we do not know the measurement outcome, we must use ρ_x^O. If we do know the outcome, we should use ρ_ψ. We can use the density operator ρ_x^O, but ρ_ψ encapsulates more of the information we know. If we were to use ρ_x^O, the outcome of the measurement must be kept track of separately, and since ρ_x^O allows for more possibilities, using it means performing unnecessary calculations involving possibilities that did not happen. The same distinction arises when sampling from a probability distribution; before the sample is taken, or if the sample is taken but the outcome is unknown, the best model for the sample is the probability distribution itself. But once the outcome is known, the sample is best modeled by the known value. Appendix A discusses such relations between the classical and quantum situations. While issues with measurement connect with the deepest issues in quantum mechanics, the distinction between these two models for measurement outcomes is not one of these issues. The deeper questions involve when and how a measurement outcome becomes known and by whom. We do not elaborate on these quantum mechanical issues here.

10.4 Transformations of Quantum Subsystems and Decoherence

Density operators were introduced to enable us to better discuss quantum subsystems. In the preceding section, we used it fruitfully to gain insight into entanglement. So far we have only used it to discuss static situations. We turn now to dynamics. In the first two parts of the book, we discussed quantum systems modeled by pure states that are acted upon by unitary operators. As we saw in section 10.1, to discuss quantum subsystems, we needed to expand from considering only pure states to considering density operators. Similarly, to discuss the dynamics of quantum subsystems, we need to expand from considering only unitary operators to a more general class of operators. Section 10.4.1 develops *superoperators*, this more general class of operators, by considering unitary operators on the entire system and looking to see what can be understood about their effect on a subsystem. Section 10.4.2 describes a decomposition that gives insight into superoperators. Section 10.4.3 discusses superoperators corresponding to measurements. Section 10.4.4 makes use of the superoperator formalism to discuss decoherence, errors caused by the interaction of the quantum system under consideration with the environment. This discussion of decoherence provides the setting for the discussion of quantum error correction in chapter 11.

10.4.1 Superoperators

This section considers the dynamics of subsystems. Section 10.1 first considers the case in which the subsystem A is the whole system ($A = X$), and then considers the general case. Here, first consider a unitary operator acting on a system X. In the original notation for pure states, the unitary operator U applied to X takes $|\psi\rangle$ to $U|\psi\rangle$. The density operator for a pure state $|\psi\rangle$ is $\rho = |\psi\rangle\langle\psi|$, so U takes ρ to $U|\psi\rangle\langle\psi|U^\dagger = U\rho U^\dagger$. The general case, in which A is a subsystem of $X = A \otimes B$, is more complicated. Suppose $|\psi\rangle \in X = A \otimes B$ and $U : X \to X$. Then the density operator $\rho_A = \mathbf{tr}_B|\psi\rangle\langle\psi|$ is sent to $\rho'_A = \mathbf{tr}_B(U|\psi\rangle\langle\psi|U^\dagger)$. When $U = U_A \otimes U_B$, ρ'_A can be deduced from just ρ_A and U, and it will be $\rho'_A = U_A \rho_A U_A^\dagger$. For a general unitary operator U, however, it is not possible to deduce ρ'_A from only U and ρ_A; the density operator ρ'_A depends on the original state $|\psi\rangle$ of the whole system. Two examples illustrate this point.

Example 10.4.1 Let $X = A \otimes B$, where A and B are both single-qubit systems. Suppose $\rho_A = |0\rangle\langle 0|$, and $U = C_{not}$ where B is the control qubit and A the target:

$$U = |00\rangle\langle 00| + |11\rangle\langle 01| + |10\rangle\langle 10| + |01\rangle\langle 11|.$$

The density operator ρ_A for subsystem A is consistent with many possible states of the entire system X, including $|\psi_0\rangle = |00\rangle$, $|\psi_1\rangle = |01\rangle$, and $|\psi_2\rangle = \frac{1}{\sqrt{2}}|0\rangle(|0\rangle + |1\rangle)$. What is the density operator ρ'_A for system A after U has been applied? If the state of the entire system is $|\psi_0\rangle = |00\rangle$, then $\rho'_A = |0\rangle\langle 0|$. But if it were $|\psi_1\rangle = |01\rangle$, then $\rho'_A = |1\rangle\langle 1|$, or if it were $|\psi_2\rangle = \frac{1}{\sqrt{2}}|0\rangle(|0\rangle + |1\rangle)$, then $\rho'_A = \frac{1}{2}I$.

In fact, the resulting mixed state ρ'_A may have no relation with the initial mixed state ρ_A.

Example 10.4.2 Consider the unitary operator

$$U_{Switch} = |00\rangle\langle 00| + |10\rangle\langle 01| + |01\rangle\langle 10| + |11\rangle\langle 11|$$

acting on single-qubit systems A and B. The transformation exchanges the states of the two systems. Suppose system A is originally in state $\rho_A = |\psi\rangle\langle\psi|$ and system B is in state $|0\rangle\langle 0|$. After applying U, the resulting state of system A is $|0\rangle\langle 0|$ no matter what $|\psi\rangle$ is.

Let $\mathcal{D}_A$ be the set of all density operators for subsystem A. When initially subsystem A is not entangled with subsystem B, and subsystem B is in state $|\phi_B\rangle$, a unitary operator $U : X \to X$ induces a transformation $S_U^{\phi_B} : \mathcal{D}_A \to \mathcal{D}_A$. Specifically, the unitary transformation

$$U : X \to X$$

$$|\psi\rangle \mapsto U|\psi\rangle$$

induces

$$S_U^{\phi_B} : \mathcal{D}_A \to \mathcal{D}_A$$

$$\rho_A \mapsto \rho_A',$$

where $\rho_A = \mathbf{tr}_B |\psi\rangle\langle\psi|$ and $\rho_A' = \mathbf{tr}_B U|\psi\rangle\langle\psi|U^\dagger$. Induced transformations such as $S_U^{\phi_B}$ are called *superoperators*.

Superoperators are linear: the effect of a superoperator S on any density operator ρ that is a probabilistic mixture of other density operators, $\rho = \sum_i p_i \rho_i$, is the sum of the superoperator applied to each of the components:

$$S : \rho \mapsto \sum_i p_i S(\rho_i).$$

10.4.2 Operator Sum Decomposition

Given a superoperator $S : \mathcal{D}_A \to \mathcal{D}_A$, it would be handy to describe it just in terms of system A and formalisms we already have for operators on A. General superoperators, however, are not of the form $U\rho U^\dagger$ for some unitary operator. They are not even reversible in general: from example 10.4.2, for $U = U_{Switch}$ and $|\phi\rangle = |0\rangle$, S_U^ϕ takes $\rho_A = |\psi\rangle\langle\psi|$ to $\rho_A' = |0\rangle\langle 0|$ for all $|\psi\rangle$. Furthermore, most superoperators are not even of the form $A\rho A^\dagger$ for some linear operator A. However, it turns out that every superoperator is the sum of operators of this form; for every superoperator S, there exist linear operators $A_1, \ldots, A_K$ such that

$$S(\rho) = \sum_{i=1}^{K} A_i \rho A_i^\dagger.$$

Such a representation is known as the *operator sum decomposition* for S. The operator sum decomposition for a given superoperator S is not, in general, unique.

To obtain an operator sum decomposition for S_U^ϕ, let $\{|\beta_i\rangle\}$ be a basis for B and let $A_i : A \to A$ be the operator $A_i = \langle\beta_i|U|\phi\rangle$ defined in equation 10.5 of box 10.2. Then

$$S_U^\phi(\rho) = \mathbf{tr}_B(U(\rho \otimes |\phi\rangle\langle\phi|)U^\dagger)$$

$$= \sum_{i=1}^{K} \langle\beta_i|U(\rho \otimes |\phi\rangle\langle\phi|)U^\dagger|\beta_i\rangle$$

$$= \sum_{i=1}^{K} \langle\beta_i|U|\phi\rangle\rho\langle\phi|U^\dagger|\beta_i\rangle$$

$$= \sum_{i=1}^{K} A_i \rho A_i^\dagger.$$

To see how the third line follows from the second, first consider the pure state case $\rho = |\psi\rangle\langle\psi|$, from which the general case ρ, a mixture of pure states, follows.

For a given superoperator there are many possible operator sum decompositions; the operator sum decomposition depends on which basis is used. The next two examples give the operator sum decomposition in the standard basis for the operators of examples 10.4.1 and 10.4.2.

Example 10.4.3   *Operator sum decomposition for C_{not} and $|\phi\rangle = \frac{1}{\sqrt{2}}(|0\rangle + |1\rangle)$. The C_{not} operator U of example 10.4.1 can be written $U = X \otimes |1\rangle\langle 1| + I \otimes |0\rangle\langle 0|$. Suppose that initially the two systems are unentangled and system A is in state $\rho' = |\psi\rangle\langle\psi|$ and B is in state $\rho = |\phi\rangle\langle\phi|$.*

$$S_u^\phi(\rho) = \mathbf{tr}_B(U(\rho \otimes |\phi\rangle\langle\phi|)U^\dagger)$$

$$= A_0\rho A_0^\dagger + A_1\rho A_1^\dagger$$

where $A_0 = \langle 0|U|\phi\rangle$ and $A_1 = \langle 1|U|\phi\rangle$. Then, using the definition of A_i found in equation 10.5 of box 10.2,

$$A_0|\psi\rangle = \sum_{i=0}^{1} \langle\alpha_i|\langle 0|U|\psi\rangle|\phi\rangle |\alpha_i\rangle$$

$$= \langle 0|\langle 0|(X \otimes |1\rangle\langle 1| + I \otimes |0\rangle\langle 0|)|\psi\rangle|\phi\rangle |0\rangle$$

$$+ \langle 1|\langle 0|(X \otimes |1\rangle\langle 1| + I \otimes |0\rangle\langle 0|)|\psi\rangle|\phi\rangle |1\rangle$$

$$= (\langle 0|\langle 0|(X \otimes |1\rangle\langle 1|)|\psi\rangle|\phi\rangle + \langle 0|\langle 0|(I \otimes |0\rangle\langle 0|)|\psi\rangle|\phi\rangle)|0\rangle$$

$$+ (\langle 1|\langle 0|(X \otimes |1\rangle\langle 1|)|\psi\rangle|\phi\rangle + \langle 1|\langle 0|(I \otimes |0\rangle\langle 0|)|\psi\rangle|\phi\rangle)|1\rangle.$$

Because $\langle 0|1\rangle = 0$, the first and third terms are zero, so

$$A_0|\psi\rangle = \langle 0|\langle 0|I \otimes |0\rangle\langle 0|)|\psi\rangle|\phi\rangle |0\rangle + \langle 1|\langle 0|I \otimes |0\rangle\langle 0|)|\psi\rangle|\phi\rangle |1\rangle$$

$$= \langle 0|\psi\rangle\langle 0|\phi\rangle|0\rangle + \langle 1|\psi\rangle\langle 0|\phi\rangle|1\rangle$$

$$= \langle 0|\phi\rangle|\psi\rangle.$$

Since $|\phi\rangle = \frac{1}{\sqrt{2}}(|0\rangle + |1\rangle)$,

$$A_0|\psi\rangle = \frac{1}{\sqrt{2}}|\psi\rangle,$$

so

$$A_0 = \frac{1}{\sqrt{2}}I.$$

Similar reasoning shows that

$$\begin{aligned}
A_1|\psi\rangle &= \langle 0|\langle 1|X \otimes |1\rangle\langle 1|)|\psi\rangle|\phi\rangle\ |0\rangle + \langle 1|\langle 1|X \otimes |1\rangle\langle 1|)|\psi\rangle|\phi\rangle\ |1\rangle \\
&= \langle 0|X|\psi\rangle\langle 1|\phi\rangle\ |0\rangle + \langle 1|X|\psi\rangle\langle 1|\phi\rangle\ |1\rangle \\
&= \langle 1|\phi\rangle(X|\psi\rangle) \\
&= \frac{1}{\sqrt{2}}X|\psi\rangle,
\end{aligned}$$

so

$$A_1 = \frac{1}{\sqrt{2}}X.$$

Example 10.4.4 *Operator sum decomposition for U_{Switch} and $|\phi\rangle = |0\rangle$.* Let

$$U_{Switch} = |00\rangle\langle 00| + |10\rangle\langle 01| + |01\rangle\langle 10| + |11\rangle\langle 11|$$

and $|\phi\rangle = |0\rangle$.

$$\begin{aligned}
S_u^\phi(\rho) &= \mathbf{tr}_B(U\rho \otimes |\phi\rangle\langle\phi|U^\dagger) \\
&= A_0\rho A_0^\dagger + A_1\rho A_1^\dagger
\end{aligned}$$

where $A_0 = \langle 0|U|\phi\rangle$ and $A_1 = \langle 1|U|\phi\rangle$.

$$\begin{aligned}
A_0|\psi\rangle &= \sum_{i=0}^{1}\langle\alpha_i|\langle 0|U|\psi\rangle|\phi\rangle\ |\alpha_i\rangle \\
&= \langle 00||\phi\rangle|\psi\rangle\ |0\rangle + \langle 10||\phi\rangle|\psi\rangle\ |1\rangle \\
&= \langle 0|\phi\rangle\langle 0|\psi\rangle\langle 0| + \langle 1|\phi\rangle\langle 1|\psi\rangle\langle 1|.
\end{aligned}$$

Since $|\phi\rangle = |0\rangle$,

$$A_0|\psi\rangle = \langle 0|\psi\rangle|0\rangle$$

and

$A_0 = |0\rangle\langle 0|$.

Similar reasoning gives

$A_1 = |0\rangle\langle 1|$.

Each term $A_i \rho A_i^\dagger$ in the operator sum decomposition is Hermitian and positive, but generally does not have trace one. Since $\mathbf{tr}(A_i \rho A_i^\dagger) \geq 0$, the operator $\frac{A_i \rho A_i^\dagger}{\mathbf{tr}(A_i \rho A_i^\dagger)}$ is Hermitian, positive, and has trace one, and therefore is a density operator. Furthermore, since the trace of a Hermitian, positive operator is zero only if the operator is zero, and $1 = \mathbf{tr}(S_U^\phi(\rho)) = \sum_{i=1}^K \mathbf{tr}(A_i \rho A_i^\dagger)$, $S_U^\phi(\rho)$ is a probabilistic mixture of the operators $\frac{A_i \rho A_i^\dagger}{\mathbf{tr}(A_i \rho A_i^\dagger)}$:

$$S_U^\phi(\rho) = \sum p_i \frac{A_i \rho A_i^\dagger}{\mathbf{tr}(A_i \rho A_i^\dagger)} \tag{10.8}$$

where $p_i = \mathbf{tr}(A_i \rho A_i^\dagger)$ and we have ignored any zero terms.

Operator sum decompositions for superoperators S on subsystem A of system $X = A \otimes B$, and their dependence on the basis chosen for B can be understood in terms of measurement. It is not a coincidence that equation 10.8 is reminiscent of the equation

$$\rho' = \sum_j p_j \frac{P_j \rho P_j^\dagger}{\mathbf{tr}(P_j \rho P_j^\dagger)}$$

that encapsulates the possible outcomes of measurement of ρ by operator O with associated projectors P_j. Let A_i be the operator obtained in the operator sum decomposition for S_U^ϕ when using basis $\{|b_i\rangle\}$ for B. Suppose that after $U : A \otimes B \to A \otimes B$ was applied to ρ, subsystem B were measured with respect to the projectors $P_i = |b_i\rangle\langle b_i|$ for the $K = 2^k$ basis elements $|b_i\rangle$ for B. The best description of subsystem A after this measurement is a probabilistic mixture of mixed states $\rho' = \sum_i p_i \rho_i$ where

$$\rho_i = \mathbf{tr}_B \left(\frac{(I \otimes P_i) U (\rho \otimes |\phi\rangle\langle\phi|) U^\dagger (I \otimes P_i^\dagger)}{\mathbf{tr}\left((I \otimes P_i) U (\rho \otimes |\phi\rangle\langle\phi|) U^\dagger (I \otimes P_i^\dagger) \right)} \right)$$

and

$$p_i = \mathbf{tr}\left((I \otimes P_i) U (\rho \otimes |\phi\rangle\langle\phi|) U^\dagger (I \otimes P_i^\dagger) \right).$$

Since

$$\mathbf{tr}_B \left((I \otimes |\beta_i\rangle\langle\beta_i|) U \rho \otimes |\phi\rangle\langle\phi| U^\dagger (I \otimes |\beta_i\rangle\langle\beta_i|) \right) = \langle\beta_i| U \rho \otimes |\phi\rangle\langle\phi| U^\dagger |\beta_i\rangle,$$

the density operator $\rho' = \sum_i p_i \rho_i$ is identical to the density operator $S_U^\phi(\rho)$.

10.4.3 A Relation Between Quantum State Transformations and Measurements

Section 10.3.1 showed that the density operator representing the probabilistic mixture of outcomes of a measurement O with associated projectors $\{P_j\}$ of a system A initially represented by the mixed state ρ is

$$\rho' = \sum_j p_j \frac{P_j \rho P_j^\dagger}{\mathbf{tr}(P_j \rho P_j^\dagger)}.$$

For any measurement O, the map

$$S_O : \mathcal{D}_A \to \mathcal{D}_A$$

$$\rho \mapsto \rho'$$

can also be obtained in a different way, as the superoperator coming from a unitary transformation on a larger system. More specifically, for any observable O of system A, there is a larger system $X = A \otimes B$, a unitary operator $U : X \to X$, and a state $|\psi\rangle$ of B such that $S_U^\phi = S_O$.

To prove this statement, suppose O has M distinct eigenvalues. Let B be a system of dimension M with basis $\{|\beta_i\rangle\}$, and suppose that B is initially in the state $|\phi\rangle = |\beta_0\rangle$. Let U be any unitary operator on $X = A \otimes B$ that maps

$$|\psi\rangle |\beta_0\rangle \mapsto \sum_{i=1}^M P_i |\psi\rangle |\beta_i\rangle.$$

Then for $\rho = |\psi\rangle\langle\psi|$,

$$S_U^\phi(\rho) = \mathbf{tr}_B(U(\rho \otimes |\phi\rangle\langle\phi|)U^\dagger)$$

$$= \sum_{i=1}^M A_i \rho A_i^\dagger$$

$$= \sum_{i=1}^M A_i |\psi\rangle\langle\psi| A_i^\dagger$$

where $A_i = \langle\beta_i|U|\phi\rangle$. Since $|\phi\rangle = |\beta_0\rangle$,

$$A_i|\psi\rangle = \sum_j \langle\alpha_j|\langle\beta_i|U|\psi\rangle|\beta_0\rangle \, |\alpha_j\rangle$$

$$= \sum_j \langle\alpha_j|\langle\beta_i| \left(\sum_{k=1}^M P_k|\psi\rangle|\beta_k\rangle \right) |\alpha_j\rangle$$

$$= \sum_j \langle\alpha_j|P_i|\psi\rangle \, |\alpha_j\rangle$$

$$= P_i|\psi\rangle.$$

So

$$S_U^\phi(\rho) = \sum_{i=1}^{M} P_i |\psi\rangle\langle\psi| P_i^\dagger$$

$$= \sum_{i=1}^{M} p_i \frac{P_i |\psi\rangle\langle\psi| P_i^\dagger}{\mathbf{tr}(P_i |\psi\rangle\langle\psi| P_i^\dagger)}$$

where $p_i = \mathbf{tr}(P_i |\psi\rangle\langle\psi| P_i^\dagger)$. There is debate within the quantum physics community as to the extent to which this relationship between unitary operators and measurement clarifies various issues in the foundations of quantum mechanics. We do not elaborate on these issues here.

10.4.4 Decoherence

In practice, it is impossible to isolate a quantum computer completely from its environment. Because all physical qubits interact with their environment, the computational qubits of a quantum computer are properly viewed as a subsystem of a larger system consisting of the computation qubits and their environment. By an *environment* we mean a subsystem over which we have no control: we cannot gain information from it by measurement or apply gates to it.

In some cases, the effect of an environmental interaction on the computational subsystem is reversible by transformations on the subsystem alone. But in other cases, *decoherence* occurs. In decoherence, information about the state of the computational subsystem is lost to the environment. Such errors are serious because the environment is beyond our computational control. The next two chapters develop quantum error correction and fault-tolerant techniques to counteract errors due to decoherence as well as other sorts of errors, such as those stemming from imperfections in the implementations of quantum gates. This section lays a foundation for that discussion by setting up an error model for errors due to interaction with the environment.

The operator sum decomposition provides a means for describing the effect on the computational subsystem of an interaction with another subsystem in terms of operations on the computational subsystem alone. Using the operator sum decomposition, the effect on the computational subsystem of any interaction with the environment can be viewed as a mixture of K errors resulting in the K mixed states $\frac{A_i \rho A_i^\dagger}{\mathbf{tr}(A_i \rho A_i^\dagger)}$.

Common error models suppose that the environment interacts separately with different parts of the computational subsystem. For example, a common error model consists of errors that are both local and Markov:

- *local*   each qubit interacts only with its own environment, and

- *Markov*   the state of a qubit's environment, and its interaction with the qubit, is independent of the state of the environment in previous time intervals.

More precisely, under a *local* error model, the errors to which an n-qubit system is subjected can be modeled by interaction with an environment $E = E_1 \otimes \cdots \otimes E_n$ such that the environment E_i

interacts with only the ith qubit of X; the errors can be modeled by unitary transforms of the form $U = U_1 \otimes \cdots \otimes U_n$ such that U_i acts on E_i and the ith qubit of X is given by superoperators of the form $S_U = S_{U_1} \otimes \cdots \otimes S_{U_n}$, where S_{U_i} acts on only the ith qubit of X.

A reasonable way to think of the Markov condition is that each qubit's environment is renewed (or replaced) at each computational time step. More concretely, under a local and Markov error model, the computational subsystem X at a given time t interacts with an environment $E^t = E_1^t \otimes \cdots \otimes E_n^t$ in such a way that the only interactions are between E_i^t and the ith qubit of system X, and the current state of the environment E, and its interaction with X, is independent of the state of the environment at any previous time s. Most of the quantum error correcting codes and fault-tolerant techniques discussed in chapters 11 and 12 are designed to handle local and Markov errors. Techniques to handle other error models have been developed, some of which are briefly described in section 13.3.

10.5   References

Jozsa and Linden [167] show that any quantum algorithm that achieves exponential speedup over classical algorithms must entangle an increasing number of qubits. Their proof applies only to algorithms run in isolation in which the state is always in a pure state. The results of section 10.4 show that any mixed state algorithm can be viewed as a pure state algorithm on a larger system. The result of Jozsa and Linden still applies in this more general setting, except that the entanglement could involve noncomputational qubits of the larger system; it is not required to be between the computational qubits.

Efficient classical simulations of certain quantum systems have be found by Vidal and others [278, 204]. Meyer discusses the lack of entanglement throughout the Bernstein-Vazirani algorithm and related results [213].

Bennett and Shor's "Quantum information theory" [38] discusses various entanglement measures for mixed states of bipartite systems, including some examples and a distillation protocol. It is generally a good overview of topics in quantum information theory, including a number of interesting topics we will not cover in this book. Bruss's "Characterizing entanglement" [69] is an excellent fifteen-page overview of many of the most significant results about entanglement to date. Myhr's master's thesis, "Measures of entanglement in quantum mechanics" [215], gives a readable and more detailed and account of many of these results.

Nielsen's majorization results is found in Nielsen [217]. The SLOCC classification of 3-qubit states was first described by Dür, Vidal, and Cirac in [107]. Briegel and Raussendorf define persistency of entanglement and maximal connectedness in [65], as well as introducing cluster states.

10.6   Exercises

Exercise 10.1.   Show that the definition of the partial trace is basis independent.

Exercise 10.2. Show that $\text{tr}_B(O_A \otimes O_B) = O_A \, \text{tr}(O_B)$.

Exercise 10.3.

a. Find the density operators for the whole system and both qubits of $|\Psi^-\rangle = \frac{1}{\sqrt{2}}(|00\rangle - |11\rangle)$.

b. Find the density operators for the whole system and both qubits of $|\Phi^+\rangle = \frac{1}{\sqrt{2}}(|01\rangle + |10\rangle)$.

Exercise 10.4. *Distinguishing pure and mixed states.*

a. Show that a density operator ρ represents a pure state if and only if $\rho^2 = \rho$. In other words, ρ is a projector.

b. What can be said about the rank of the density operator of a pure state?

Exercise 10.5. We showed that any density operator can be viewed as a probability distribution over a set of orthogonal states. Show by example that some density operators have multiple associated probability distributions, so that in general the probability distribution associated to a density operator is not unique.

Exercise 10.6. *Geometry of Bloch regions.*

a. Show that the Bloch region, the set S of mixed states of an n-qubit system, can be parametrized by $2^{2n} - 1$ real parameters.

b. Show that S is a convex set.

c. Show that the set of pure states of an n qubit system can be parametrized by $2^{n+1} - 2$ real parameters, and therefore the set of density matrices corresponding to pure states can be parametrized in this way also.

d. Explain why for $n > 2$ the boundary of the set of mixed states must consist of more than just pure states.

e. Show that the extremal points, those that are not convex linear combinations of other points, are exactly the pure states.

f. Characterize the non-extremal states that are on the boundary of the Bloch region.

Exercise 10.7. Give a geometric interpretation for $R(\theta)$ and $T(\phi)$ of Section 5.4.1 by determining their behavior on the set of mixed states viewed as points of the Bloch sphere.

Exercise 10.8. *The Schmidt decomposition.* Every $m \times n$ matrix M, with $m \leq n$, has a singular value decomposition $M = UDV$ where D is an $m \times n$ diagonal matrix with non-negative real entries, and U and V are $m \times m$ and $n \times n$ unitary matrices.
Let $|\psi\rangle \in A \otimes B$, where A has dimension m and B has dimension n, with $m \leq n$. Let $\{|i\rangle\}$ be a basis for A and $\{|j\rangle\}$ be a basis for B, then for some choice of $m_{ij} \in \mathbf{C}$

$$|\psi\rangle = \sum_{i=0}^{m-1} \sum_{j=0}^{n-1} a_{ij} |i\rangle |j\rangle.$$

Let M be the $m \times n$ matrix with entries a_{ij}. Use the singular value decomposition (SVD) for M to find sets of orthonormal unit vectors $\{|\alpha_i\rangle\} \in A$ and $\{|\beta_j\rangle\} \in B$ such that

$$|\psi\rangle = \sum_{i=0}^{m-1} \lambda_i |\alpha_i\rangle |\beta_j\rangle$$

where λ_i is non-negative. The λ_i are called the *Schmidt coefficients*, and K, the number of λ_i, is called the *Schmidt rank* or *Schmidt number* of $|\psi\rangle$.

Exercise 10.9. *Singular value decomposition.* Let A be an $n \times m$ matrix.

a. Let $|u_j\rangle$ be unit length eigenvectors of $A^\dagger A$ with eigenvalues λ_j. Explain how we know that λ_j is real and non-negative for all j.

b. Let U be the matrix with $|u_j\rangle$ as its columns. Show that U is unitary.

c. For all eigenvectors with non-zero eigenvalues define $|v_i\rangle = \frac{A|x_i\rangle}{\sqrt{\lambda_i}}$. Let V be the matrix with $|v_i\rangle$ as columns. Show that V is unitary.

d. Show that $V^\dagger A U$ is diagonal.

e. Conclude that $A = VDU^\dagger$ for some diagonal D. What is D?

Exercise 10.10. For $|\psi\rangle \in A \otimes B$, show that $|\psi\rangle$ is unentangled if and only if $S(\text{tr}_B \rho) = 0$, where $\rho = |\psi\rangle\langle\psi|$.

Exercise 10.11.

a. Show that the states $\frac{1}{\sqrt{2}}(|01\rangle + |10\rangle)$ and $\frac{1}{\sqrt{2}}(|00\rangle - i|11\rangle)$ are maximally entangled.

b. Write down two other maximally entangled states.

Exercise 10.12. What is the maximum possible amount of entanglement, as measured by the von Neumann entropy, over all pure states of a bipartite quantum system $A \otimes B$ where A has dimension n and B has dimension m with $n \geq m$?

Exercise 10.13. Claim: LOCC cannot convert an unentangled state to an entangled one.

a. State the claim in more precise language.

b. Prove the claim.

Exercise 10.14. Show that the four Bell states $|\Psi\rangle^\pm$ and $|\Phi\rangle^\pm$ are all LOCC equivalent.

Exercise 10.15.

a. Show that any two-qubit state can be converted to $|00\rangle$ via LOCC.

b. Show that any n-qubit state can be converted to a state unentangled with respect to the tensor decomposition into the n qubits.

Exercise 10.16. Show that the vector of ordered eigenvalues λ^ψ for the density operator of any unentangled state $|\psi\rangle$ of a bipartite system majorizes the vectors for any other state of the bipartite system.

Exercise 10.17. *Maximally entangled bipartite states.* Let $|\psi\rangle$ be a state of the form

$$|\psi\rangle = \frac{1}{\sqrt{m}} \sum_{i=1}^{m} |\phi_i^A\rangle \otimes |\phi_i^B\rangle$$

where the $\{|\phi_i^A\rangle\}$ and $\{|\phi_i^B\rangle\}$ are orthonormal sets. Show that the vector λ^ψ is majorized by λ^ϕ for all states $|\phi\rangle \in A \otimes B$.

Exercise 10.18. Classify all two-qubit states up to SLOCC equivalence.

Exercise 10.19. Show that $|GHZ_3\rangle$ can be converted via SLOCC to any A-BC decomposable state.

Exercise 10.20. Show that the states $|GHZ_n\rangle$ are maximally connected.

Exercise 10.21. Show that the states $|W_n\rangle$ are not maximally connected.

Exercise 10.22.

a. If $|\psi\rangle$ has persistency n and $|\phi\rangle$ has persistency m, what is the persistency of $|\psi\rangle \otimes |\phi\rangle$?

b. Show by induction that the persistency of $|W_n\rangle$ is $n - 1$. (Hint: You may want to use (a).)

Exercise 10.23.

a. Check that each of the cluster states of examples 10.2.7, 10.2.8, and 10.2.9 is stabilized by the operators of equation 10.7.

b. Find the cluster state for the 1×5 lattice.

c. Find the cluster state for the 2×2 lattice.

Exercise 10.24. *Maximal connectedness of cluster states.*

a. Show by induction that for the qubits corresponding to the ends of the chain in the cluster state $|\phi_n\rangle$ for the $1 \times n$ lattice, there is a sequence of single-qubit measurements that place these qubits in a Bell state.

b. Show that for any two qubits q_1 and q_2 in a graph state, there exists a sequence of single-qubit measurements that leave these qubits as the end qubits of a cluster state of a $1 \times r$ lattice. Conclude that graph states are maximally connected.

Exercise 10.25. *Persistency of cluster states.* For the cluster state $|\phi_N\rangle$ corresponding to the $1 \times N$ lattice for N even, give a sequence of $N/2$ single-qubit measurements that result in a completely unentangled state.

Exercise 10.26. Show that if $\{|x_i\rangle\}$ is the set of possible states resulting from a measurement and p_i is the probability of each outcome, then $\rho = \sum p_i |x_i\rangle\langle x_i|$ is Hermitian, trace 1, and positive.

Exercise 10.27. For initial mixed state $\rho_A \otimes \rho_B$, find the mixed state of A after the transformation $U = |00\rangle\langle00| + |10\rangle\langle01| + |01\rangle\langle10| + |11\rangle\langle11|$ has been applied.

Exercise 10.28. Suppose that subsystem $A = A_1 \otimes A_2$ and that $U : A \otimes B \rightarrow A \otimes B$ behaves as the identity on A_1. In other words, suppose $U = I \otimes V$ where I acts on A_1 and V acts on $A_2 \otimes B$. Show that for any state $|\phi\rangle$ of system B, the superoperator S_U^ϕ can be written as $I \otimes S$ for some superoperator S on subsystem A_2 alone.

Exercise 10.29.

a. Give an alternative operator sum decomposition for example 10.4.3.

b. Give an alternative operator sum decomposition for example 10.4.4.

c. Give a general condition for two sets of operators $\{A_i\}$ and $\{A_j'\}$ to give operator sum decompositions for the same superoperator.

Exercise 10.30.

a. Describe a strategy for determining which sequence was sent in the game of section 10.3 if both qubits are received. More specifically, you receive a sequence of pairs of qubits. Either all pairs are randomly chosen from $\{|00\rangle, |11\rangle\}$ or all pairs are in the state $\frac{1}{\sqrt{2}}(|00\rangle + |11\rangle)$. Describe a strategy for determining which sequence was sent.

b. For each sequence, write down the density operator representing that sequence.

11   Quantum Error Correction

For practical quantum computers to be built, techniques for handling environmental interactions that muddle the quantum computations are required. Shor's algorithms, while universally acclaimed, were initially thought by many to be of only theoretical interest; estimates suggested that unavoidable interactions with the environment were many orders of magnitude too strong to be able to run Shor's factoring algorithm on a number that was of practical interest, and no one had any idea as to how to perform error correction for quantum computation. Given the impossibility of copying an unknown quantum state, a straightforward application of classical methods to the quantum case is not possible, and it was far from obvious what else to do. Results such as the no-cloning theorem made many experts believe that robust quantum computation might be impossible. It turns out, however, that an elegant and surprising use of classical techniques forms the foundation of sophisticated quantum error correction techniques. Quantum error correction is now one of the most extensively developed areas of quantum computation. It was the discovery of quantum error correction, as much as of Shor's algorithms, that turned quantum information processing into a significant field in its own right.

In the classical world, error correcting codes are primarily used in data transmission. Quantum systems, however, are difficult to isolate sufficiently from environmental interactions while retaining the ability to perform computations. In any quantum system used to perform quantum information processing, the effects of interaction with the environment are likely to be so pervasive that quantum error correction will be used at all times.

We begin in section 11.1 with a few simple examples to give a sense for the workings of quantum error correction, particularly purely quantum aspects such as how quantum superpositions of both errors and states are handled. A general framework for quantum error correction is given in section 11.2. This framework has similarities to the framework for classical codes but is considerably more complicated. Quantum error correcting codes must handle the infinite variety of single-qubit states and the peculiarly quantum ways in which qubits can interact with each other. In section 11.3, Calderbank-Shor-Steane (CSS) codes are presented. Then, in section 11.4, the more general class of stabilizer codes is described. Most of the specific quantum error correcting codes we consider are designed to correct all errors on k or fewer qubits. Such codes work well for systems subject to independent single-qubit, or few-qubit, errors. While that sort of error behavior is expected in many situations, other reasonable error models exist. Throughout the chapter we pretend that quantum error correction can be carried out perfectly. Chapter 12 discusses fault-tolerant methods that enable quantum error correction to work even when carried out imperfectly. Other approaches to robust quantum computation are discussed in chapter 13.

11.1   Three Simple Examples of Quantum Error Correcting Codes

Classical error correcting codes map message words into a code space, consisting of longer words, in a redundant way that allows detection and correction of errors. Quantum error correcting codes embed the vector space of message states, called words, into a subspace of a larger vector space, the code space. A quantum algorithm that logically operates on n-qubits is implemented as an algorithm operating on the much larger m-qubit system in which the n-qubits are encoded. To detect and correct an error, computation into ancilla qubits is performed and the ancilla are measured. Error correcting transformations are applied according to the result of that measurement. To preserve superpositions, the encoding and measurements must be carefully designed so that these measurements give information only about what error occurred and not about the encoded state of the computation.

To give a general sense for quantum error correction, particularly its use of measurement and its ability to correct superpositions of correctable errors, we first describe a simple code that corrects only single-qubit bit-flip errors, then a code that corrects only single-qubit phase errors, and finally a code that corrects all single-qubit errors.

11.1.1   A Quantum Code That Corrects Single Bit-Flip Errors

A single-qubit bit-flip error applies X to one of the qubits of the quantum computer. The following simple code is a quantum version of the classical [3, 1] repetition code, which will be described more formally in section 11.2. It detects and corrects any of the three single bit-flip errors

$$\{X_2 = X \otimes I \otimes I, X_1 = I \otimes X \otimes I, X_0 = I \otimes I \otimes X\},$$

where X_i means the tensor product of X applied to the ith qubit with the identity on all other qubits.

In brief, the [3, 1] repetition code encodes each bit in three bits as

$$0 \rightarrow 000$$

$$1 \rightarrow 111.$$

Decoding is done by *majority rules*

$$\left.\begin{matrix} 000 \\ 001 \\ 010 \\ 100 \end{matrix}\right\} \mapsto 0$$

$$\left.\begin{matrix} 011 \\ 101 \\ 110 \\ 111 \end{matrix}\right\} \mapsto 1.$$

To implement majority rules, first determine if an error has occurred by comparing the first bit with each of the other bits. More formally, to make the comparisons, use two additional bits, called *ancilla*, to hold the computation of $b_2 \oplus b_1$ and $b_2 \oplus b_0$ respectively. This computation is called the *syndrome computation*. The *syndrome* values of $b_2 \oplus b_1$ and $b_2 \oplus b_0$ determine which error correcting transformation should be applied, as shown in table 11.1. The first line of the table says that if $b_2 \oplus b_1$ and $b_2 \oplus b_0$ are both zero, do nothing. The second lines says that if $b_2 = b_1$ but $b_2 \neq b_0$, flip b_0 so that it agrees with b_2 and b_1, the majority. Similarly, if $b_2 \neq b_1$ and $b_2 = b_0$, flip b_1. Finally, if $b_2 \neq b_1$ and $b_2 \neq b_0$, then b_1 and b_0 must agree, so flip b_2 to make it agree with the majority. No matter what happened previously, this procedure results in a codeword. However, if more than one error has occurred it will correct to the wrong word. For example, if the original string was 000 and two bit-flip errors occur, one on the first qubit and one on the third, the resulting string, 101, will be "corrected" to 111 under this procedure. The [3, 1] repetition code can correct only single bit-flip errors. More powerful codes, such as the [n, 1] repetition codes that encode one bit in n bits and decode by majority rules, can correct more errors.

Both classical and quantum error correction spread the information we want to protect across several qubits so that individual errors have less of an effect. The [3, 1] repetition code encodes 0 and 1 as the bit strings 000 and 111 respectively. In the quantum setting, let C_{BF} be the subspace spanned by $\{|000\rangle, |111\rangle\}$. This quantum code encodes the state $|0\rangle$ in the state $|000\rangle$ and $|1\rangle$ in the state $|111\rangle$. Linearity of the code and these relations define a general encoding c_{BF} of single-qubit

Table 11.1
Syndrome and corresponding error correcting transformations for the classical [3, 1] repetition code.

$b_2 \oplus b_1$	$b_2 \oplus b_0$	Error correcting transformation
0	0	identity
0	1	flip b_0
1	0	flip b_1
1	1	flip b_2

states into the subspace C_{BF} of the state space for a three-qubit system:

$$c_{BF} : \quad |0\rangle \otimes |00\rangle \rightarrow |000\rangle$$
$$|1\rangle \otimes |00\rangle \rightarrow |111\rangle,$$

so $a|0\rangle + b|1\rangle$ maps to $a|000\rangle + b|111\rangle$. In general, for a quantum code, we use the notation $|\tilde{0}\rangle$ for the encoding of $|0\rangle$ and likewise for other states. For this code, $|\tilde{0}\rangle = |000\rangle$ and $|\tilde{1}\rangle = |111\rangle$.

The set of states $a|\tilde{0}\rangle + b|\tilde{1}\rangle = a|000\rangle + b|111\rangle$ is a two-dimensional vector space, so it may be considered a qubit in its own right. It is called the *logical qubit* to distinguish it from the three *computation qubits* whose tensor product forms the entire eight-dimensional code space. States such as $|101\rangle$ that are not logical qubit values are not legitimate computational states. Legitimate states, the possible values of the logical qubits, are called *codewords*. On a logical qubit $a|000\rangle + b|111\rangle$, single bit-flip errors no longer take legitimate computational states to legitimate computational states, but to states that are not codewords. For example, a bit-flip error on the first qubit results in the state $a|100\rangle + b|011\rangle$, which is not a codeword because it is not in C_{BF}. The goal of an error correction scheme is to detect non-codeword states and transform them back to codewords.

To detect an error, we compute the XOR of the first and second qubits into one ancilla qubit, and the XOR of the first and third qubits into another. More formally,

$$U_{BF} : |x_2, x_1, x_0, 0, 0\rangle \rightarrow |x_2, x_1, x_0, x_2 \oplus x_1, x_2 \oplus x_0\rangle.$$

The transformation U_{BF} is called the *syndrome extraction operator* and has quantum circuit

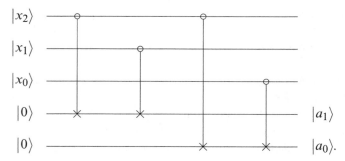

The ancilla qubits are then measured in the standard basis, and the error *syndrome* is obtained.

The use of the syndrome parallels that of the classical [3, 1] repetition code. In addition to correcting all single bit-flip errors, the code must not corrupt correct states, so for convenience we also consider $I \otimes I \otimes I$ as an "error" we can correct. The information we gain from measuring the ancilla enables us to choose the right transformation to apply to correct the error. Since $X = X^{-1}$, the correcting transformation in this case is the same as the error transformation that occurred. The following table gives the transformation to apply given the measurement of the ancilla:

Bit flipped	Syndrome	Error correction
none	$\lvert 00 \rangle$	none
0	$\lvert 11 \rangle$	$X_2 = I \otimes I \otimes X$
1	$\lvert 10 \rangle$	$X_1 = I \otimes X \otimes I$
2	$\lvert 01 \rangle$	$X_0 = X \otimes I \otimes I.$

Because of its close parallel with the classical [3, 1] code, it is not surprising that this procedure corrects any single bit-flip errors on the encoded standard basis states $\lvert \tilde{0} \rangle = \lvert 000 \rangle$ and $\lvert \tilde{1} \rangle = \lvert 111 \rangle$. In addition, it corrects single bit-flip errors on superpositions of codewords.

Example 11.1.1 *Correcting a bit-flip error on a superposition.* A general superposition $\lvert \psi \rangle = a \lvert 0 \rangle + b \lvert 1 \rangle$ is encoded as

$$\lvert \tilde{\psi} \rangle = a \lvert \tilde{0} \rangle + b \lvert \tilde{1} \rangle = a \lvert 000 \rangle + b \lvert 111 \rangle.$$

Suppose $\lvert \tilde{\psi} \rangle$ is subject to the single bit-flip error $X_2 = X \otimes I \otimes I$, resulting in

$$X_2 \lvert \tilde{\psi} \rangle = a \lvert 100 \rangle + b \lvert 011 \rangle.$$

Applying the syndrome extraction operator U_{BF} to $X_2 \lvert \tilde{\psi} \rangle \otimes \lvert 00 \rangle$ results in the state

$$U_{BF}((X_2 \lvert \tilde{\psi} \rangle) \otimes \lvert 00 \rangle) = a \lvert 100 \rangle \lvert 11 \rangle + b \lvert 011 \rangle \lvert 11 \rangle$$

$$= (a \lvert 100 \rangle + b \lvert 011 \rangle) \lvert 11 \rangle$$

Measuring the two ancilla qubits yields $\lvert 11 \rangle$, and the state is now

$$(a \lvert 100 \rangle + b \lvert 011 \rangle) \otimes \lvert 11 \rangle.$$

The error can be removed by applying the inverse error operator X_2, corresponding to the measured syndrome $\lvert 11 \rangle$, to the first three qubits. Doing so reconstructs the original encoded state

$$\lvert \psi \rangle = a \lvert \tilde{0} \rangle + b \lvert \tilde{1} \rangle = a \lvert 000 \rangle + b \lvert 111 \rangle).$$

The intuition behind why this procedure does not irreparably disturb the quantum state, even though it includes measurement, is that measurement of the ancilla by the syndrome extraction operator tells us nothing about individual computational qubit states, only about what errors occurred. If the syndrome extraction operator is applied to a codeword $a \lvert \tilde{0} \rangle + b \lvert \tilde{1} \rangle$, the result of the measurement of the ancilla will be the syndrome 00 regardless of whether the codeword is $\lvert \tilde{0} \rangle$, $\lvert \tilde{1} \rangle$, or some superposition of the two. Similarly, if error $X_2 = X \otimes I \otimes I$ has occurred, the syndrome will be in state $\lvert 11 \rangle$ regardless of whether the computational qubits are in state $\lvert 100 \rangle$, $\lvert 011 \rangle$, or a superposition of the two. Thus measuring the ancilla qubits gives no information about the states of the computational qubits, but it does give information about what error has occurred. Measuring the ancilla qubits gives information about the error without disturbing the computation, even when the initial state is a superposition $a|000\rangle + b|111\rangle$.

Unlike in the classical case, linear combinations of quantum errors are also possible. This same procedure also corrects linear combinations of bit-flip errors.

Example 11.1.2 *Correcting a linear combination of bit-flip errors.* Suppose the state $|0\rangle$ has been encoded as $|\tilde{0}\rangle = |000\rangle$ and an error $E = \alpha X \otimes I \otimes I + \beta I \otimes X \otimes I$, a linear combination of the two single bit-flip errors X_2 and X_1, occurs, yielding

$$E|\tilde{0}\rangle = \alpha|100\rangle + \beta|010\rangle.$$

Applying the syndrome extraction operator U_{BF} to $(E|\tilde{0}\rangle) \otimes |00\rangle$ results in the state

$$U_{BF}((E|\tilde{0}\rangle) \otimes |00\rangle) = \alpha|100\rangle|11\rangle + \beta|010\rangle|10\rangle.$$

Measuring the two auxiliary qubits of this state yields either $|11\rangle$ or $|10\rangle$. If the measurement produces the former, the state is now $|100\rangle$. The measurement has the almost magical effect of causing all but one summand of the error to disappear. The remaining part of the error can be removed by applying the inverse error operator $X_2 = X \otimes I \otimes I$, corresponding to the measured syndrome $|11\rangle$. Doing so reconstructs the original encoded state $|\tilde{0}\rangle = |000\rangle$. If instead the syndrome measurement yields $|10\rangle$, we would apply X_1 to $|010\rangle$ to recover the original state $|\tilde{0}\rangle = |000\rangle$.

While linear combinations of single bit-flip errors can be corrected in this way, multiple bit-flip errors cannot be corrected by this code. The distinction between linear combinations of single bit-flip errors and multiple bit-flip errors is that in the former case any term in the superposition representing a computational state contains only one error, but in the second case a single term may contain multiple errors that will be misinterpreted by the syndrome.

In the classical case, the [3, 1] code corrects all possible single bit errors. The quantum code C_{BF}, while based on the [3, 1] code, does not correct all single-qubit errors. In the classical case, bit flips are the only possible errors; in the quantum case, there is an infinite continuum of possible single-qubit errors. The code C_{BF} does not even detect, let alone correct, phase errors.

Example 11.1.3 *Undetected phase error.* Suppose the quantum state $|+\rangle$, encoded as

$$|\tilde{+}\rangle = \frac{1}{\sqrt{2}}(|000\rangle + |111\rangle),$$

is subjected to a phase error $E = Z \otimes I \otimes I$. The state $|\tilde{+}\rangle$ becomes the error state

$$E|\tilde{+}\rangle = \frac{1}{\sqrt{2}}(|000\rangle - |111\rangle).$$

The syndrome extraction operator U_{BF} applied to $E|\tilde{+}\rangle|00\rangle$ results in $E|\tilde{+}\rangle|00\rangle$, so no error is detected, let alone corrected.

It is easy to construct a code that corrects all single-qubit phase-flip errors, but does not correct single-qubit bit-flip errors. The next section describes such a code. To obtain a code that corrects all single-qubit errors requires more cleverness. It turns out that, by carefully combining codes that correct bit-flip and phase-flip errors, a code correcting all single-qubit errors can be constructed. Such a code is given in section 11.1.3.

11.1.2 A Code for Single-Qubit Phase-Flip Errors

Consider the three single-qubit phase-flip errors Z_2, Z_1, Z_0 of a three-qubit system, where

$$\{Z_2 = Z \otimes I \otimes I, Z_1 = I \otimes Z \otimes I, Z_0 = I \otimes I \otimes Z\}.$$

Phase-flip errors, Z_i, in the standard basis are bit-flip errors $X = HZH$ in the Hadamard basis $\{|+\rangle, |-\rangle\}$, and vice versa. This observation suggests that appropriate modifications to the bit-flip code C_{BF} of section 11.1.1 will result in a code C_{PF} that corrects phase-flip errors instead. To obtain the logical qubits for the code C_{PF}, apply the Walsh-Hadamard transformation $W^{(3)} = H \otimes H \otimes H$ to the logical qubits of the code C_{BF}; the logical qubits for C_{PF} are $|\tilde{0}\rangle = |+++\rangle$ and $|\tilde{1}\rangle = |---\rangle$.

The phase-flip error Z_2 sends $|+++\rangle$ to $|-++\rangle$ and $|---\rangle$ to $|+--\rangle$. To detect such errors, the syndrome extraction operator U_{PF} for C_{PF} can be obtained from U_{BF} by changing basis from the standard basis to the Hadamard basis. Since, in the Hadamard basis, phase flips appear as bit flips, applying U_{BF} from code C_{BF} detects the error. Once the syndrome has been obtained by measuring the ancilla qubits in the standard basis, the error can be corrected by applying the bit-flip operator corresponding to the syndrome for code C_{BF} and then applying W to change back to the original basis. Instead, because $HX = ZH$, the error may be corrected by first applying W, and then the appropriate error correction transformation from the following table:

Bit shifted	Syndrome	Error correction
none	$|00\rangle$	none
0	$|11\rangle$	$Z_2 = Z \otimes I \otimes I$
1	$|10\rangle$	$Z_1 = I \otimes Z \otimes I$
2	$|01\rangle$	$Z_0 = I \otimes I \otimes Z.$

Thus $U_{PF} = W U_{BF} W$, with implementation

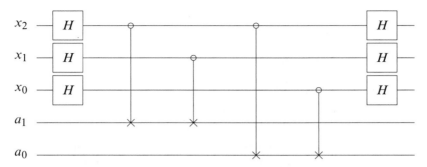

The code C_{PF} corrects all single-qubit relative phase errors, not just Z, because any single-qubit phase error is a linear combination of Z and I up to an irrelevant global phase factor:

$$\begin{pmatrix} 1 & 0 \\ 0 & e^{\mathbf{i}\phi} \end{pmatrix} = e^{\mathbf{i}\frac{\phi}{2}} \left(\cos\frac{\phi}{2} I - \mathbf{i}\sin\frac{\phi}{2} Z \right).$$

The code C_{PF} does not correct bit-flip errors, let alone general single-qubit errors.

11.1.3 A Code for All Single-Qubit Errors

Section 11.2.11 shows that a quantum error correcting code C that can correct all X_i and all Z_i errors can also correct all Y_i errors. Section 11.2.9 shows that any superposition (linear combination) of correctable errors is correctable. Section 11.2.9 also shows that the Pauli errors I, X, Y, and Z form a basis for all single-qubit errors. So if we can design a code that corrects all X_i and Z_i errors, the code will actually correct all single-qubit errors.

To construct such a code, it is natural to try to combine C_{BF} and C_{PF}. First encoding a qubit using C_{PF} and then encoding each resulting qubit using C_{BF} leads to the nine-qubit code

$$|0\rangle \rightarrow |\tilde{0}\rangle = \frac{1}{\sqrt{8}}(|000\rangle + |111\rangle) \otimes (|000\rangle + |111\rangle) \otimes (|000\rangle + |111\rangle),$$

$$|1\rangle \rightarrow |\tilde{1}\rangle = \frac{1}{\sqrt{8}}(|000\rangle - |111\rangle) \otimes (|000\rangle - |111\rangle) \otimes (|000\rangle - |111\rangle),$$

known as Shor's nine-qubit code. For convenience, we often write these states as

$$|0\rangle \rightarrow |\tilde{0}\rangle = \frac{1}{\sqrt{8}}(|000\rangle + |111\rangle)^{\otimes 3}$$

$$|1\rangle \rightarrow |\tilde{1}\rangle = \frac{1}{\sqrt{8}}(|000\rangle - |111\rangle)^{\otimes 3}.$$

To perform error correction, first use U_{BF} on each block of three qubits to correct for possible X errors in each block separately. At this point the bit values of the state are correct — in any term of the superposition, the three qubits of each block now have the same bit value — but the relative phases may be wrong. To correct phase errors, a variant of U_{PF} is used, essentially an expansion of U_{PF} to nine qubits instead of three. More details are given in section 11.3.

The term *code*, in both the classical and quantum setting, refers to the set of codewords. The mapping of the original strings or states into the codewords is not of great importance; a different mapping allows exactly the same set of errors to be corrected. Moreover, the encoding map is not generally implemented. The mapping $a|0\rangle + b|1\rangle$ to $a|\tilde{0}\rangle + b|\tilde{1}\rangle$ should be viewed as an abstract mapping; we do not start with qubits of the form $a|0\rangle + b|1\rangle$ and then encode them. Rather, we define the logical qubits of a system in this way, and we design gates and interpret measurements in terms of these logical qubits. For example, for Shor's code, instead of computing directly on n single qubits, each qubit is encoded in 9 qubits, totaling $9n$ qubits altogether. All quantum computation takes place on the n logical qubits, each consisting of nine qubits. It is on the 2^n-dimensional subspace containing the *logical qubits*, not on the full 2^{9n}-dimensional space, that we compute. Error correction returns states to this subspace, and it is on this 2^n-dimensional subspace, not on the full 2^{9n}-dimensional space, that we need a universal set of gates. Sections 11.2.8 and 11.4.4, and then much of chapter 12, concern the design of such gates.

Later sections describe codes that correct multiple-qubit errors and codes that correct all single-qubit errors using fewer than nine qubits. Before discussing those codes, we need to develop more systematic ways of thinking about and describing codes.

11.2 Framework for Quantum Error Correcting Codes

As section 10.4.4 explained, errors on the computational system due to interactions with the environment are linear, but not necessarily unitary. Because unitary transformations are invertible, if we can figure out what unitary error has occurred, we can correct it. But general errors may not have inverse transformations, so if such an error occurs, even if we have been able to determine which error has occurred, it is not obvious how to correct it. At first glance we might guess that such errors cannot be corrected without access and control over the part of the environment that interacted with the system. It is true that these errors cannot be corrected by applying unitary quantum transformations to the computational system alone. By measuring the system, however, or by entangling the system with auxiliary qubits, nonunitary errors can be corrected.

When a system has been subjected to decoherence under which it undergoes a nonunitary transformation, information about the original state of the system has been lost. For example, decoherence could swap a qubit in the environment with a qubit of the computational system, resulting in a complete loss of information about that qubit, except what can be deduced from other qubits. If the qubit's state was completely uncorrelated with the other qubits' states, all information about the state of that qubit is lost. The idea behind any sort of scheme for protecting information stored in quantum states is to embed the quantum states we care about in highly correlated states of a larger quantum system. To correct against general quantum errors, this correlation must be quantum; these states must be highly entangled states.

The art of designing quantum error correcting codes is to choose the embedding of k logical qubits into an n-qubit system in such a way that measurements can correct the most common errors to which the system is likely to be subjected. Generally, this embedding is taken to be linear: it is given by a linear map between the 2^k-dimensional vector space of the logical system and the 2^n-dimensional vector space of the larger system. We consider only linear codes here. Quantum codes have been designed for many types of errors. The most frequently considered family of errors consists of all errors on t or fewer qubits. We concentrate on this family of errors after presenting a general framework for quantum error correction. As physical implementations of quantum computers are developed it will be possible to determine to which sorts of errors a given physical device is most subject and to design error correcting codes or other forms of error protection to guard most efficiently and effectively against those errors.

Linear quantum codes are closely related to classical block codes. For each concept in quantum error correction, we first review related concepts from classical codes. For this reason, this section alternates between short subsections describing classical error correction and subsections describing quantum error correction.

This exposition is most suitable for readers who have some familiarity with classical error correcting codes; readers new to error correcting codes may wish to read all of the classical sections first to get a feel for the general strategies employed in error correction. Both classical and quantum error correction rely heavily on group theory. Boxes containing brief reviews of groups, subgroups, and Abelian groups can be found in section 8.6.1 and section 8.6.2. A few more boxes are interspersed throughout this chapter. Readers new to group theory will need to study the relevant sections of a text devoted to group theory. Suggested texts are given in the reference section at the end of this chapter.

This section describes a general nonconstructive framework for linear quantum error correcting codes, specifying properties that all linear quantum error correcting codes must satisfy. This framework pays no attention to whether or how a code can be efficiently implemented. This issue is crucial to whether the code is useful or not and will be dealt with more carefully later in this chapter and in chapter 12.

11.2.1 Classical Error Correcting Codes

A classical $[n, k]$ *block code* C is a size 2^k subset of the 2^n possible n-bit strings. The set of n-bit strings is a group, written $\mathbf{Z}_2^n$, under bitwise addition modulo 2. If the 2^k size subset C is a subgroup of $\mathbf{Z}_2^n$, then the code is said to be an $[n, k]$ *linear* block code. When a code is used,

Box 11.1
Group Homomorphisms

A *homomorphism* f from a group G to a group H is a map $f : G \rightarrow H$ that satisfies, for any elements g_1 and g_2 of G,

$$f(g_1 \circ g_2) = f(g_1) \circ f(g_2).$$

The product used on the right-hand side is the product for group G, while on the left-hand side it is the product for group H. An *isomorphism* from group G to group H is a homomorphism that is both one-to-one and onto. If there is an isomorphism between H and G, H and G are *isomorphic*, written $H \cong G$.

The *kernel* of a homomorphism $f : G \rightarrow H$ is the set of elements of G that are mapped to the identity element e_H of H.

a specific *encoding function* $c : \mathbf{Z}_2^k \rightarrow \mathbf{Z}_2^n$ is chosen, where c is an isomorphism between $\mathbf{Z}_2^k$, the *message space*, the set of all k-bit strings, and C, the code space: $c : \mathbf{Z}_2^k \rightarrow C \subset \mathbf{Z}_2^n$. In general, for any code C, there are many possible encoding functions. It may seem odd that the code is defined purely in terms of the subgroup C, not in terms of an encoding function. The reason for this convention is that no matter which encoding function is chosen, exactly the same set of errors can be corrected.

To encode a length mk message, each of the m blocks of length k are separately encoded using c to obtain a ciphertext of length mn. For this reason these codes are called block codes. The encoding function c can be represented by an $n \times k$ *generator matrix* G that takes a message word, an element of $\mathbf{Z}_2^k$ viewed as a length k column vector, to a codeword, an element of $C \subset \mathbf{Z}_2^n$: the generator matrix G multiplied with a message word gives the corresponding codeword. The k columns of G form a linearly independent set of binary words.

Example 11.2.1 *The* [3, 1] *repetition code.* The [3, 1] repetition code is defined to be the subset $C = \{000, 111\}$ of all 3-bit strings. This subset is a subgroup of $\mathbf{Z}_2^3$ under bitwise addition modulo 2.

The standard encoding function sends

$$0 \rightarrow 000$$

$$1 \rightarrow 111$$

and the associated generator matrix is

$$G = \begin{pmatrix} 1 \\ 1 \\ 1 \end{pmatrix},$$

which acts on bit strings viewed as column vectors:

$$\begin{pmatrix} 0 \\ 0 \\ 0 \end{pmatrix} = \begin{pmatrix} 1 \\ 1 \\ 1 \end{pmatrix} (0)$$

$$\begin{pmatrix} 1 \\ 1 \\ 1 \end{pmatrix} = \begin{pmatrix} 1 \\ 1 \\ 1 \end{pmatrix} (1).$$

A more interesting code is the [7, 4] Hamming code. A widely used quantum code, the Steane code, is built using special properties of the [7, 4] Hamming code. The Steane code will be introduced in section 11.3.3 and is a member of some major code families, including CSS codes and stabilizer codes, which are the subjects of sections 11.3 and 11.4 respectively.

Example 11.2.2 *The [7, 4] Hamming code.* The [7, 4] Hamming code C encodes 4-bit strings, elements of $\mathbf{Z}_2^4$, in 7-bit strings, elements of $\mathbf{Z}_2^7$. The code C is the subgroup of $\mathbf{Z}_2^7$ generated by {1110100, 1101010, 1011001, 1111111}. The reasoning behind this construction will become clear in section 11.2.5. One encoding function for C sends

$$1000 \mapsto 1110100$$

$$0100 \mapsto 1101010$$

$$0010 \mapsto 1011001$$

$$0001 \mapsto 1111111$$

These relations, together with linearity, fully define the encoding. The generator matrix G' for this encoding is

$$G' = \begin{pmatrix} 1 & 1 & 1 & 0 & 1 & 0 & 0 \\ 1 & 1 & 0 & 1 & 0 & 1 & 0 \\ 1 & 0 & 1 & 1 & 0 & 0 & 1 \\ 1 & 1 & 1 & 1 & 1 & 1 & 1 \end{pmatrix}^T.$$

An alternative encoding function sends

$$1000 \mapsto 1000111$$

$$0100 \mapsto 0100110$$

$$0010 \mapsto 0010101$$

$$0001 \mapsto 0001011$$

with generator matrix G

$$G = \begin{pmatrix} 1 & 0 & 0 & 0 \\ 0 & 1 & 0 & 0 \\ 0 & 0 & 1 & 0 \\ 0 & 0 & 0 & 1 \\ 1 & 1 & 1 & 0 \\ 1 & 1 & 0 & 1 \\ 1 & 0 & 1 & 1 \end{pmatrix}.$$

11.2.2 Quantum Error Correcting Codes

A $[[n, k]]$ *quantum block code* C is a 2^k-dimensional subspace C of the vector space V associated with the state space of an n-qubit system. The double square brackets are used to distinguish $[[n, k]]$ quantum codes from $[n, k]$ classical codes. View W, the k-qubit message space, as the subspace of V that has as basis the subset of the standard basis consisting of all strings in which the first $n - k$ elements are 0. Any unitary transformation $U_C : V \rightarrow V$ that takes W to C is a possible encoding operator for code C. In most cases we do not care how U_C behaves on states outside W, so frequently when we define an encoding operator U_C we will specify only its behavior on W and not on all of V. Elements $|w\rangle \in W$ are called *message words*, and elements of C are called *codewords* in analogy with the classical case. This terminology should not be taken too literally; neither message words in W nor codewords in C are bit strings, but rather quantum states of k and n qubits, respectively.

Just as in the classical case, it is the subspace C, not the encoding function, that defines the code; the same set of errors can be corrected no matter which encoding function is used. Given an encoding function and any state represented by $|w\rangle \in W$, the image $U_C(|w\rangle) = |\tilde{w}\rangle$ of $|w\rangle$ is an n-qubit state referred to as the *logical k-qubit state* corresponding to $|w\rangle$.

Example 11.2.3 *The bit-flip code revisited.* The code C is the subspace spanned by $\{|000\rangle, |111\rangle\}$. The standard encoding operator is

$$U_C : |0\rangle \mapsto |000\rangle$$
$$|1\rangle \mapsto |111\rangle.$$

So $|\tilde{0}\rangle = |000\rangle$ and $|\tilde{1}\rangle = |111\rangle$.

Strictly speaking, we should write

$$U_C : |000\rangle \mapsto |000\rangle$$
$$|001\rangle \mapsto |111\rangle$$

and define U_C on the rest of V, but we will generally define encoding functions in this way, since we do not care how the encoding behaves on states outside W, and the function definition is easier to read if we leave off the initial prefix string of zeros.

Example 11.2.4 *The Shor code revisited.* The Shor code is a $[[9, 1]]$ code, where C is the two-dimensional subspace spanned by

$$\frac{1}{\sqrt{8}}(|000\rangle + |111\rangle)^{\otimes 3}$$

and

$$\frac{1}{\sqrt{8}}(|000\rangle - |111\rangle)^{\otimes 3}.$$

The standard encoding operator used with this code sends

$$|0\rangle \rightarrow |\tilde{0}\rangle = \frac{1}{\sqrt{8}}(|000\rangle + |111\rangle)^{\otimes 3}$$

$$|1\rangle \rightarrow |\tilde{1}\rangle = \frac{1}{\sqrt{8}}(|000\rangle - |111\rangle)^{\otimes 3},$$

but any other function mapping $|0\rangle$ and $|1\rangle$ to two orthogonal vectors within the subspace C would also be a legitimate encoding function.

In practice it is not necessary to implement the encoding and decoding functions. At the beginning of a computation we simply construct the valid starting state, and at the end we interpret the classical information obtained from measurement to deduce information about the final logical state. Sections 11.2.8 and 11.4.4, and then much of chapter 12, discuss how to compute directly on the encoded data.

11.2.3 Correctable Sets of Errors for Classical Codes

A classical error may be viewed as an n-bit string $e \in \mathbf{Z}_2^n$ that acts on code words through bitwise addition $\oplus$, flipping a subset of the code bits. Any code C corrects some sets of errors and not others. A set of errors $\mathcal{E}$ is said to be *correctable* by code C if, for a w in $\mathbf{Z}_2^n$, there is at most one error that could result in w: for all $e_1, e_2 \in \mathcal{E}$ and $c_1, c_2 \in C$,

$$e_1 \oplus c_1 \neq e_2 \oplus c_2. \tag{11.1}$$

This condition is called the *disjointness condition* for classical error correction. Usually $\mathcal{E}$ is taken to be a group under bitwise addition modulo 2, so $\mathcal{E}$ contains the identity element, the non-error $00\cdots 0$. The disjointness condition for $e_1 = 00\cdots 0$ means that a correctable error cannot take a codeword to a different codeword. For any code C, there are many possible sets of correctable errors. Some correctable sets of errors are better than others from a practical point of view.

Example 11.2.5 *Correctable error sets for the* [3, 1] *repetition code.* The set $\mathcal{E} = \{000, 001, 010, 100\}$ is a correctable set of errors for the [3, 1] repetition code C. The set $\mathcal{E}' = \{000, 011, 101, 110\}$ is also a correctable set of of errors for C. The union of $\mathcal{E}$ and $\mathcal{E}'$ is not a correctable set for C.

11.2.4 Correctable Sets of Errors for Quantum Codes

For classical error correction, it suffices to consider bit-flip errors, a simple discrete set of errors. For quantum error correction, neither the encoded states nor the possible errors form a discrete set. For this reason, specifying correctable sets of errors for a quantum code C is more complicated than for a classical code. Fortunately, it is simpler than we might at first fear.

Let $B_C = \{|c_1\rangle, \ldots, |c_k\rangle\}$ be a (orthonormal) basis for C. A finite set $\mathcal{E} = \{E_1, E_2, \ldots, E_L\}$ of unitary transformations $E_i : V \rightarrow V$ is said to be a *correctable set of errors* for code C if there exists a matrix M with entries m_{ij} such that

$$\langle c_a | E_i^\dagger E_j | c_b \rangle = m_{ij} \delta_{ab} \tag{11.2}$$

for all $|c_a\rangle, |c_b\rangle \in C$ and $E_i, E_j \in \mathcal{E}$. The next few paragraphs clarify the meaning and motivation for this definition.

Just as in the classical case, there are many possible sets of correctable errors for a code C. Furthermore, there is no maximal correctable set, but some sets are more useful than others from a practical point of view. To perform error correction, one set of correctable errors is chosen, and the error correction procedures are designed with respect to that set. In the quantum case, the set of errors corrected by these procedures is much larger than the original correctable set $\mathcal{E}$; section 11.2.9 shows that if there is a procedure for a code C that corrects a set of errors $\mathcal{E} = \{E_1, E_2, \ldots, E_L\}$, then any superposition or mixture of errors in $\mathcal{E}$ can also be corrected by code C. It is this property that enables the correction of the general errors, discussed in section 10.4, that can be modeled as probabilistic mixtures of linear transformations. Since unitary errors E are easily corrected by applying the inverse transform $E^\dagger$, the errors of a correctable set have a clear error correction procedure once the error is known. The next two paragraphs give intuitive justification for the *Correctable Error Set Condition* (equation 11.2).

Just as in the classical case, there is no hope of correctly recovering from a set of errors $\mathcal{E}$ that contains a pair of error transformations that take two different codewords to the same state. The quantum case has a stronger requirement along these lines: any two distinct errors in $\mathcal{E}$ must take orthogonal codewords to orthogonal states. The reason for this requirement is that in order to determine which error is likely to have occurred, we need to make measurements, and two states can be distinguished with certainty if and only if the two states are orthogonal. This condition guarantees that the images of two different codewords under errors in $\mathcal{E}$ are distinguishable if the original codewords are distinguishable. This condition is written

$$\langle c|E_i^\dagger E_j|c'\rangle = 0 \tag{11.3}$$

for all $E_i, E_j \in \mathcal{E}$, and all $|c\rangle, |c'\rangle \in C$ such that $\langle c|c'\rangle = 0$. This *orthogonality condition* is the analog of the disjointness condition, equation. 11.1, for classical error correction.

In the quantum case, in order for error correction not to destroy the quantum computation, an additional condition is needed. Measurements made to determine the error must not give any information about the logical state, since otherwise superpositions may be destroyed, making the quantum computation useless. For this reason, we require

$$\langle c_a|E_i^\dagger E_j|c_a\rangle = \langle c_b|E_i^\dagger E_j|c_b\rangle \tag{11.4}$$

for all $|c_a\rangle, |c_b\rangle \in C$ and $E_i, E_j \in \mathcal{E}$. This requirement means that for every pair of indices i and j, there is a value m_{ij} such that

$$\langle c_a|E_i^\dagger E_j|c_a\rangle = m_{ij}.$$

Putting conditions 11.3 and 11.4 together results in the original equation 11.2:

$$\langle c_a|E_i^\dagger E_j|c_b\rangle = m_{ij}\delta_{ab}$$

for all $|c_a\rangle, |c_b\rangle \in C$ and $E_i, E_j \in \mathcal{E}$, where a significant part of the meaning of this formula is that m_{ij} is independent of a and b.

Condition 11.2 holds if

$$\langle c_a|E_i^\dagger E_j|c_b\rangle = 0 \tag{11.5}$$

for all $|c_a\rangle, |c_b\rangle \in C$ and $E_i, E_j \in \mathcal{E}$ such that $i \neq j$, but this condition is stronger than necessary. If two different errors E_1 and E_2 take a state $|\psi\rangle$ to the same state $|\psi'\rangle$, no matter which error occurred, applying $E_1^\dagger$ (or equally well $E_2^\dagger$) corrects the error. Condition 11.5 holds for many quantum codes, but not for some important codes. A code that does not satisfy this condition is called a *degenerate code* for error set $\mathcal{E}$. Shor's code is degenerate, for example: a relative phase error acting on the first qubit will have the same effect as a relative phase error acting on the second qubit. The existence of degenerate codes complicates matters. There is no classical analog for degenerate quantum codes.

The unitarity of the E_i means that E_iC has dimension 2^k for all errors E_i. Since there can be at most 2^{n-k} mutually orthogonal subspaces of dimension 2^k in a space of dimension 2^n, the maximum size of a set $\mathcal{E}$ of correctable errors for a nondegenerate code is 2^{n-k}. For degenerate codes, the size of a maximal set of correctable errors can be greater than 2^{n-k}.

Example 11.2.6 *The bit-flip code revisited.* The set of errors $\mathcal{E} = \{E_{ij}\}$ with

$$E_{00} = I \otimes I \otimes I, \; E_{01} = X \otimes I \otimes I, \; E_{10} = I \otimes X \otimes I, \; E_{11} = I \otimes I \otimes X$$

is a correctable error set for the bit-flip code.

The set of errors $\mathcal{E}' = \{E'_{ij}\}$ with

$$E'_{00} = I \otimes I \otimes I, \; E'_{01} = I \otimes X \otimes X, \; E'_{10} = X \otimes I \otimes X, \; E'_{11} = X \otimes X \otimes I$$

is a different correctable error set for the bit-flip code. In this case, the code corrects all two-qubit flip errors, but none of the single bit-flip errors. Of course, this set of correctable errors is of little practical value, since single bit-flip errors are generally more likely than pairs of bit-flip errors. But it is conceivable that in certain physical implementations, bit-flip errors are more likely to appear in pairs.

11.2.5 Correcting Errors Using Classical Codes

Let C be a classical $[n, k]$ linear block code, and suppose $\mathcal{E}$ is a correctable set of errors for C. Suppose $w = e \oplus c$ for some codeword $c \in C$ and error $e \in \mathcal{E}$. We wish to correct w to c. To find e and c, it is helpful to consider cosets of the code C.

This paragraph shows that there is a unique error associated with each coset. Let H be the set of cosets of C in $\mathbf{Z}_2^n$. An error $e \in \mathcal{E}$ changes a code word c into $e \oplus c$, an element of some coset of C. Given errors $e_1 \neq e_2$ and codewords c_1 and c_2, by disjointness condition 11.1, $e_1 \oplus c_1$ and $e_2 \oplus c_2$ are in two different cosets. To see this, suppose $e_1 \oplus c_1$ and $e_2 \oplus c_2$ were in the same coset. Then there would exist a $c_3 \in C$ such that $e_1 \oplus c_1 \oplus c_3 = e_2 \oplus c_2$.

Box 11.2
Cosets

Given a subgroup $H < G$, for each $a \in G$, the set $aH = \{ah | h \in H\}$ is called a (left) *coset* of H in G. (Right cosets are analogously defined, but we do not need to consider them here, so we will simply refer to left cosets as cosets.)

For a and b in G, either $aH = bH$ or $aH \cap bH = \emptyset$, so the cosets partition G. Thus, the order of a subgroup must divide the order of the group and, similarly, the number of distinct cosets must divide the order of the group. The *index* of H in G is the number of distinct cosets of H in G, and is denoted by $[G : H]$.

For example, let $G = \mathbf{Z}_n$, and let $H = m\mathbf{Z}_n$ be the set of multiples of m for some integer m dividing n. The order of G is n, the order of H is n/m, and the number of distinct cosets is $[G : H] = |G|/|H| = m$.

If $K < H < G$, then $[G : K] = |G|/|K| = (|G|/|H|)(|H|/|K|) = [G : H][H : K]$.

But $c_1 \oplus c_3$ is in C, which violates the disjointness condition 11.1 that says that two distinct correctable errors cannot take two codewords to the same word. Thus, knowing to which coset the word $e \oplus c$ belongs, tells us which error e has occurred. Let us make this more precise.

Because $\mathbf{Z}_2^n$ is Abelian, the set of all cosets forms a group, H. It is of size 2^{n-k}. Since H is Abelian and nontrivial, and all elements of H have order 2, H is isomorphic to $\mathbf{Z}_2^{n-k}$. Let $\sigma : H \to \mathbf{Z}_2^{n-k}$ be an isomorphism. The map

$$h : \mathbf{Z}_2^n \to \mathbf{Z}_2^{n-k} \cong H$$

$$w \mapsto \sigma(w \oplus C)$$

sends all elements of C to the zero element of $\mathbf{Z}_2^{n-k}$; the kernel of h is C. The element $h(w)$ characterizes each coset since $h(w) = h(w')$ if and only if w and w' are in the same coset. By the previous paragraph, there is a unique error $e \in \mathcal{E}$ associated with this coset. Since $h(w)$ characterizes the coset, it also characterizes this error. For this reason, $h(w)$ is called the *error syndrome*, or simply *syndrome*.

More concretely, h can be realized by an $(n-k) \times n$ matrix P. To construct a concrete P, find $n-k$ linearly independent elements p_i of $\mathbf{Z}_2^n$ such that $p_i \cdot c = 0 \bmod 2$ for all $c \in C$, and take these as the rows of the matrix:

$$P = \begin{pmatrix} p_1^T \\ \vdots \\ p_{n-k}^T \end{pmatrix}$$

For a given code C, there are many possible matrices P (just as there are many possible isomorphisms σ). The matrix P, acting on $w \in \mathbf{Z}_2^n$ viewed as a column vector, produces an $n-k$ length binary column vector Pw, the *syndrome*, that characterizes the coset of C containing w. Each of these $n-k$ values is the inner product (mod 2) of w with a row of P. For this reason, the rows p_i are called *parity checks* and P is called a *parity check matrix* for code C. The parity check matrix P distinguishes between distinct correctable errors e_i and e_j since $P(e_i) \neq P(e_j)$. If G is a generator matrix for the codewords of C, and P is an arbitrary $(n-k) \times n$ matrix, the $(n-k) \times k$ product matrix PG is 0 if and only if P is a parity check matrix for C. The code C is both the image of $\mathbf{Z}_2^k$ in $\mathbf{Z}_2^n$ under G, and the kernel of P, the set of elements of $\mathbf{Z}_2^n$ sent to $00 \cdots 0$ under P.

Hamming codes are among the simplest classical codes and are used as the basis for many quantum codes. There is a Hamming code C_n for every integer $n \geq 2$. A parity check matrix for Hamming codes has columns consisting of all the non-zero n-bit strings. Since the parity check matrix for the Hamming code C_n is a $n \times (2^n - 1)$ matrix, the generator matrix for C_n is therefore a $(2^n - 1) \times (2^n - n - 1)$ matrix, and the Hamming code C_n is a $[2^n - 1, 2^n - n - 1]$ code. All Hamming codes correct single bit-flip errors.

Example 11.2.7 *The Hamming code C_2*. The $[3, 1]$ repetition code is also the Hamming code C_2, the code with parity check matrix

$$P = \begin{pmatrix} 0 & 1 & 1 \\ 1 & 0 & 1 \end{pmatrix}.$$

A different parity check matrix for the same code is

$$P' = \begin{pmatrix} 1 & 1 & 0 \\ 1 & 0 & 1 \end{pmatrix}.$$

The matrix P' has form $(A|I)$. By exercise 11.2, $\left(\frac{I}{A}\right)$ is a generator matrix for the code. The generator matrix obtained in this way from P' is

$$G = \begin{pmatrix} 1 \\ 1 \\ 1 \end{pmatrix}.$$

The code C_2 is called a repetition code, since $0 \mapsto 000$ and $1 \mapsto 111$.

Example 11.2.8 *The Hamming code C_3*. The Hamming code C_3 is a $[7, 4]$ code. Section 11.3.3 uses C_3 to define the quantum Steane code.

A parity check matrix for the $[7, 4]$ Hamming code is

$$P' = \begin{pmatrix} 0 & 0 & 0 & 1 & 1 & 1 & 1 \\ 0 & 1 & 1 & 0 & 0 & 1 & 1 \\ 1 & 0 & 1 & 0 & 1 & 0 & 1 \end{pmatrix};$$

its columns are exactly the seven non-zero 3-bit strings. Our next task is to find a generator matrix G' for C. Since each row of P' contains an even number of 1s, each row is orthogonal to itself. Furthermore, these elements are orthogonal to each other, that is, $PP^T = 0$, so we may take as the first three columns of G' the transposes of the rows of P'. We need to find one other vector orthogonal to and linearly independent of these columns. The vector

$$\begin{pmatrix} 1 & 1 & 1 & 1 & 1 & 1 & 1 \end{pmatrix}^T$$

satisfies both conditions. So a generator matrix for the $[7, 4]$ Hamming code is

$$G' = \begin{pmatrix} 0 & 0 & 0 & 1 & 1 & 1 & 1 \\ 0 & 1 & 1 & 0 & 0 & 1 & 1 \\ 1 & 0 & 1 & 0 & 1 & 0 & 1 \\ 1 & 1 & 1 & 1 & 1 & 1 & 1 \end{pmatrix}^T.$$

Alternatively, the [7, 4] Hamming code can be defined in terms of a more convenient parity check matrix of the form $(A|I)$,

$$P = \begin{pmatrix} 1 & 1 & 1 & 0 & 1 & 0 & 0 \\ 1 & 1 & 0 & 1 & 0 & 1 & 0 \\ 1 & 0 & 1 & 1 & 0 & 0 & 1 \end{pmatrix}.$$

By exercise 11.2, a generator matrix corresponding to a parity check matrix of this form is $(\frac{I}{A})$,

$$G = \begin{pmatrix} 1 & 0 & 0 & 0 \\ 0 & 1 & 0 & 0 \\ 0 & 0 & 1 & 0 \\ 0 & 0 & 0 & 1 \\ 1 & 1 & 1 & 0 \\ 1 & 1 & 0 & 1 \\ 1 & 0 & 1 & 1 \end{pmatrix}.$$

11.2.6 Diagnosing and Correcting Errors Using Quantum Codes

This section describes a procedure for correcting errors handled by nondegenerate quantum codes. Let C be an $[[n, k]]$ quantum code that is nondegenerate with respect to a correctable error set $\mathcal{E} = \{E_i\}$, where $0 \leq i < M$. Suppose $|w\rangle = E_s|v\rangle$ for some $E_s \in \mathcal{E}$ and $|v\rangle \in C$. Because C is nondegenerate with respect to $\mathcal{E}$, the subspaces E_iC and E_jC are orthogonal for all $i \neq j$, so $E_s|v\rangle$ is the only way to obtain $|w\rangle$ from a codeword in C and an error in $\mathcal{E}$: the elements E_s and $|v\rangle$ are unique. Thus, if we can determine in which subspace E_sC the state $|w\rangle$ lives, from among the M subspaces $\{E_iC\}$, we can correct the error by applying $E_s^\dagger$ to $|w\rangle$. To make this determination, we must measure the state $|w\rangle$. The standard model of quantum computation allows only single-qubit measurements in the standard basis. Any other measurement can be carried out by computing into ancilla qubits and measuring each of these in the standard basis, but only some measurements can be efficiently carried out in this way. This section presents a general framework. Later sections of this chapter and the next consider implementation issues with respect to specific codes.

The aim of the measurement is to determine in which error subspace the state $|w\rangle$ lies. Let $W_i = E_iC$, and

$$W = \bigoplus_{i=0}^{M-1} W_i.$$

Let $W^\perp$ be the possibly empty subspace of the computational space V orthogonal to W; vectors in $W^\perp$ are orthogonal to all codewords and also to all states that are images of codewords under a correctable error $E_i \in \mathcal{E}$. For notational convenience, define $W_M = W^\perp$. Since $|w\rangle$ is the result of an error E_i applied to a codeword, by definition of W_M, $|w\rangle$ does not lie in W_M. Since the W_i are mutually orthogonal, there is an observable O with eigensubspaces exactly the W_i.

Let P_i be the projector onto the subspace W_i. Let $m = \lceil \log_2 M \rceil$, and let U_P be a unitary operator on $n + m$ qubits such that

$$U_P : |w\rangle|0\rangle \mapsto \sum_{j=0}^{M-1} b_j|w_j\rangle|j\rangle, \tag{11.6}$$

where $|w\rangle = \sum_{j=0}^{M-1} b_j|w_j\rangle$ is written in terms of its components $b_j|w_j\rangle = P_j|w\rangle$. Measuring the m auxiliary qubits in the standard basis gives the *error syndrome*, the subspace index j. By definition of W_M, the index M cannot occur. After measurement, the state of the first n qubits is in the subspace $W_j = E_j C$, so applying the operator $E_j^\dagger$ corrects the error. The operator U_P is called a *syndrome extraction operator* since it plays a similar role to the syndrome in classical error correction. The notation U_P is meant to suggest a unitary operator that plays the role of the parity check matrix P in classical error correction. Since the labels for the subspaces can be arbitrarily chosen, many different unitary operators can serve as a syndrome extraction operator for a given code C and error set $\mathcal{E}$.

Measuring a single qubit l of the m auxiliary qubits on its own corresponds to a binary observable with two 2^{n-1}-dimensional eigensubspaces, the subspace spanned by all of the W_i for which the lth bit of the binary representation of its index i is 0, and the subspace spanned by all of the W_i for which the lth bit of the binary representation of its index i is 1. In this way, the syndrome extraction operator can be viewed as a set of m observables.

Example 11.2.9 *The bit-flip code revisited.* Consider the bit-flip code C and the set of correctable errors $\mathcal{E} = \{E_{ij}\}$ with

$$E_{00} = I \otimes I \otimes I, \; E_{01} = X \otimes I \otimes I, \; E_{10} = I \otimes X \otimes I, \; E_{11} = I \otimes I \otimes X.$$

More simply, $E_{00} = I$, $E_{01} = X_2$, $E_{10} = X_1$, and $E_{11} = X_0$, where X_i is the operator X applied to the ith qubit. The orthogonal subspaces corresponding to this error set are $W_{00} = E_{00}C$, $W_{01} = E_{01}C$, $W_{10} = E_{10}C$ and $W_{11} = E_{11}C$ with bases $B_{00} = \{|000\rangle, |111\rangle\}$, $B_{01} = \{|100\rangle, |011\rangle\}$, $B_{10} = \{|010\rangle, |101\rangle\}$ and $B_{11} = \{|001\rangle, |110\rangle\}$, respectively. The operator

$$U_P : |x_2, x_1, x_0, 0, 0\rangle \rightarrow |x_2, x_1, x_0, b_1 = x_1 \oplus x_0, b_0 = x_2 \oplus x_0\rangle$$

serves as a syndrome extraction operator for C with error set $\mathcal{E}$. Measuring bit b_1 in the standard basis distinguishes between the eigenspaces spanned by subspaces $\{W_{00}, W_{01}\}$ and $\{W_{10}, W_{11}\}$ respectively. Similarly, measuring b_0 distinguishes between errors in the spaces spanned by $\{W_{00}, W_{10}\}$ and $\{W_{01}, W_{11}\}$. Measuring b_1 and b_0 as i and j projects the state into $W_{ij} = E_{ij}C$.

The error can be corrected by applying $E_{ij}^{\dagger}$. If, for example, measuring the ancilla b_1 and b_0 yields 0 and 1 respectively, we apply the transformation X_2.

Measuring b_0 (resp. b_1) directly, without the use of ancilla bits, can be done using the observable $Z \otimes I \otimes Z$ (resp. $I \otimes Z \otimes Z$). Compare the classical parity check matrix

$$P = \begin{pmatrix} 1 & 0 & 1 \\ 0 & 1 & 1 \end{pmatrix}$$

for the [3, 1] code with the array

$$\begin{array}{ccc} Z & I & Z \\ I & Z & Z \end{array},$$

where the factors of the two observables have been placed in the rows. In the classical case, the parity check matrix multiplied by a word will be 0 if the word is a codeword. At least one of the rows of the parity check matrix when multiplied by a non-codeword will be non-zero. In the quantum case a codeword is in the $+1$-eigenspace for all the observables, and non-codewords are in the -1-eigenspace for at least one of the observables. The stabilizer codes of section 11.4 exploit this connection.

Before turning to another example, we use this example to illustrate an alternative to the syndrome measurement. The use of ancilla qubits in quantum error correction to correct general errors is one of the most elegant and surprising aspects of quantum computing. By computing information into ancilla qubits and measuring them, nonunitary errors can be converted to unitary errors. When the result of the measurement tells us which unitary error remains, we can correct it by applying the inverse unitary operator. Alternatively, and equivalently, instead of measuring the ancilla after computing into them, a controlled operation from the ancilla qubits to the computational qubits can correct the error. In general, for $\mathcal{E} = \{E_s\}$, Instead of measuring after applying U_P to the computational system and the ancilla, apply the following controlled operation with the ancilla as the control bits:

$$V_P = \sum_s E_s^{\dagger} \otimes |s\rangle\langle s|.$$

In this way, errors can be corrected without measurement.

Example 11.2.10 *Bit-flip code C_{BF} correction by controlled operations.* After applying U_{BF} in example 11.1.2, instead of measuring, a controlled operation V_P from the ancilla qubits to the computational qubits can be performed, one that applies each of the three error correction transformations when the ancilla qubits are in the corresponding state:

$$V_P = I \otimes |00\rangle\langle 00| + X_2 \otimes |01\rangle\langle 01| + X_1 \otimes |10\rangle\langle 10| + X_0 \otimes |11\rangle\langle 11|.$$

The circuit for this controlled operation is

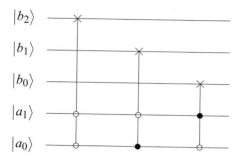

Suppose an error $E = \alpha X_2 + \beta X_1$ has occurred. Applying this circuit to the state

$$U_{BF}(E|\tilde{0}\rangle \otimes |00\rangle) = \alpha|100\rangle|11\rangle + \beta|010\rangle|10\rangle$$

results in

$$\alpha|000\rangle|11\rangle + \beta|000\rangle|10\rangle = |000\rangle(\alpha|11\rangle + \beta|10\rangle).$$

We may wish to measure the two ancilla qubits in order to transform them to $|00\rangle$ so that they can be reused in a later error correction step, but measurement is not required to achieve quantum error correction.

Example 11.2.11 *The phase-flip code revisited.* Recall that the relative phase code of section 11.1.2 is dual to the bit-flip code of example 11.2.9 through the transformation $W = H \otimes H \otimes H$. Applying W to all states and replacing all transformations T used in example 11.2.9 with WTW results in an error correction procedure for the relative phase code. Since $X = HZH$, the observables corresponding to the syndrome operator U'_P are $X \otimes I \otimes X$ and $I \otimes X \otimes X$, which have corresponding array

$$
\begin{array}{ccc}
X & I & X \\
I & X & X
\end{array}
,
$$

which is related to the classical parity check matrix

$$P = \begin{pmatrix} 1 & 0 & 1 \\ 0 & 1 & 1 \end{pmatrix}.$$

In this case, errors can be corrected without measurement using

$$V'_P = I \otimes |00\rangle\langle00| + Z_2 \otimes |01\rangle\langle01| + Z_1 \otimes |10\rangle\langle10| + Z_0 \otimes |11\rangle\langle11|.$$

11.2.7 Quantum Error Correction Across Multiple Blocks

Just as the classical $[n, k]$ block codes of section 11.2.1 encode length mk bit-strings as a length mn bit-string by encoding each of the m blocks of k bits, a quantum $[[n, k]]$ code C encodes mk logical qubits in mn computational qubits by encoding each of the m blocks of k logical bits using C. A logical superposition such as

$$|\psi\rangle = \sum_i \sum_j \alpha_{ij}(|w_i\rangle \otimes |w_j\rangle)$$

is encoded as

$$|\tilde{\psi}\rangle = \sum_i \sum_j \alpha_{ij}(|c_i\rangle \otimes |c_j\rangle),$$

where $|c_i\rangle = U_C|w_i\rangle$ and U_C is an encoding function for C. Quantum block codes must be able to correct errors on such superpositions. Furthermore, if C can correct errors $E_i \in \mathcal{E}$, then C applied blockwise must be able to correct errors of the form $E_{i_1} \otimes \cdots \otimes E_{i_m}$ on the encoded state. The rest of this section illustrates, in the two-block case, quantum error correction on superpositions and across multiple blocks.

Suppose the encoded state $|\tilde{\psi}\rangle = \sum_i \sum_j \alpha_{ij}(|c_i\rangle \otimes |c_j\rangle)$ were subject to error $E_a \otimes E_b$, where E_a and E_b are both correctable errors for code C. Applying the syndrome extraction operator U_P for C to each block separately, measuring the ancilla for each block, and applying the appropriate correcting operators will restore the state $|\tilde{\psi}\rangle$:

$$U_P \otimes U_P((E_a \otimes E_b|\tilde{\psi}\rangle) \otimes |0\rangle|0\rangle) = \sum_{ij} \alpha_{ij}(U_P(E_a|c_i\rangle|0\rangle) \otimes (U_P(E_b|c_j\rangle|0\rangle)))$$

$$= \sum_{ij} \alpha_{ij}(E_a|c_i\rangle|a\rangle \otimes E_b|c_j\rangle|b\rangle),$$

where we have reordered the qubits for clarity. Measurement of the two ancilla yields $|a\rangle$ and $|b\rangle$ respectively, with the computation qubits in state $|\phi\rangle = \sum_{ij} \alpha_{ij}(E_a|c_i\rangle \otimes E_b|c_j\rangle)$. The syndrome $|a\rangle|b\rangle$ indicates that the error can be corrected by applying $E_a^{\dagger} \otimes E_b^{\dagger}$. Applying $E_a^{\dagger} \otimes E_b^{\dagger}$ does indeed correct the error:

$$E_a^{\dagger} \otimes E_b^{\dagger}|\phi\rangle = \sum_i \sum_j \alpha_{ij}(|c_i\rangle \otimes |c_j\rangle) = |\tilde{\psi}\rangle.$$

11.2.8 Computing on Encoded Quantum States

For error correcting codes to be useful for quantum computation, we must still be able to perform computation on the states after they are encoded. Let $C \subset V$ be an $[[n, k]]$ quantum code and let U_C be an encoding function $U_C : W \to C$. In order to perform general computation on encoded states, for any unitary operator $U : W \to W$, we must find an analogous unitary operator $\tilde{U}$ acting on the encoded states, one that for all $|w\rangle \in W$ sends $U_C(|w\rangle)$ to $U_C(U|w\rangle)$. Because we do not care how $\tilde{U}$ behaves outside C, there are many unitary operators acting on V that have this property. For a given unitary operator $\tilde{U}$, there are many ways to implement it in terms of basic gates. Furthermore, for a given $U : W \rightarrow W$ with two logical analogs $\tilde{U} : V \rightarrow V$ and $\tilde{U}' : V \rightarrow V$, one of $\tilde{U}$ and $\tilde{U}'$ may be more efficiently implementable than the other, and some implementations have better robustness properties than others.

One such operator can be constructed using the encoding operator. Let U_C be the unitary coding function that sends $|w\rangle \otimes |0\rangle$ to $|\tilde{w}\rangle$. The transformation $U_C^\dagger$ sends a valid codeword $|\tilde{w}'\rangle$ to $|w'\rangle \otimes |0\rangle$. The operator $\tilde{U} = U_C(U \otimes I)U_C^\dagger$ acts as desired on the code space; $\tilde{U}$ is the logical equivalent to U on the encoded states. In general, however, this construction yields a $\tilde{U}$ with poor robustness properties: after applying $U_C^\dagger$, the state is unencoded, making it extremely vulnerable to any errors that occur during this time. Chapter 12 takes a careful look at how logical operations are best implemented on encoded states.

11.2.9 Superpositions and Mixtures of Correctable Errors Are Correctable

Section 10.4 showed that general errors E can be modeled as probabilistic mixtures of linear transformations, and that these linear error transformations A_i are not necessarily unitary:

$$E : \rho \mapsto \sum_{i=1}^{K} A_i \rho A_i^\dagger.$$

This section shows that errors that are non-zero complex linear combinations of elements of a correctable error set $\mathcal{E}$ for a code C can be corrected by this code. The term *set of correctable errors* refers to a set of errors the code can correct via a unitary transformation, but the set of errors the code corrects is much larger: all linear combinations of such errors. Measurement is used to project a linear combination of errors onto one of the correctable errors, and it is also used to detect which error remains after measurement so that the corresponding unitary error transformation can be applied. As in the classical case, there are many possible maximal sets of correctable errors for a given code, and some of these distinct maximal sets of correctable errors generate distinct subspaces.

Let error $E = \sum_{i=0}^{m} \alpha_i E_i$ be a probabilistic mixture of errors, a linear combination of errors E_i from a correctable set $\mathcal{E}$ such that $\sum_i |\alpha_i|^2 = 1$. The error E may or may not be unitary, so we consider the general case and show that, if E takes a codeword $|c\rangle$, with density operator $\rho = |c\rangle\langle c|$, to a mixed state $\rho' = E\rho E^\dagger$, we can correct for the error. The mixed state ρ' can be written

$$\rho' = \sum |\alpha_i|^2 E_i |c\rangle\langle c| E_i^\dagger.$$

Since the $E_i|c\rangle$ are mutually orthogonal and $\sum_i |\alpha_i|^2 = 1$, ρ' has trace 1 and is a mixed state. Thus ρ' is a probability distribution over the orthogonal pure states $E_i|c\rangle$. Consider the observable $O = \sum_i \lambda_i P_i$, where the λ_i are distinct and P_i is the projector onto the subspace $E_i C$. Using the definitions in section 10.3, measurement with O results in the state $P_i \rho' P_i^\dagger = E_i|c\rangle\langle c|E_i^\dagger$ with probability $|\alpha_i|^2$. Thus, after measurement, we have a pure state $E_i|c\rangle$. The measurement result, λ_i, tells us in which subspace $E_i C$ the state resides. Applying $E_i^\dagger$ corrects the state.

11.2.10 The Classical Independent Error Model

In both the quantum and classical case, a general error correction strategy consists of three parts: detecting non-codewords, determining the most likely error, and applying a transformation to correct that error. Determining the most likely error requires an error model. A common family of error models for classical computation is the *independent error model* in which each bit has probability $p \leq 1/2$ of flipping. In this model, the chance of any of the single bit-flip errors $100\cdots 0, 010\cdots 0, \ldots, 000\cdots 1$ is $p(1-p)^{n-1}$, the chance of the two-bit error $110\cdots 0$ occurring is probability $p^2(1-p)^{n-2}$, and the chance of no error occurring is $(1-p)^n$.

This error model guides the error correction strategy. Since under this model no error is more likely than any error, if a codeword is received, our best bet is to assume that no error occurred. Suppose w is a non-codeword we wish to correct. Let c be an element of C that is closest to w in the Hamming distance. If the closest element is unique, the most likely error to have occurred under the independent error model is $e = c \oplus w$. Let w' be another element of the coset containing w, so $w' = w \oplus k$ for some $k \in C$. If c is the closest element in C to w, then $c' = c \oplus k$ must be the closest element in C to w'. The most likely error resulting in w' is also e, because

$$w' \oplus c' = w \oplus k \oplus c \oplus k = w \oplus c = e.$$

Thus, all elements of a coset are equally close to C in the Hamming distance. By definition of c, the most likely error e is the element of the coset with the lowest Hamming weight.

Once the syndrome computation tells us the coset, we correct by applying the lowest weight element e of that coset. If the actual error was a different one, we have "corrected" to the wrong word, but no better strategy exists. In particular, if we receive a codeword, we do nothing. In general, error correcting codes cannot correct errors that take codewords to codewords. Furthermore, if there is more than one closest element to w in C, it is unclear how best to correct the error. For this reason, when working under the independent error model, the set of correctable errors is usually taken to be $\mathcal{E}_t$, the set of all words of Hamming weight t or less, where t is as large as possible without introducing ambiguity or, equivalently, violating the disjointness condition (equation 11.1) for a set of correctable errors.

The minimum Hamming distance between any pair of codewords is called the *distance of the code*. An $[n, k, d]$ code is one that uses n-bit words to encode k-bit message words and has distance d. For each codeword c, let

$$e_t(c) = \{v | d_H(v, c) \leq t\}$$

be the set of words no more than Hamming distance t away from c. The set $e_t(c)$ contains exactly words v obtained from c by an error of weight at most t. If the sets $e_t(c)$ are disjoint for all pairs of code words c and c', the code can correct any weight t error by mapping words in $e_t(c)$ to the codeword c. The sets $e_t(c)$ are disjoint if and only if $d \geq 2t + 1$. So an $[n, k, d]$-code can correct at most all errors with weight less than or equal to $t = \lfloor \frac{d-1}{2} \rfloor$. For a distance d code C, the maximum possible t satisfies $2t + 1 \leq d$, since otherwise two codewords could be mapped to the same error word under two different t-bit errors, and the disjointness condition would not hold.

11.2.11 Quantum Independent Error Models

For a given quantum code C, some sets of correctable errors for C are better than others from a practical point of view. As in the classical case, which correctable sets are better depends on which errors are more probable. Because there is a richer class of quantum errors, there is a greater variety of quantum error models to choose from. The most common quantum error models assume, as the classical independent error model does, that errors on separate qubits occur independently and that, with probability p, a given qubit is subject to an error. The error model we describe is motivated by the local and Markov assumptions discussed in section 10.4.4.

Because unitary errors are easily corrected by applying the inverse transformation, sets of correctable errors are chosen to contain only unitary error transformations. It is particularly common to choose a correctable set of errors containing only elements of the generalized Pauli group $\mathcal{G}_n$. The *generalized Pauli group* $\mathcal{G}_n$ consists of n-fold tensor products of Pauli group elements: all elements of $\mathcal{G}_n$ are of the form

$$\mu A_1 \otimes A_2 \otimes \cdots \otimes A_n,$$

where $A_i \in \{I, X, Y, Z\}$ and $\mu \in \{1, -1, \mathbf{i}, -\mathbf{i}\}$. The commutation relations, the relations between group products $g_i g_j$ and $g_j g_i$, for the Pauli group imply that every element of $\mathcal{G}_n$ can be written as

$$\mu(X^{a_1} \otimes \cdots \otimes X^{a_n})(Z^{b_1} \otimes \cdots \otimes Z^{b_n}),$$

where the a_i and b_i are binary values.

Section 10.4.4 showed that any error can be expressed as a mixture of linear transformations

$$\frac{A_i}{\sqrt{\mathbf{tr}(A_i \rho A_i^\dagger)}}.$$

The generalized Pauli group $\mathcal{G}_n$ forms a basis for the vector space of linear transformations acting on the vector space associated with an n-qubit system. Thus, a general error E on an n-qubit quantum register can be expressed as linear combination $\sum_j e_j E_j$ where $E_j \in \mathcal{G}_n$. All linear transformations arising in an operator sum decomposition can be written not only in terms of unitary operators, but also in terms of generalized Pauli operators. By results of section 11.2.9, a mixture of errors, each of which is corrected by a procedure, is also corrected by that procedure.

We write X_i for the transform that applies X to the ith qubit and leaves the others alone:

$$\underbrace{I \otimes \cdots \otimes I}_{i} \otimes X \otimes \underbrace{I \otimes \cdots \otimes I}_{n-i-1}.$$

The meaning of Y_i and Z_i is similar. The weight of a Pauli error is the number of nonidentity terms in its tensor product expression. The weight of an error is defined only for Pauli errors, not for general errors.

The generalized Pauli group has a number of convenient properties. For example, the stabilizer codes of section 11.4 make heavy use of the fact that any two elements g_1 and g_2 in the Pauli group either commute ($g_1 g_2 = g_2 g_1$) or anticommute ($g_1 g_2 = -g_2 g_1$). Another convenient property is that if the set of all single-qubit bit-flip and phase-flip errors X_i and Z_i for all i is a correctable set $\mathcal{E}$ for a code C, then $\mathcal{E}$ can be expanded to contain the Y_i errors for all i. The orthogonality condition for $\mathcal{E}$ and C says that if X_i and Z_i are correctable errors, then for all i, the following four expressions are zero:

$$\langle c_1 | X_i^\dagger Z_i | c_2 \rangle = \langle c_1 | Z_i^\dagger X_i | c_2 \rangle = \langle c_1 | I Z_i | c_2 \rangle = \langle c_1 | I X_i | c_2 \rangle = 0.$$

To show that the Y_i are compatible correctable errors, it suffices to show that for all i and j and for all orthonormal $|c_1\rangle \neq |c_2\rangle \in C$,

$$\langle c_1 | X_j^\dagger Y_i | c_2 \rangle = 0$$

$$\langle c_1 | Z_j^\dagger Y_i | c_2 \rangle = 0$$

$$\langle c_1 | I Y_i | c_2 \rangle = 0,$$

and for all $j \neq i$

$$\langle c_1 | Y_j^\dagger Y_i | c_2 \rangle = 0.$$

These equalities follow immediately from multiplication in the Pauli group. For example, because

$$X_i^\dagger Y_i = -X_i^\dagger X_i Z_i = -I Z_i,$$

$$\langle c_1 | X_i^\dagger Y_i | c_2 \rangle = -\langle c_1 | I Z_i | c_2 \rangle = 0$$

Thus, any code that corrects all bit-flip errors X and all phase-flip errors Z also corrects all Y errors.

Let t be the maximum weight for which the set of Pauli group elements of weight t or less satisfies correctable error set condition (equation. 11.2). Any nondegenerate $[[n, k]]$-quantum code cannot correct errors of more than weight t. Section 11.2.4 showed that the maximum number of elements in a correctable set for a nondegenerate code is 2^{n-k}. The number of elements of weight t is $3^t \binom{n}{t}$. Thus, any nondegenerate code that corrects all errors with weight t or less must satisfy the quantum Hamming bound

$$\sum_{i=0}^{t} 3^i \binom{n}{i} \leq 2^{n-k}.$$

A nondegenerate code that obtains equality in the quantum Hamming bound is called a *perfect code*. The classical Hamming Bound is discussed in box 11.3. The quantum Hamming bound does not apply to degenerate codes. All classical codes satisfy the classical Hamming bound. That the quantum Hamming bound does not apply to all codes provides an example of how the existence of degenerate codes complicates the quantum picture.

Just as in the classical case, the term *perfect* should not be taken to imply that perfect codes are necessarily the best ones to use in practice. The quantum Hamming bound quantifies the best trade-off in terms of code expansion (ratio of size of encoded state to original message state) and the strength of the error correction in terms of the number of single-qubit errors the code can correct. A third quantity is also of great practical interest: the efficiency with which errors can be detected. There are many codes that come close to the quantum Hamming bound but that do not have efficient error detection schemes, as measured in terms of the number of gates needed for syndrome extraction and the number of qubits that need to be measured. Both in the quantum and the classical cases, significant structure must be in place in order for efficient error detection schemes to be possible. The design of classical, as well as quantum, error correction schemes with efficient error detection and good trade-offs between data expansion and strength is a continuing area of research. Stabilizer codes provide this structure; for this reason, nearly all

Box 11.3
The Classical Hamming Bound

For any [n, k, d]-code, there are $\binom{n}{t}$ weight t errors, so the cardinality of $E_t(c)$ is

$$|E_t(c)| = \sum_{i=0}^{t} \binom{n}{i}.$$

Since there are 2^k codewords, the sets $E_t(c)$ can be disjoint only if $|E_t(c)| 2^k \leq 2^n$. Thus, any [n, k] code that corrects all errors of weight t or less must satisfy the following bound:

$$\sum_{i=0}^{t} \binom{n}{i} \leq 2^{n-k}.$$

This condition is called the (classical) *Hamming bound*. A code for which equality holds is called a *perfect code*, since it uses the minimum size n to encode k-bit message words in such a way that all weight t errors can be corrected. This bound on t is independent of d.

quantum error correction codes are stabilizer codes. CSS codes, a subset of stabilizer codes, have the advantage that they can be built from pairs of classical codes that are related to each other in a special way.

11.3 CSS Codes

Shor's code encodes a single qubit into three qubits to correct bit-flip errors and then re-encodes the resulting logical qubits to correct phase-flip errors. Recall from section 11.1.2 that $X_i = H Z_i H$, so bit-flip errors X_i are closely related to phase-flip errors Z_i; bit-flip errors in the standard basis $\{|0\rangle, |1\rangle\}$ are phase-flip errors in the Hadamard basis $\{|+\rangle, |-\rangle\}$ and vice versa. Calderbank and Shor, and separately Steane, recognized that by using this relation they could construct quantum codes from pairs of classical codes that satisfy a certain duality relation. These codes, called CSS codes after their founders, have a number of advantages. For example, by encoding only once to correct both phase- and bit-flip errors, the number of qubits required to correct t qubit errors can be reduced: the most famous CSS code, Steane's [[7, 1]] code requires seven qubits to correct all single qubits, as opposed to the nine qubits needed in Shor's code.

11.3.1 Dual Classical Codes

Two classical codes C_1 and C_2 are *dual* to each other, $C_1 = C_2^{\perp}$, if a generator matrix for one is the transpose of a parity check matrix for the other: $G_1 = P_2^T$. Two sets of words V and W are said to be orthogonal if for all $v \in V$ and $w \in W$, the inner product, $v^{\dagger} w = 0 \bmod 2$, where v and w are viewed as vectors. Let C and $C^{\perp}$ be dual codes with generator matrices and parity check matrices $\{G, P\}$ and $\{G^{\perp}, P^{\perp}\}$, respectively. The codewords $C^{\perp}$ are orthogonal to the codewords of C because $G^{\perp} = P^T$; $v \in C^{\perp}$ and $w \in C$ means that there exist x and y such that $v = G^{\perp} x$ and $w = Gy$, so

$$v^{\dagger} w = (G^{\perp} x)^T Gy = (P^T x)^T Gy = x^T P Gy = 0.$$

Example 11.3.1 The dual code to the [7, 4] Hamming code is the is the [7, 3] code $C^{\perp}$ with generator matrix

$$G^{\perp} = P^T = \begin{pmatrix} 1 & 1 & 1 & 0 & 1 & 0 & 0 \\ 1 & 1 & 0 & 1 & 0 & 1 & 0 \\ 1 & 0 & 1 & 1 & 0 & 0 & 1 \end{pmatrix}^T$$

and parity check matrix

$$P^{\perp} = G^T = \begin{pmatrix} 1 & 1 & 1 & 0 & 1 & 0 & 0 \\ 1 & 1 & 0 & 1 & 0 & 1 & 0 \\ 1 & 0 & 1 & 1 & 0 & 0 & 1 \\ 1 & 1 & 1 & 1 & 1 & 1 & 1 \end{pmatrix}.$$

Since the rows of P are a subset of those of $P^\perp$, it follows that C contains its own dual: $C^\perp \subset C$. The eight codewords of $C^\perp$ are the linear combinations of the columns of $G^\perp$:

$$C^\perp = \{0000000, 1110100, 1101010, 0011110, 1011001, 0101101, 0110011, 1000111\}.$$

The sixteen codewords of C are those of $C^\perp$ plus those obtained by adding 1111111 to all of the codewords of $C^\perp$.

For any $[n, k]$ classical code C,

$$\sum_{c \in C}(-1)^{c \cdot x} = \begin{cases} 2^k & \text{if } x \in C^\perp \\ 0 & \text{otherwise.} \end{cases} \tag{11.7}$$

This identity may be established by relating it to the identity

$$\sum_{y=0}^{N-1}(-1)^{y \cdot x} = \begin{cases} 0 & \text{for } x \neq 0 \\ N = 2^n & \text{for } x = 0 \end{cases}$$

from box 7.1. Because $x \cdot Gy = G^T x \cdot y$, the inner product of the two n-bit strings x and Gy is equal to the inner product of the two k-bit strings $G^T x$ and y, so

$$\sum_{c \in C}(-1)^{c \cdot x} = \sum_{y=0}^{2^k-1}(-1)^{Gy \cdot x}$$

$$= \sum_{y=0}^{2^k-1}(-1)^{y \cdot G^T x}$$

$$= \begin{cases} 2^k & \text{if } G^T x = 0 \\ 0 & \text{otherwise} \end{cases}$$

Identity 11.7 follows, since $G^T x = P^\perp x = 0$ precisely when $x \in C^\perp$.

11.3.2 Construction of CSS Codes from Classical Codes Satisfying a Duality Condition

Identity 11.7 enables the construction of states that are superpositions of codewords from a classical code C when viewed in the standard basis and are superpositions of dual codewords $w \in C^\perp$ when viewed in the Hadamard basis. More precisely, we construct states $|\psi_g\rangle$ that are superpositions of codewords from C, and show that they have amplitude only in the states $|h_i\rangle$ where $i \in C^\perp$ and the $|h_i\rangle$ are elements of the n-qubit Hadamard basis:

$$|h_i\rangle = W|i\rangle = H \otimes \cdots \otimes H|i\rangle.$$

After constructing these states, this section shows how this property enables the correction of both phase-flip and bit-flip errors.

Box 11.4
Quotient Groups

If $H < G$ and the *conjugacy condition*, $gHg^{-1} = H$, holds for all elements of $g \in G$, then the cosets of H form a group, called a *quotient group* G/H of the group G. Let us be more precise. Let $S = g_1 H$ and $T = g_2 H$ be cosets of H in G. Then $ST = g_1 H g_2 H = g_1 g_2 H g_2^{-1} g_2 H = g_1 g_2 H$ is another coset $R = g_1 g_2 H$ of H in G.

Let $f : G \to H$ be a homomorphism. Let K be the set of elements of G that are sent to the identity in H. Then, K is a subgroup of G that satisfies the conjugacy condition, so the cosets of K form a quotient group G/K. If f is onto, the quotient group G/K is isomorphic to H.

Let C_1 and $C_2^{\perp}$ be $[n, k_1]$ and $[n, k_2]$ classical codes respectively, and suppose both codes correct t errors. Furthermore, suppose $C_2^{\perp} \subset C_1$. There are $2^{k_1 - k_2}$ distinct cosets of $C_2^{\perp}$ in C_1; every $c \in C_1$ defines a coset $c \oplus C_2^{\perp} = \{c \oplus c' | c' \in C_2^{\perp}\}$ and $c \oplus C_2^{\perp} = d \oplus C_2^{\perp}$ if and only if $c \oplus d \in C_2^{\perp}$. The set of cosets forms a group, the quotient group $G = C_1 / C_2^{\perp}$. Since $C_1 \equiv \mathbf{Z}_2^{k_1}$ and $C_2^{\perp} \equiv \mathbf{Z}_2^{k_2}$, the quotient group $G \equiv \mathbf{Z}_2^{k_1 - k_2}$. For each element $g \in G$, define a quantum state

$$|\psi_g\rangle = \frac{1}{\sqrt{2^{k_2}}} \sum_{c \in C_2^{\perp}} |c_g \oplus c\rangle,$$

where c_g is any element of C_1 contained in the coset of $C_2^{\perp}$ labeled by g. The $2^{k_1 - k_2}$-dimensional subspace spanned by the $|\psi_g\rangle$ for all $g \in G$ defines a $[[n, k_1 - k_2]]$ quantum code C, the CSS code $CSS(C_1, C_2)$.

This paragraph shows that $|\psi_g\rangle$, when viewed in the Hadamard basis, only has amplitude in the codewords of C_2. The components of $|\psi_g\rangle$ in the Hadamard basis are

$$\langle h_i | \psi_g \rangle | h_i \rangle = \langle i | W | \psi_g \rangle W | i \rangle.$$

Therefore it suffices to show that $W|\psi_g\rangle$ is a superposition of codewords $|c\rangle \in C_2$: recall from section 7.1.1 that

$$W|y\rangle = \frac{1}{\sqrt{N}} \sum_{x=0}^{N-1} (-1)^{y \cdot x} |x\rangle.$$

So

$$W|\psi_g\rangle = \frac{1}{\sqrt{2^{k_2}}} \sum_{c \in C_2^{\perp}} \frac{1}{\sqrt{2^n}} \sum_{x=0}^{N-1} (-1)^{(c_g \oplus c) \cdot x} |x\rangle$$

$$= \frac{1}{\sqrt{2^{n+k_2}}} \sum_{x=0}^{N-1} (-1)^{x \cdot c_g} \sum_{c \in C_2^{\perp}} (-1)^{x \cdot c} |x\rangle$$

$$= \frac{1}{\sqrt{2^{n+k_2}}} \sum_{x \in (C_2^\perp)^\perp} (-1)^{x \cdot c_g} (2^{k_2}) |x\rangle$$

$$= \frac{1}{\sqrt{2^{n-k_2}}} \sum_{x \in C_2} (-1)^{x \cdot c_g} |x\rangle,$$

where line 3 follows from line 2 by identity 11.7.

Now we turn to how the error correction is carried out. Since each $|\psi_g\rangle$ is a linear combination of codewords in C_1, a quantum version of the syndrome for C_1 can be used to correct all t bit-flip errors. More specifically, each row of the parity check matrix P_1 tests whether the sum (mod 2) of the corresponding set of bits is even or odd. If a row of the parity check matrix reads $b = b_{n-1} b_2 \ldots b_1$, then the observable that makes the analogous check on quantum states is $Z^{b_{n-1}} \otimes \cdots \otimes Z^{b_1}$, the operator with a Z in every place the parity check has a 1 and an I every place the parity check has a 0. More generally, for any single-qubit unitary transformation Q, let Q^b be the tensor product $Q^{b_{n-1}} \otimes \cdots \otimes Q^{b_1} \otimes Q^{b_0}$. Let $b \in P$ mean that b appears as a row in P. To realize these observables in terms of single-qubit measurement, each row $b \in P_1$ corresponds to a component of a quantum circuit on $n + 1$ qubits, the n computational qubits plus an ancilla qubit. The component has a C_{not} between the ith qubit and the ancilla wherever the ith entry of the row has a 1.

To see how the code handles phase errors, we first confirm that phase-flip errors become bit-flip errors under W. Let e be the bit string indicating the location of the phase-flip errors. Under this error, $|\psi_g\rangle$ becomes

$$\frac{1}{\sqrt{2^{k_2}}} \sum_{c \in C_2^\perp} (-1)^{e \cdot (c_g \oplus c)} |c_g \oplus c\rangle$$

which, after applying W, becomes

$$\frac{1}{\sqrt{2^{n+k_2}}} \sum_{c \in C_2^\perp} (-1)^{e \cdot (c_g \oplus c)} \sum_{x=0}^{N-1} (-1)^{x \cdot (c_g \oplus c)} |x\rangle$$

$$= \frac{1}{\sqrt{2^{n+k_2}}} \sum_{x=0}^{N-1} (-1)^{(e \oplus x) \cdot c_g} \sum_{c \in C_2^\perp} (-1)^{(e \oplus x) \cdot c} |x\rangle$$

$$= \frac{1}{\sqrt{2^{n-k_2}}} \sum_{x \oplus e \in C_2} (-1)^{(e \oplus x) \cdot c_g} |x\rangle$$

$$= \frac{1}{\sqrt{2^{n-k_2}}} \sum_{y \in C_2} (-1)^{y \cdot c_g} |y \oplus e\rangle.$$

This state differs from $W|\psi_g\rangle$ by exactly the bit-flip error corresponding to the string e.

Since applying W to $|\psi_g\rangle$ yields a linear combination of elements of C_2, and under W phase-flip errors become bit-flip errors, a quantum version of the syndrome for C_2 can be used to correct phase errors. Following the construction for bit-flip error detection, for each row in the parity matrix P_2, construct a component of a quantum circuit that has a C_{not} operator between qubit i and the ancilla qubit if and only if there is a 1 in the ith entry of the row.

A variant of this construction gives an even more direct connection between the quantum syndrome computation and the parity check matrices. The following two circuits, involving a computational qubit and an ancilla, have the same effect on states $|\psi\rangle|0\rangle$ where $|\psi\rangle$ is any single-qubit state.

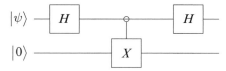

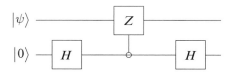

Measuring the ancilla in either case results in the same state with the same probability. So instead of applying the Walsh-Hadamard transformation to the computational qubits and then C_{not} from computational qubits to the ancilla, simply apply a Hadamard gate to the ancilla qubit and use it to control phase flips on the computational qubits. A similar argument implies that to correct bit flips we can apply a Hadamard transformation to the ancilla and use it to control bit flips on the computational qubits.

Studying stabilizer codes, a generalization of CSS codes, will illuminate why correct computational states remain undisturbed by the syndrome computation. Altogether, the CSS code has

$$n - k_1 + n - k_2 = 2n - k_1 - k_2$$

observables, $n - k_1$ that contain only Z and I terms and correct bit-flip errors, and $n - k_2$ that contain only X and I terms and correct phase-flip errors. Instead of constructing CSS codes starting with superpositions of classical codewords, we could have begun the construction with the observables corresponding to the parity check matrices for the codes C_1 and C_2. This approach will be pursued in section 11.4 on stabilizer codes.

11.3.3 The Steane Code

Steane's $[[7, 1]]$ code C is based on the $[7, 4]$ Hamming code. We revisit this code multiple times, first in section 11.4 as an example of stabilizer codes, and then in chapter 12 as the running example illustrating the design of fault-tolerant procedures.

Recall from example 11.3.1 that

$$C^\perp = \{0000000, 1110100, 1101010, 0011110, 1011001, 0101101, 0110011, 1000111\},$$

and that C contains sixteen codewords, those of $C^\perp$ plus those obtained by adding 1111111 to all of the codewords of $C^\perp$. Since C contains its own dual, the conditions for the CSS construction are satisfied by taking $C_1 = C$ and $C_2 = C$. Following the CSS construction,

$$|0\rangle \rightarrow |\tilde{0}\rangle = \frac{1}{\sqrt{8}} \sum_{c \in C^\perp} |c\rangle$$

$$= \frac{1}{\sqrt{8}} (|0000000\rangle + |1110100\rangle + |1101010\rangle + |0011110\rangle +$$

$$|1011001\rangle + |0101101\rangle + |0110011\rangle + |1000111\rangle)$$

and

$$|1\rangle \rightarrow |\tilde{1}\rangle = \frac{1}{\sqrt{8}} \sum_{c \in C,\, c \notin C^\perp} |c\rangle$$

$$= \frac{1}{\sqrt{8}} (|1111111\rangle + |0001011\rangle + |0010101\rangle + |1100001\rangle +$$

$$|0100110\rangle + |1010010\rangle + |1001100\rangle + |0111000\rangle).$$

A syndrome extraction operator U_P for the Steane code is based on a parity check matrix P for the $[7, 4]$ Hamming code,

$$P = \begin{pmatrix} 1 & 1 & 1 & 0 & 1 & 0 & 0 \\ 1 & 1 & 0 & 1 & 0 & 1 & 0 \\ 1 & 0 & 1 & 1 & 0 & 0 & 1 \end{pmatrix}.$$

The six observables for the Steane code are (a circuit for S_1 is shown in figure 11.1):

$$S_1 = Z \otimes Z \otimes Z \otimes I \otimes Z \otimes I \otimes I$$
$$S_2 = Z \otimes Z \otimes I \otimes Z \otimes I \otimes Z \otimes I$$
$$S_3 = Z \otimes I \otimes Z \otimes Z \otimes I \otimes I \otimes Z$$
$$S_4 = X \otimes X \otimes X \otimes I \otimes X \otimes I \otimes I \qquad (11.8)$$
$$S_5 = X \otimes X \otimes I \otimes X \otimes I \otimes X \otimes I$$
$$S_6 = X \otimes I \otimes X \otimes X \otimes I \otimes I \otimes X$$

We postpone discussion of how to compute on the encoded states until after developing stabilizer codes, a general class of codes that contains CSS codes.

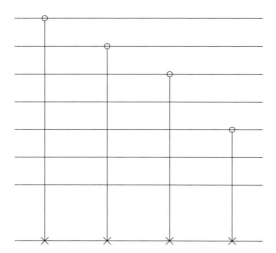

Figure 11.1
One of six component circuits for a syndrome extraction operator U_P for the Steane code.

11.4 Stabilizer Codes

The stabilizer code construction generalizes from the construction of CSS codes described in section 11.3. The construction begins by recognizing that certain $[[n, k]]$ codes, 2^k-dimensional subspaces of a 2^n-dimensional space, can be defined in terms of the set of operators that *stabilize* the subspace.

Example 11.4.1 The Steane code is stabilized by the six observables S_1, S_2, S_3, S_4, S_5, S_6 of equations 11.8; the states $|\tilde{0}\rangle$ and $|\tilde{1}\rangle$ of the Steane code are $+1$-eigenvectors of all six observables.

Section 11.4.1 explains how codes are defined by their stabilizers. It looks at the case of binary observables serving as stabilizers for a code C. All of the observables used in the CSS code construction have only two eigenvalues, -1 and $+1$. Section 11.4.1 uses properties of these observables to determine conditions on a set of correctable errors for C. Section 11.4.2 further restricts from binary observables to elements of the generalized Pauli group, yielding more specific conditions on a set of correctable errors. This setup prepares for a full development of stabilizer code error correction in section 11.4.3. Section 11.4.4 explains how computation is done on the logical qubits of a stabilizer code using a new code, the $[[5, 1]]$ stabilizer code, as a running example.

11.4.1 Binary Observables for Quantum Error Correction

A subspace W of a vector space V is *stabilized* by an operator $S : V \to V$ if for all $|w\rangle \in W$, $S|w\rangle = |w\rangle$. In other words, W is stabilized by S if $|w\rangle$ is a $+1$-eigenstate of S for all $|w\rangle \in W$.

The *stabilizer* of a subspace $W \subset V$ is the set of all operators that stabilize W. Let S be the set of all binary observables on V with only $+1$ and -1 as eigenvalues. Any set of observables $\{S_i\} \subset S$ defines a subspace C, the largest subspace stabilized by all elements S_i. Sometimes the code C is an attractive quantum error correcting code, while in other cases C is not. For example, for some sets of observables, C is simply the zero vector. Our next task is to understand for what sets of observables we get an interesting code by learning how to determine a correctable set of errors for C from the set of observables that defines C.

Suppose S stabilizes $|v\rangle$ and T anticommutes with S; in other words, $ST = -TS$. Then,

$$ST|v\rangle = -TS|v\rangle = -T|v\rangle,$$

so $T|v\rangle$ is a -1-eigenvector for S. If a code C is stabilized by S, then for all $|v\rangle \in C$, the state $T|v\rangle$ cannot be a codeword of C. That $T|v\rangle$ is not a codeword can be detected by measurement with S. This fact enables us to express the condition on a set $\mathcal{E}$ of unitary errors, equation 11.2 of section 11.2.4

$$\langle c_a | E_i^\dagger E_j | c_b \rangle = m_{ij} \delta_{ab}, \tag{11.9}$$

in terms of the set of stabilizers. Let C be the code defined defined by the r stabilizers $S_1, \ldots, S_r$. Suppose that for all pairs E_i and E_j of distinct elements, either $E_i^\dagger E_j$ stabilizes C or there is at least one S_l that anticommutes with $E_i^\dagger E_j$. The next paragraph shows that such a $\mathcal{E}$ is a correctable set of errors for C.

Box 11.5
Stabilizers and Groups Acting on Sets

A group G *acts* on a set S if for all elements $g, g_1, g_2 \in G$ and element $s \in S$

- it is meaningful to talk about applying g to s to obtain another element gs of S,
- the identity e of G takes any s to itself, $es = s$, and
- $(g_1 g_2)s = g_1(g_2 s)$.

A group may act on a set in many different ways, so it is important to define which action is being talked about. We give some examples.

- For any $H < G$, the group G acts on the set of cosets of H in a canonical way: an element $g_1 \in G$ acts on a coset $g_2 H$ taking it to the coset $(g_1 g_2)H$.
- The group of unitary operators $U : V \to V$ acts on the vector space V, viewed as a set, by sending $|v\rangle \in V$ to the vector $U|v\rangle$.

For any $s \in S$, the set of group elements that *stabilize* s is a subgroup,

$$H_s = \{g \in G | gs = s\},$$

called the *stabilizer* of s.

If $E_i^\dagger E_j$ stabilizes C,

$$\langle c_a | E_i^\dagger E_j | c_b \rangle = \langle c_a | c_b \rangle = \delta_{ab}.$$

On the other hand, suppose $E_i^\dagger E_j$ anticommutes with a stabilizer S_l. Then

$$\langle c_a | E_i^\dagger E_j | c_b \rangle = \langle c_a | E_i^\dagger E_j S_l | c_b \rangle = -\langle c_a | S_l E_i^\dagger E_j | c_b \rangle = -\langle c_a | E_i^\dagger E_j | c_b \rangle,$$

from which it follows that

$$\langle c_a | E_i^\dagger E_j | c_b \rangle = 0.$$

Since the E_i are unitary,

$$\langle c_a | E_i^\dagger E_i | c_b \rangle = \delta_{ab}$$

for all i, a, and b. These equations show that $\mathcal{E}$ satisfies the quantum error condition, equation 11.2. If, for some i and j, the transformation $E_i^\dagger E_j$ stabilizes C, the code C is *degenerate* with respect to $\mathcal{E}$. Otherwise, if for all $i \neq j$ each $E_i^\dagger E_j$ anticommutes with at least one S_l, then the code C is *nondegenerate*.

11.4.2 Pauli Observables for Quantum Error Correction

The observations of section 11.4.1 suggest a general mechanism for constructing a code C with correctable error set $\mathcal{E}$ from a set of operators satisfying certain relations. Because of the generalized Pauli group's commutation relations, it is relatively easy to find sets of generalized Pauli operators satisfying these relations. Because $Y^\dagger = -Y$, $X^\dagger = X$ and $Z^\dagger = Z$, any element of the generalized Pauli group $\mathcal{G}_n$ that contains an even number of Y terms and arbitrarily many X and Z terms is Hermitian, and so can be viewed as an observable.

Let S be an Abelian subgroup of $\mathcal{G}_n$ that does not contain $-I$. All elements of the generalized Pauli group $\mathcal{G}_n$ square to either $\pm I$. Since S is a subgroup that does not contain $-I$, all elements of S square to I which means they can only have ± 1 as eigenvalues. Because S is an Abelian group in which all elements square to the identity, S must be isomorphic to $\mathbf{Z}_2^k$ for some k. Let $S_1, \ldots, S_r$ be generators for S. Let C be the subspace stabilized by S:

$$C = \{ |v\rangle \in V \,|\, S_a |v\rangle = |v\rangle, \, \forall S_a \in S \}.$$

The next paragraph shows that C has dimension 2^{n-r}.

Let C_i be the subspace stabilized by the first i stabilizers:

$$C_i = \{ |v\rangle \in V \,|\, S_j |v\rangle = |v\rangle, \, \forall 0 < j \leq i \}.$$

Because all nonidentity elements S_α of $\mathcal{G}_n$ have trace 0, and $+1$ and -1 are the only eigenvalues, the $+1$ eigenspace of S_α must have half the dimension of V. Thus, the subspace C_1 stabilized by S_1 must have dimension half that of V: the subspace C_1 has dimension 2^{n-1}. For all i, the operator $P_i = \frac{1}{2}(I + S_i)$ is a projector onto the $+1$-eigenspace of S_i, so $C_1 = P_1 V$. Since $S_2 P_1 = \frac{1}{2}(I + S_1) S_2$ has trace zero, exactly half of C_1 is in the $+1$-eigenspace of S_2. Thus C_2

has dimension 2^{n-2}. Since $C = C_r$, induction yields dim $C = 2^{n-r}$. To find C explicitly, for any element $S_\alpha \in S$, the set of elements $\{S_\alpha S_\beta | S_\beta \in S\} = S$ because S is a group. Thus,

$$\frac{1}{\sqrt{|S|}} \sum_{S_\alpha \in S} S_\alpha |\psi\rangle$$

is stabilized by S, where $|\psi\rangle$ is any n-qubit state.

Let $\mathcal{E} \in \mathcal{G}_n$ be a set of errors $\{E_i\}$ such that for all i and j, either $E_i^\dagger E_j$ is in the stabilizer of S or anticommutes with at least one element of S. In other words,

$$E_i^\dagger E_j \notin Z(S) - S,$$

where $Z(S)$ is the centralizer of S, the subgroup of $\mathcal{G}_n$ that contains elements that commute with all elements of S. As per section 11.4.1, if $E_i^\dagger E_j$ stabilizes C, then $\langle c_a | E_i^\dagger E_j | c_b \rangle = \delta_{ab}$, and if $E_i^\dagger E_j$ anticommutes with an S_l, then $\langle c_a | E_i^\dagger E_j | c_b \rangle = 0$ unless $i = j$ and $a = b$. Thus, any $\mathcal{E}$ such that all $E_i, E_j \in \mathcal{E}$ satisfy $E_i^\dagger E_j \notin Z(S) - S$ is a correctable set of errors for code C. Of particular interest is the maximal t such that all errors E_i and E_j on t or fewer qubits satisfy $E_i^\dagger E_j \notin Z(S) - S$.

The distance d of a stabilizer code is the minimum weight of an element in $Z(S) - S$. A $[[n, k, d]]$ quantum code represents k-qubit message words in n-qubit codewords and has distance d. We use double brackets to distinguish quantum from classical codes. An $[[n, k, d]]$-quantum code is able to correct all errors of weight t or less if $d \geq 2t + 1$.

11.4.3 Diagnosing and Correcting Errors

Let C be a stabilizer code with stabilizers S given in terms of an independent generating set $S_1, \ldots, S_r$. Because S is Abelian, measurements by different S_i do not affect each other; the probability that a state $|v\rangle \in V$ is measured and determined to be in the -1-eigenspace of S_i is the same no matter what other S_j have been measured before. Measurement of all r observables S_i distinguishes 2^r subspaces $\{V_e\}$ of V, each of dimension 2^{n-r}. Each subspace has a unique signature e, a length r bit-string whose i^{th} bit e_i indicates whether V_e is in the $+1$ or -1-eigenspace of S_i:

$$V_e = \bigcap_i (-1)^{e_i}\text{-eigenspace of } S_i.$$

Any error $E \in \mathcal{G}_n$ either commutes or anticommutes with each S_i. The discussion of stabilizer codes started with the observation that for any $|v\rangle$ stabilized by S_i, the state $E|v\rangle$ is in the $+1$-eigenspace of S_i if E and S_i commute, and in the -1-eigenspace of S_i if they anticommute. Since both EC and V_e have dimension 2^{n-r}, the subspace $EC = V_e$ for some e. Recall from section 11.4.1 that if $\mathcal{E}$ is a correctable error set for code C, then for all E_i and E_j in $\mathcal{E}$, either $E_i^\dagger E_j$ anticommutes with S or is in S. If $E_i^\dagger E_j$ anticommutes with S, then $E_i C$ and $E_j C$ are orthogonal subspaces. If $E_i^\dagger E_j$ is in S, then $E_i |v\rangle = E_j |v\rangle$ for all $|v\rangle \in C$, and $E_i C = E_j C$. In the first case, measurement by the r observables S_i distinguishes $E_i C$ from $E_j C$. In the second case, while the measurement cannot determine whether error E_i or error E_j has occurred, it is not necessary to know; applying either $E_i^\dagger$ or $E_j^\dagger$ returns the state to the correct original. Every E_i in $\mathcal{E}$ is associated

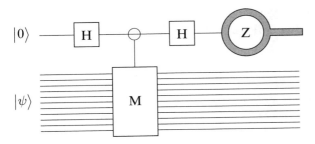

Figure 11.2
Indirect measurement for operators M that are both Hermitian and unitary. Measurement of the ancilla qubit in the standard basis yields the same state on the quantum register with the same probability as would have been obtained by direct measurement of the register with M.

with a unique signature e. When the S_i are measured and an r bit string e is obtained, applying $E_i^\dagger$ for any of the E_i with signature e will return the state to the correct one no matter which error $E_j \in \mathcal{E}$ occurred.

Let M be any Hermitian unitary operator on n qubits. Because M is both Hermitian and unitary, an indirect measurement may be performed with an additional ancilla qubit and the circuit of figure 11.2, where a gray circle means "measure according to the Hermitian operator encircled." In this case, measure the ancilla qubit with operator Z, a measurement in the standard basis. The remainder of this section explains how this circuit achieves an indirect measurement according to M.

Because M is both unitary and Hermitian, its only possible eigenvalues are $+1$ and -1. The circuit uses the fact that, for any $|\psi\rangle$, the state $c(|\psi\rangle + M|\psi\rangle)$ is a $+1$-eigenvector of M, and the state $c'(|\psi\rangle - M|\psi\rangle)$ is a -1-eigenvector of M, where c and c' are the normalization factors $c = 1/\||\psi\rangle + M|\psi\rangle\|$ and $c' = 1/\||\psi\rangle - M|\psi\rangle\|$: if we write $|\psi\rangle$ in terms of eigenstates for M, we see that the -1-eigenvectors cancel in $|\psi\rangle + M|\psi\rangle$, leaving the $+1$-eigenvectors. Let P^+ be the projection onto the $+1$-eigenspace of M, so $\frac{1}{2}(|\psi\rangle + M|\psi\rangle) = P^+|\psi\rangle$, and P^- the projection onto the -1-eigenspace of M, so $\frac{1}{2}(|\psi\rangle - M|\psi\rangle) = P^-|\psi\rangle$. Direct measurement of $|\psi\rangle$ according to M yields $\frac{P^+|\psi\rangle}{|P^+|\psi\rangle|}$ with probability $\langle\psi|P^+|\psi\rangle$ and $\frac{P^-|\psi\rangle}{|P^-|\psi\rangle|}$ with probability $\langle\psi|P^-|\psi\rangle$.

This paragraph shows that the circuit of figure 11.2 yields these same states with the same probability. Prior to measurement, the state is

$$\frac{1}{\sqrt{2}}(|+\rangle|\psi\rangle + |-\rangle M|\psi\rangle) = \frac{1}{2}((|0\rangle + |1\rangle)|\psi\rangle + (|0\rangle - |1\rangle)M|\psi\rangle)$$

$$= \frac{1}{2}(|0\rangle(|\psi\rangle + M|\psi\rangle) + |1\rangle(|\psi\rangle - M|\psi\rangle))$$

$$= \frac{1}{2}(\frac{1}{c}|0\rangle(c(|\psi\rangle + M|\psi\rangle)) + \frac{1}{c'}|1\rangle(c'(|\psi\rangle - M|\psi\rangle))).$$

Measurement of the ancilla qubit with Z yields 0 with probability $p_+ = \langle\psi|P^+|\psi\rangle$ and results in the n-qubit state $c(|\psi\rangle + M|\psi\rangle)$. Similarly, the measurement yields 1 with probability $p_- =$

$\langle\psi|P^-|\psi\rangle$ resulting in $c'(|\psi\rangle - M|\psi\rangle)$ as the state of the n qubit register. As an alternative argument, since $M = P^+ - P^-$ and $I = P^+ + P^-$,

$$c(|\psi\rangle + M|\psi\rangle) = c((P^+ + P^-)|\psi\rangle + (P^+ - P^-)|\psi\rangle) = cP^+|\psi\rangle.$$

Thus, the circuit of figure 11.2 has the same effect on the n computational qubits as a direct measurement by M. To measure each of the S_i for $1 \le i \le m$, we need m such circuits and m ancilla qubits. Measurement of these qubits yields the string e.

11.4.4 Computing on Encoded Stabilizer States

For stabilizer codes, certain operators U, including those in the Pauli group, have logically equivalent operators $\tilde{U}$ that are simple to obtain. Chapter 12 shows that stabilizer codes also have good error handling properties. Again we work backward; instead of defining an encoding function and then finding logical operators for that encoding, we find candidate logical operators and then use them to define the encoding. In particular, we find logical single-qubit Pauli operators $\tilde{Z}_1, \ldots, \tilde{Z}_k$ and use them to define an encoding function.

In general, a state can be defined in terms of operators for which it is an eigenstate; for example, all of the standard basis elements for a k-qubit system can be defined in terms of whether they are $+1$ of -1-eigenvectors for each of the operators $Z_1, \ldots, Z_k$. As another example, section 10.2.4 defined cluster states in terms of the operators that stabilize them. Here, the encoding function takes any standard basis vector $|b_1 \ldots b_k\rangle$ to the unique state in the code C that is a $(-1)^{b_i}$-eigenstate of $\tilde{Z}_i$ for all i. The rest of this section describes this program in more detail, developing the five-qubit code as a running example.

Example 11.4.2 The set of observables

$S_0 = X \otimes Z \otimes Z \otimes X \otimes I$

$S_1 = Z \otimes Z \otimes X \otimes I \otimes X$

$S_2 = Z \otimes X \otimes I \otimes X \otimes Z$

$S_3 = X \otimes I \otimes X \otimes Z \otimes Z,$

defines a $[[5, 1]]$ code. The four observables are independent, so each of the four observables divides the 2^5-dimensional code space into two eigenspaces, leaving a space of $2^5/2^4 = 2^1$ dimensions of codewords. This code satisfies the quantum Hamming bound, and so is a perfect code.

Consider an element $A \in Z(S)$. For any $|v\rangle \in C$, the state $A|v\rangle$ is also in C:

$$S_i A|v\rangle = A S_i|v\rangle = A|v\rangle,$$

so $A|v\rangle$ is a $+1$-eigenstate of all the S_i. If A is in S, then $A|v\rangle = |v\rangle$, but for all A in $Z(S) - S$, A acts nontrivially on C. If $A_1 = A_2 S_a$ for some $S_a \in S$, then A_1 and A_2 behave in the same way on C. All the elements of the quotient group $Z(S)/S$ act in distinct ways on C. To understand how they act on C, we need to know more about the structure of the centralizer $Z(S)$.

The symplectic view of $\mathcal{G}_n$ illuminates the structure of $Z(S)$. Recall from section 11.2.11 that any element of $\mathcal{G}_n$ can be written uniquely as

$$\mu(X^{a_1} \otimes \cdots \otimes X^{a_n})(Z^{b_1} \otimes \cdots \otimes Z^{b_n}),$$

so to each element of $\mathcal{G}_n$ there is an associated $2n$-bit string

$$(a|b) = a_1 \ldots a_n b_1 \ldots b_n.$$

Moreover $h : \mathcal{G}_n \to \mathbf{Z}_2^{2n}$ is a group homomorphism where

$$(a|b) \cdot (a'|b') = (a \oplus a'|b \oplus b').$$

The homomorphism h is four-to-one and loses the phase information contained in μ. Since S does not contain $-I$, and therefore does not contain $\mathbf{i}I$ or $-\mathbf{i}I$, no two elements of S map to the same string, so on S the homomorphism h is one-to-one. The elements $S_1, \ldots, S_r$ of $\mathcal{G}_n$ are independent, meaning that none of them can be written as a product of the others, if and only if the corresponding bit strings $(a|b)$ are linearly independent.

Two elements g and g' of $\mathcal{G}_n$ commute if and only if

$$ab' + a'b = 0 \bmod 2,$$

where ab' is the the usual inner product, the sum of bitwise multiplication of the corresponding bits, and $(a|b) = h(g)$ and $(a'|b') = h(g')$. The expression $ab' + a'b \bmod 2$ is called the *symplectic inner product*.

Example 11.4.3 The stabilizer group generated by the four observables of example 11.4.2 has sixteen elements S_α:

$$
\begin{aligned}
S_i S_i &= I = I \otimes I \otimes I \otimes I \otimes I & S_0 &= X \otimes Z \otimes Z \otimes X \otimes I \\
S_1 &= Z \otimes Z \otimes X \otimes I \otimes X & S_2 &= Z \otimes X \otimes I \otimes X \otimes Z \\
S_3 &= X \otimes I \otimes X \otimes Z \otimes Z & S_0 S_1 &= -Y \otimes I \otimes Y \otimes X \otimes X \\
S_0 S_2 &= -Y \otimes Y \otimes Z \otimes I \otimes Z & S_0 S_3 &= -I \otimes Z \otimes Y \otimes Y \otimes Z \\
S_1 S_2 &= -I \otimes Y \otimes X \otimes X \otimes Y & S_1 S_3 &= -Y \otimes Z \otimes I \otimes Z \otimes Y \\
S_2 S_3 &= -Y \otimes X \otimes X \otimes Y \otimes I & S_0 S_1 S_2 &= -X \otimes X \otimes Y \otimes I \otimes Y \\
S_0 S_1 S_3 &= -Z \otimes I \otimes Z \otimes Y \otimes Y & S_0 S_2 S_3 &= -Z \otimes Y \otimes Y \otimes Z \otimes I \\
S_1 S_2 S_3 &= -X \otimes Y \otimes I \otimes Y \otimes X & S_0 S_1 S_2 S_3 &= I \otimes X \otimes Z \otimes Z \otimes X.
\end{aligned}
$$

The following table shows the error syndrome for the code defined by these observables. For single-qubit errors X, Y, and Z on qubits 0 through 4, the corresponding column shows the result of measurement with observable S_i after that single error occurred on a codeword. The $+$ and $-$ indicate whether the result of measurement with S_i on that qubit is $+1$ and -1 respectively. The results of the four measurements identify the error uniquely. Counting $+$ as 0 and $-$ as 1, the last row shows a unique decimal value, coming from measurement of all four observables.

	bit 0			bit 1			bit 2			bit 3			bit 4		
	X	Z	Y	X	Z	Y	X	Z	Y	X	Z	Y	X	Z	Y
S_0	+	−	−	−	+	−	−	+	−	+	−	−	+	+	+
S_1	−	+	−	−	+	−	+	−	−	+	+	+	+	−	−
S_2	−	+	−	+	−	−	+	+	+	+	−	−	−	+	−
S_3	+	−	−	+	+	+	+	−	−	−	+	−	−	+	−
	6	9	15	3	4	7	1	10	11	8	5	13	12	2	14

Let $S_1, \ldots, S_r$ be an independent generating set for S. Form the $r \times 2n$ binary matrix

$$M = \begin{pmatrix} (a|b)_1 \\ (a|b)_2 \\ \vdots \\ (a|b)_r \end{pmatrix}$$

with the $(a|b)_i = h(S_i)$ as the rows. Because the S_i are independent, so are the rows of M, and so M has rank r. The matrix M acts on a $2n$-bit string $(a|b)$, viewed as a column vector $\left(\frac{b}{a}\right)$, to produce a length r vector in which the ith entry is the symplectic inner product of $(a|b)$ with $(a|b)_i$. The matrix M has a kernel of dimension $2n - r$. Elements of this kernel correspond to elements of $\mathcal{G}_n$ that commute with all elements of the stabilizer. Thus, there are $4 \cdot 2^{2n-r}$ elements in $Z(S)$, where the factor of 4 comes from the four possible values of μ; these values of μ will not be relevant for the remaining discussion because elements of S are uniquely determined by the corresponding string $(a|b)$. For an $[[n, k]]$ stabilizer code, the size of the stabilizer subgroup is 2^{n-k}. So for an $[[n, k]]$ code, there are $2^{2n-r} = 2^{n+k}$ elements in $Z(S)$.

Take $\tilde{Z}_1$ to be any element of $Z(S)$ that is independent of $S_1, \ldots, S_r$. Form the $(r+1) \times 2n$ binary matrix M_1 by adding as an additional row to M the $2n$-bit string corresponding to $\tilde{Z}_1$. The matrix M_1 has full rank $r + 1$. Let C_1 be the size $2^{2n-(r+1)} = 2^{n+k-1}$ set of binary strings, viewed as column vectors $\left(\frac{b}{a}\right)$, that are in the kernel of M_1. Let $\tilde{Z}_2$ be any element of $Z(S)$ that corresponds to a bit string in C_1. We can continue this process k times to obtain operators $\tilde{Z}_1, \ldots \tilde{Z}_k$ that commute with each other and with all elements of S. The kernel of M_k will be S.

Consider unencoded standard basis vectors for a moment. The k-qubit state $|00\ldots0\rangle$ is the unique $+1$-eigenstate of the $Z_1, \ldots, Z_k$. More generally, the standard basis vector $|b_1 \ldots b_k\rangle$ is the unique state that is a $(-1)^{b_i}$-eigenstate of Z_i for all i. For any k-bit string $b_1 \ldots b_k$, there is a unique element of the code C that is a $(-1)^{b_i}$-eigenstate of Z_i for all i; the argument that there is a unique element is similar to the argument that established the dimension of C. We define an encoding function U_C for the code C that takes standard basis elements to elements of C that have analogous eigenstate relations with the logical versions $\tilde{Z}_i$ of the Z_i. A k-qubit state $\sum_{x=0}^{2^k-1} a_x |x\rangle$ is encoded as follows:

$$U_C : \sum_{x=0}^{2^k-1} a_x |x\rangle \rightarrow \sum_{x=0}^{2^k-1} a_x |\tilde{x}\rangle \tag{11.10}$$

where $|\tilde{x}\rangle$ is the unique element of C that is in the $(-1)^{x_i}$-eigenspace of $\tilde{Z}_i$ for all $0 \le i \le k$.

The $(r + k) \times 2n = n \times 2n$ matrix M_k has full rank; therefore for any i, there is a bit string $(a|b)$ that, when viewed as a column vector $\left(\frac{b}{a}\right)$, yields the n-bit string e_i which has a 1 in the ith place and 0 elsewhere: in particular, there is a $2n$-bit string $(a|b)$ that satisfies

$$M_k \left(\frac{b}{a}\right) = e_1.$$

Let $\tilde{X}_1$ be the element of $Z(S)$ with bit string $(a|b)$ that yields e_1 when multiplied by M_k. Construct M_{k+1} by adding as a row to M_k the bit string corresponding to $\tilde{X}_1$. Let $\tilde{X}_2$ be such that its bit string $(a|b)$ satisfies

$$M_{k+1} \left(\frac{b}{a}\right) = e_2.$$

We can continue in this way until we obtain $\tilde{X}_1, \ldots, \tilde{X}_k$. By construction, $\tilde{X}_i$ anticommutes with $\tilde{Z}_i$, and commutes with all of S, all $\tilde{X}_j$, and all the $\tilde{Z}_j$ for $j \ne i$.

Example 11.4.4 For the $[[5, 1]]$ code of example 11.4.2, the binary matrix corresponding to the independent generating set $\{S_i\}$ is

$$M = \begin{pmatrix} 1 & 0 & 0 & 1 & 0 & 0 & 1 & 1 & 0 & 0 \\ 0 & 0 & 1 & 0 & 1 & 1 & 1 & 0 & 0 & 0 \\ 0 & 1 & 0 & 1 & 0 & 1 & 0 & 0 & 0 & 1 \\ 1 & 0 & 1 & 0 & 0 & 0 & 0 & 0 & 1 & 1 \end{pmatrix}.$$

The bit string $(a|b) = (11111|00000)$ is independent of the rows $m \in M$ and satisfies $Mb = 0$, so we may take

$$\tilde{Z} = Z \otimes Z \otimes Z \otimes Z \otimes Z$$

and, since $(b|a)$ is orthogonal to $(a|b)$ and all rows of M, we may take

$$\tilde{X} = X \otimes X \otimes X \otimes X \otimes X.$$

Let $|\tilde{e}_i\rangle$ be the unique state in C that is a -1-eigenstate of $\tilde{Z}_i$ but a $+1$-eigenstate for all the $\tilde{Z}_j$ with $j \ne i$. For $j \ne i$,

$$\tilde{Z}_j \tilde{X}_i |\tilde{e}_i\rangle = \tilde{X}_i \tilde{Z}_j |\tilde{e}_i\rangle = \tilde{X}_i |\tilde{e}_i\rangle,$$

so $\tilde{X}_i |\tilde{e}_i\rangle$ is a $+1$-eigenstate of $\tilde{Z}_j$ for $j \ne i$. For $\tilde{Z}_i$,

$$\tilde{Z}_i \tilde{X}_i |\tilde{e}_i\rangle = -\tilde{X}_i \tilde{Z}_i |\tilde{e}_i\rangle = -\tilde{X}_i |\tilde{e}_i\rangle,$$

so $\tilde{X}_i |\tilde{e}_i\rangle$ is in the $+1$-eigenstate of $\tilde{Z}_i$ as well. This calculation suggests that $\tilde{X}_i$ is the logical analog of X_i for C with encoding U_C of equation 11.10. A full proof is straightforward.

Example 11.4.5 For the $[[5, 1]]$ code of example 11.4.2, the $+1$-eigenspace of $\tilde{Z}$ is spanned by the set of standard basis states with an even number of 1s. Thus, we may take $|\tilde{0}\rangle$ to be

$$|\tilde{0}\rangle = \frac{1}{\sqrt{|S|}} \sum_{S_\alpha \in S} S_\alpha |00000\rangle$$

$$= \frac{1}{4}(|00000\rangle + |10010\rangle + |00101\rangle + |01010\rangle$$
$$+ |10100\rangle - |10111\rangle - |11000\rangle - |00110\rangle$$
$$- |01111\rangle - |10001\rangle - |11110\rangle - |11101\rangle$$
$$- |00011\rangle - |01100\rangle - |11011\rangle + |01001\rangle),$$

and $|\tilde{1}\rangle$ to be

$$|\tilde{1}\rangle = \frac{1}{\sqrt{|S|}} \sum_{S_\alpha \in S} S_\alpha |11111\rangle,$$

a superposition of all basis vectors with an odd number of ones.

The construction of logical versions of other single-qubit gates and multiple-qubit gates for a stabilizer code C is more complicated. Chapter 12 looks at this issue in more detail and provides constructions for a universal approximating set of logical gates for the Steane code.

11.5 CSS Codes as Stabilizer Codes

Let C_1 and C_2 be $[n, k_1]$ and $[n, k_2]$ classical codes respectively, and suppose both correct t errors. Furthermore, suppose $C_2^\perp \subset C_1$. These codes satisfy the condition required for the construction of an $[[n, k_1 - k_2]]$ CSS code. This section describes an alternative to the CSS code construction of section 11.3, one that uses the stabilizer viewpoint.

Let P_1 (resp. P_2) be the parity check matrix for the code C_1 (resp. C_2). For each row of P_1, viewed as a bit string $b = b_1 \ldots b_n$, construct an observable

$$X^b = X^{b_1} \otimes \cdots \otimes X^{b_n}.$$

These $n - k_1$ observables are independent, since the rows of P_1 are linearly independent. For each row of P_2, also construct an observable

$$Z^b = Z^{b_1} \otimes \cdots \otimes Z^{b_n}.$$

These $n - k_2$ observables are also independent, and, since X and Z are independent, the entire set of $2n - k_1 - k_2$ observables is independent. The group S generated by these observables does not contain $-I$, so S defines a stabilizer code if and only if it is Abelian.

This paragraph shows that the CSS condition implies that S is Abelian. Since X and I commute, all elements of $\{X^a | a \in P_1\}$ commute. Similarly, all elements of $\{Z^b | b \in P_2\}$ commute. Since X and Z anticommute, the group elements X^a and Z^b commute if and only if $a \cdot b$ is even. So the elements of S commute if $ab^T = 0 \bmod 2$ for all rows a of P_1 and rows b of P_2. This equality holds if $P_1 P_2^T = 0 \bmod 2$. Because $C_2^\perp \subset C_1$, the generator matrix $G_2^\perp$ satisfies $P_1 G_2^\perp = 0$. Since $G_2^\perp = P_2^T$, we have $P_1 P_2^T = 0$. Therefore, S is Abelian and the subspace C, stabilized by S, is a stabilizer code.

The code $CSS(C_1, C_2)$ of section 11.3.2 is stabilized by S. Since S has $n - k_1 + n - k_2$ independent generators, it stabilizes a subset of dimension

$$n - (2n - k_1 - k_2) = k_1 + k_2 - n.$$

Since $CSS(C_1, C_2)$ has dimension $k_1 + k_2 - n$, it is the stabilizer code for S.

Example 11.5.1 *The Steane code revisited.* The parity check matrix

$$P = \begin{pmatrix} 1 & 1 & 1 & 0 & 1 & 0 & 0 \\ 1 & 1 & 0 & 1 & 0 & 1 & 0 \\ 1 & 0 & 1 & 1 & 0 & 0 & 1 \end{pmatrix}$$

defines the [7, 4] Hamming code. The Steane code takes the [7, 4] Hamming code as both C_1 and C_2. To obtain stabilizers for the Steane code, define an operator in $\mathcal{G}_7$ for each row in the parity check matrix that has a Z in every place a 1 occurs and an I for every 0:

$Z \otimes Z \otimes Z \otimes I \otimes Z \otimes I \otimes I$
$Z \otimes Z \otimes I \otimes Z \otimes I \otimes Z \otimes I$
$Z \otimes I \otimes Z \otimes Z \otimes I \otimes I \otimes Z.$

For each row in the parity check matrix, also define an operator that has an X wherever a 1 occurs:

$X \otimes X \otimes X \otimes I \otimes X \otimes I \otimes I$
$X \otimes X \otimes I \otimes X \otimes I \otimes X \otimes I$
$X \otimes I \otimes X \otimes X \otimes I \otimes I \otimes X$

These six observables stabilize exactly the Steane code C, so the Steane code is a [[7, 1]] stabilizer code.

11.6 References

Hungerford's *Abstract Algebra: An Introduction* [159] includes a chapter on classical error correction, as well as giving a thorough treatment of the group theory and linear algebra involved. Wicker's *Error Control Systems for Digital Communication and Storage* [283] discusses classical error correction and includes chapters on the relevant algebra.

The nine-qubit code of section 11.1.3 was originally proposed by Shor [251]. The seven-qubit code of section 11.3 was originally proposed by Steane [259]. The theory of stabilizer codes and fault-tolerant implementation of them are discussed at length in Daniel Gottesman's thesis [135].

11.7 Exercises

Exercise 11.1. For the code $C : \mathbf{Z}_2^1 \to \mathbf{Z}_2^3$ defined by generator matrix

$$G = \begin{pmatrix} 1 \\ 1 \\ 1 \end{pmatrix}$$

give

- the set of code words
- two distinct parity check matrices

Exercise 11.2. *Computing a parity check matrix for a code specified by a generator matrix.*
a. Show that adding a column of a generator matrix G for a code C to another column produces an alternative generator matrix G' for the same code C.

b. Show that for any $[n, k]$ code there is a generator matrix of the form $\left(\frac{A}{I} \right)$, where A is a $(n - k) \times k$ matrix and I is the $k \times k$ identity matrix.

c. Show that if $G = \left(\frac{A}{I} \right)$, then the $(n - k) \times k$ matrix $P = (I | A)$, where I is the $(n - k) \times (n - k)$ identity matrix, is a parity check matrix for the code C.

d. Show that if a parity check matrix P' has the form $(A | I)$, then $G' = \left(\frac{I}{A} \right)$ is a generator matrix for the code.

Exercise 11.3. Show that the code C_{PF} of section 11.1.2 corrects all linear combinations of single-qubit phase-flip errors $\{I, Z_2, Z_1, Z_0\}$ on any superposition $a|\tilde{0}\rangle + b|\tilde{1}\rangle$.

Exercise 11.4. Show that the code C_{PF} of section 11.1.2 does not correct bit-flip errors.

Exercise 11.5. Show that if an $[[n, k, d]]$-quantum code is able to correct all errors of weight t or less, $d \geq 2t + 1$.

Exercise 11.6. Show that all Hamming codes have distance 3 and so correct single bit-flip errors.

Exercise 11.7. Show that Shor's code is a degenerate code.

Exercise 11.8. *Alternative Steane code constructions.*

a. Find a parity check matrix of the form $(I | A)$ for the Steane code.

b. Construct an alternative circuit, based on the parity check matrix found in (a), that can serve as a syndrome extraction operator for the Steane code.

Exercise 11.9. Show that the generalized set of Pauli elements forms a basis for the linear transformations on the vector space associated with an n-qubit system.

Exercise 11.10.

a. Show that for all i and j and for all orthonormal $|c_1\rangle \neq |c_2\rangle \in C$,

$$\langle c_1 | Z_j^\dagger Y_i | c_2 \rangle = 0$$

$$\langle c_1 | I Y_i | c_2 \rangle = 0,$$

b. Show that for all $j \neq i$ and for all orthonormal $|c_1\rangle \neq |c_2\rangle \in C$,

$$\langle c_1 | Y_j^\dagger Y_i | c_2 \rangle = 0.$$

Exercise 11.11. Describe how the Shor code can be used to correct single-qubit errors without making any measurements.

Exercise 11.12. Show that if two blocks encoded according to code C are subjected to an error E that is a superposition of errors $E = E_a \otimes E_b + E_c \otimes E_d$, where E_a, E_b, E_c, and E_d are all elements of a correctable set of errors $\mathcal{E}$ for C, then E can be corrected.

Exercise 11.13. Suppose a single qubit $|\psi\rangle = a|0\rangle + b|1\rangle$ has been encoded using the Steane code and that the error $E = \frac{1}{2}X_2 + \frac{\sqrt{3}}{2}Z_3$ has occurred. Write down

a. the encoded state,

b. the state after the error has occurred,

c. for each phase of the error correction, the syndrome and the resulting state, and

d. each error correcting transformation applied and the state after each of these applications.

Exercise 11.14. Show that for a $[[n, k]]$ quantum stabilizer code there is, for any k bit string $b_1 \ldots b_k$, a unique element of the code C that is a $(-1)^{b_i}$-eigenstate of Z_i for all i.

Exercise 11.15. Show that the subspaces V_e of section 11.4.3 are of dimension 2^{n-r}.

Exercise 11.16. Show that the operators $\tilde{X}_i$, as defined for stabilizer codes in section 11.4.4, act as a logical analog of the gates X_i for the logical states obtained from the encoding c.

Exercise 11.17. Show that the $[[9, 1]]$ Shor code is a stabilizer code.

Exercise 11.18. Find alternative ways of implementing operations corresponding to X and Z on the logical qubits of the five-qubit stabilizer code.

Exercise 11.19. Let $[[n, k, d]]$ be any nondegenerate code. Such a code can correct $t = \lfloor \frac{d-1}{2} \rfloor$ errors. Show that tracing any codeword over any $n - t$ qubits results in the totally mixed state $\rho = \frac{1}{2^t} I$ on the remaining t qubits. Thus, all codewords are highly entangled states.

12   Fault Tolerance and Robust Quantum Computing

Quantum error correction by itself is not sufficient for robust quantum computation. Robust computation means that arbitrarily long computations can be run to whatever accuracy desired. The analyses of quantum error correction techniques in chapter 11 supposed that the error correcting procedures were carried out perfectly, an unrealistic assumption. Also, even if the environment interacts with the system only in ways that can be handled by the error correcting code, gates used as part of the computation may propagate errors in ways that produce errors the code cannot correct. To achieve robust quantum computation, quantum error correction must be combined with fault-tolerant techniques.

This chapter presents one approach to robust quantum computation: error correction coupled with fault-tolerant procedures. Other approaches to robust quantum computation exist, both for quantum computation in the standard circuit model and for alternative models of computation. These alternative approaches will be touched on briefly in sections 13.3 and 13.4 respectively. The chapter concludes with a threshold theorem for one error model. Threshold theorems prove that as long as the error rate is below a certain threshold, a quantum computer can run arbitrary long computations to arbitrarily high accuracy. This chapter uses a simple error model sufficient to illustrate a general approach to fault-tolerant quantum computation: the strategy is to replace a circuit with an expanded circuit that is more robust; if the original circuit's chance of failing was $O(p)$, the expanded circuit fails with only probability $O(p^2)$. Given a general method for obtaining such expanded circuits, arbitrary low probabilities of failure can be achieved by concatenation; the expanded circuit can be replaced with a yet larger and more robust circuit, and we can repeat this process, called *concatenated coding*, until the desired level of accuracy is achieved. A key feature of concatenated coding is that only polynomial resources are required to obtain exponentially low probabilities of failure.

Like quantum error correction, fault-tolerant quantum computing is a richly developed field. A variety of approaches have been developed, and threshold theorems for a variety of error models and codes have been proved. Fault-tolerant quantum computation remains an active area of research, and like quantum error correction, it will evolve as quantum information processing devices are built, more realistic error models are learned, and more sophisticated quantum computer architectures are developed.

As in chapter 11, this chapter concentrates on quantum error correcting codes that correct errors of weight t or less. The most important issue in ensuring that a circuit can always be replaced with a more robust expanded circuit is the control of *spread*; if in the course of computation a single error propagates to additional qubits before we are able to correct it, then the probability that our computation becomes subject to an error that we cannot correct is much higher. Fault-tolerant quantum computing methods aim to eliminate the propagation of errors from small numbers of qubits to a large numbers of qubits. All aspects of quantum computation must be made fault-tolerant: error correction, the gates themselves, initial state preparation, and measurement.

The sections of the chapter gradually peel away at assumptions of perfection, adding mechanisms to handle each source of errors, to arrive at a full program of fault-tolerant procedures that support robust quantum computation. The chapter uses Steane's seven-qubit code as a running example. Section 12.1 discusses the setting in which we describe fault-tolerant techniques, including the error model and when error correction steps are applied. Section 12.2 addresses fault-tolerant quantum error correction. Section 12.2 examines the design of a full set of fault-tolerant procedures for performing arbitrary computations on qubits encoded with the Steane code. Section 12.3 describes concatenated coding leading to a threshold result.

12.1   Setting the Stage for Robust Quantum Computation

To simplify the presentation, we consider only $[[k, 1]]$ quantum codes. Given a specific set of universal gates and a specific $[[k, 1]]$ quantum error correcting code, fault-tolerant techniques aim to take any circuit composed of those gates and produce a circuit on encoded qubits such that the probability of a faulty computation is reduced even though more qubits and more operations are involved. Fault-tolerant techniques for a given code address the question of how to implement logical procedures on the computational qubits, syndrome extraction operators, measurements, state preparation, and error correcting transformations in such a way that the resulting computation is more robust than the original one; roughly, if the original circuit failed with probability $O(p)$, then the expanded circuit fails with probability $O(p^2)$. Before describing fault-tolerant techniques, we need to discuss when quantum error correcting operations are applied, and how we will model the errors.

Let Q_0 be a quantum circuit for a computation we wish to make robust. We partition time into intervals in which at most one gate acts on any qubit (figure 12.1). This partitioning is not unique: for the circuit of figure 12.1, the single-qubit gate applied to the first qubit could have been placed in the second time interval instead, or the first time interval could have been split in two with, for example, the single-qubit operations performed in the first interval and the two-qubit operation in the second. Given a partitioned circuit Q_0, we will define an expanded circuit Q_1, in which every qubit expands to a block of k qubits, and each time interval expands into two parts, one in which procedures implementing the logical gates are carried out, and a second in which the syndrome is measured and error correcting transformations are applied (figure 12.2). Both of these procedures

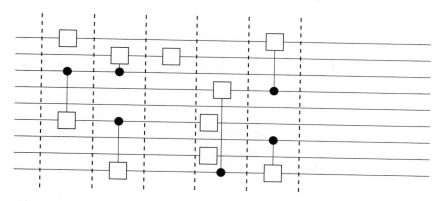

Figure 12.1
The original circuit Q_0 for a computation we wish to make robust partitioned into time intervals in which at most one gate acts on a single qubit.

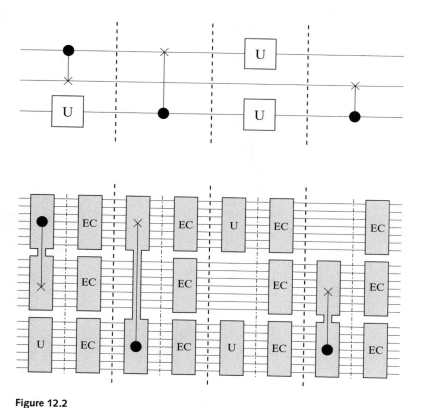

Figure 12.2
Schematic diagram showing the general structure, including the subpartitioned time intervals, for an expanded circuit for a [[7, 1]] quantum code. The expanded circuit alternates between carrying out logical procedures and performing error correction (EC).

may require ancilla qubits. The expanded circuit is then subpartitioned into time intervals in which no qubit is acted on by more than one gate. We could have chosen to apply error correction less often. We have chosen to apply it as often as possible: at the end of each logical procedure. A number of choices remain. Which procedures are used to implement the logical gates, and how the syndrome and error correcting transformations are performed, determine whether the expanded circuit Q_1 is more robust than the original circuit Q_0 or not.

For the purposes of describing fault-tolerant procedures, we use a model in which errors only take place at the beginning of time intervals. We model imperfect single-qubit gates as a single-qubit error followed by a perfect gate. Similarly, we model imperfect C_{not} gates as two single-qubit errors followed by a perfect gate. Correlations between these errors can be ignored because quantum error correction is applied separately to each block and, within our fault-tolerant procedures, we will allow C_{not} transformations only between qubits in different blocks. Errors due to interactions with the environment are modeled as occurring only at the beginning of time intervals. For our initial discussion, we use the local and Markov error model of section 10.4.4. This model means that each qubit, in each time interval, interacts only with its own environment at the beginning of each time step (figure 12.3), and that the state of the environment at the beginning of each time interval is uncorrelated with the state at all previous times. The threshold theorem discussed in section 12.3.2 uses a more general error model.

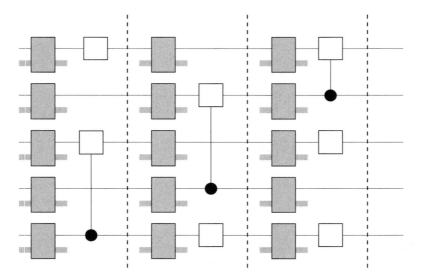

Figure 12.3
Schematic diagram of the error model with environmental interactions at the beginning of each time interval. The boxes representing the environmental interactions show each qubit interacting with its own environment of arbitrary size.

12.2 Fault-Tolerant Computation Using Steane's Code

In medicine, doctors take an oath, *"first, do no harm."* While the quantum error correction methods described in the previous chapter enable the correction of errors during the course of a quantum computation, they also increase the chance of errors, since they require more qubits and more gates. Analysis of quantum error correction in chapter 11 made the unrealistic assumption that the error correction steps were carried out perfectly. In fact, as we will see shortly, if we used the quantum error correction schemes of the last chapter exactly as described, imperfections in the process would likely introduce more errors than the process corrects. These schemes are not *fault-tolerant*. Fortunately, these schemes can be modified so that they do not introduce more errors than they correct. We illustrate fault-tolerant quantum error correction techniques by demonstrating how to make the Steane seven-qubit code fault tolerant. Fault-tolerant techniques put in safeguards so that a single error never propagates to multiple qubits, since multiple errors cannot be corrected by Steane's code. The strategy is to replace parts that fail under a single qubit error with an ensemble of parts that fails only in the presence of two or more errors, so that, if originally a part fails with probability p, the ensemble replacing it fails only with probability cp^2.

Section 12.2.1 illustrates ways in which the quantum error correction techniques of 11.3.3 fail to be fault tolerant. Section 12.2.2 shows how to perform error correction in a fault-tolerant way using the Steane code as the example. Section 12.2.3 develops fault-tolerant logical gates for the Steane code that limit the propagation of errors, and sections 12.2.4 and 12.2.5 deal with fault-tolerant measurement and fault-tolerant state preparation. Together, these procedures make the quantum computation robust against errors in the system and ancilla qubits, in the gates, in measurement, and in state preparation.

12.2.1 The Problem with Syndrome Computation

The computation of the syndrome is potentially dangerous to the computational state. Consider the first of the six parity check circuits we gave for the Steane code, the one shown in figure 12.4,

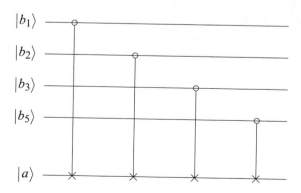

Figure 12.4
One of the six syndrome computation circuits for Steane's seven-qubit code.

which acts on the first, second, third, and fifth qubits of the encoded state and on the first ancilla qubit. Steane's code was designed to correct any single-qubit error on any one of the qubits. We want to make sure that imperfections in carrying out the error correction scheme do not make things worse. In particular, we want to make sure that a single error in carrying out the quantum error correction does not result in multiple errors on the encoded qubits. Suppose a bit-flip error occurs on an ancilla qubit leading to the "correction" of a nonexistent error. Such an occurrence is annoying but not too serious; the "correction" only introduces a single-qubit error on the coding qubits, which, as long as another error does not occur, will be corrected by the next round of error correction. (Since the ancilla qubit will be used again only if it is first reset to $|0\rangle$, its error does not propagate further.) There is a worse possibility, one that results in multiple errors on the coding qubits, something that will not be corrected by subsequent rounds of error correction even if no other error occurs. Take a minute to see if you can see the problem; spotting the problem is a good test of whether you are thinking of quantum circuits in a quantum or classical fashion.

Syndrome extraction operators for quantum codes commonly use controlled gates. On the face of it, controlled operations seems perfectly safe since how could computing *from* the computational qubits *to* the ancilla qubits adversely affect the state of the computational qubits? However, as we saw in caution 2 of section 5.2.4: the notions of *from* and *to* are basis dependent; in the Hadamard basis the control and target qubits of a C_{not} are reversed, and phase flips become bit flips and vice versa. Consider for example what happens if the coding qubits b_1, b_2, b_3, b_5 are in the state $|+\rangle$ and a ZH error occurs on the ancilla qubit before the syndrome computation has begun. The error places the ancilla qubit in the state $|-\rangle$ so that when each C_{not} is performed it acts as the control bit, with the result that all four qubits b_1, b_2, b_3, b_5 have been flipped to the $|-\rangle$ state. In this way, a single error on the ancilla qubit propagates to multiple errors on the coding qubits.

12.2.2 Fault-Tolerant Syndrome Extraction and Error Correction

The example of section 12.2.1 suggests that, to obtain fault-tolerant error correction, an ancilla qubit should be connected with at most one coding qubit. To implement Steane's code in a fault-tolerant manner, we must use a circuit of the following form:

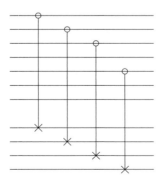

If we measure the ancilla qubits, however, we are in danger of gaining information about the quantum state of the coded qubits, not just the error, which means the measurement is likely to affect the state of the encoding qubits. For example, suppose the encoded state was $\frac{1}{\sqrt{2}}(|\tilde{0}\rangle + |\tilde{1}\rangle)$ before a single-qubit error occurs on qubit b_5. Measurement of the ancilla qubits in the standard basis will tell us that an error occurred on qubit b_5, but it will also destroy the superposition, so that the error correcting operation will "restore" the state to $|\tilde{0}\rangle$ or $|\tilde{1}\rangle$ instead of the correct state $\frac{1}{\sqrt{2}}(|\tilde{0}\rangle + |\tilde{1}\rangle)$.

The trick to avoid gaining too much information is to initialize the ancilla qubits in a state from which it is impossible to learn anything about the computational state. The four ancilla qubits replace a single qubit in the non-fault-tolerant circuit, so from measuring all four qubits of the ancilla, only one bit of information needs to be gained: the value of the corresponding syndrome operator. Exercise 5.9 suggests how to achieve this result; a carefully designed initial starting state $|\phi_0\rangle$ for the ancilla that becomes a second state $|\phi_e\rangle$ under a single-qubit bit-flip error on any one of the four qubits will yield only one bit of information. Consider

$$|\phi_0\rangle = \frac{1}{2\sqrt{2}} \sum_{d_H(x) \; even} |x\rangle,$$

where the sum is over all strings with even Hamming weight, and

$$|\phi_e\rangle = \frac{1}{2\sqrt{2}} \sum_{d_H(x) \; odd} |x\rangle.$$

Under errors in the encoded qubits that would have resulted in syndrome state $|0\rangle$ in the original syndrome computation, the ancilla remains in state $|\phi_0\rangle$. Under errors that would have resulted in $|\phi_1\rangle$, the ancilla ends up in state $|\phi_e\rangle$. These two states are distinguished by a measurement in the standard basis that yields a random even-weighted string in the no-error case and a random odd-weighted string in the error case. This measurement provides only one bit of information.

One final problem needs to be addressed before a fault-tolerant implementation of the Steane code syndrome measurement is obtained. The solution given above requires the preparation of the state $|\phi_0\rangle$. We must make sure that we can prepare $|\phi_0\rangle$ in a fault-tolerant way. Our strategy is not to use a state we prepare if it deviates too much from $|\phi_0\rangle$. In particular, we want to make sure that a faulty preparation does not produce errors in multiple coding qubits. Applying the Walsh-Hadamard transformation to the cat state $\frac{1}{\sqrt{2}}(|0000\rangle + |1111\rangle)$ produces the state $|\phi_0\rangle$. The circuit of figure 12.5 constructs the cat state in a non-fault-tolerant way. To see how to turn this construction into a fault-tolerant one, let us look at what errors may occur.

A single error in any one of the cat state qubits must not propagate to an error in more than one of the coding qubits. Bit-flip errors in the overall construction of the ancilla state, even multiple bit-flip errors resulting from a single error, are not a concern; the worst they do is cause an error in the syndrome, which results in at most a single-qubit error when the "correction" corresponding to this syndrome is carried out. Multiple phase errors resulting from a single error must be avoided, since such errors could result in multiple errors in the coding qubits. Before the final Hadamard

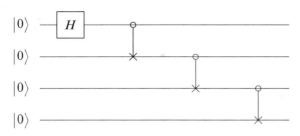

Figure 12.5
Non-fault-tolerant construction of a cat state.

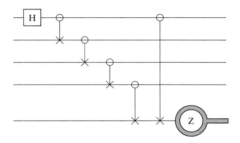

Figure 12.6
Fault-tolerant cat state ancilla preparation. The Z measurement test whether the first and fourth qubits have the same value. If this measurement fails, the state is discarded, and the state preparation is repeated until the test is passed.

transformations are applied, phase errors were bit-flip errors, so we must avoid bit-flip-error propagation in the first part of the circuit. In the circuit of figure 12.5, a bit-flip error in either the second or third qubit can propagate to the successive qubits. However, either of these bit flips would mean that the first and fourth qubit have opposite values, whereas in the error-free case the first and fourth qubits have the same value. If we insert a check for equality of these values, we can discard the state and redo the preparation if the check fails. This single-qubit test suffices. Figure 12.6 shows cat state ancilla preparation that includes this test.

12.2.3 Fault-Tolerant Gates for Steane's Code

To perform arbitrary quantum computations on the logical qubits of the Steane code, a universal set of fault-tolerant logical gates that can approximate any unitary operator on the logical qubits must be available. Even implementations of logical single-qubit gates may not be fault-tolerant, since they may propagate a single error to multiple qubits. For example, the most obvious, though far from optimal, way to carry out a logical single-qubit operation is to decode the logical qubit, apply a true single-qubit operation to the resulting single qubit, and then re-encode. Such an implementation is clearly not fault-tolerant; if an error occurs to the single qubit after the decoding, upon re-encoding the error will propagate to all seven of the encoding qubits. Furthermore, for logic gates involving more than one logical qubit, fault-tolerant implementations must not propagate a single error in one block to multiple errors in another.

For the Steane code, it is easy to find fault-tolerant implementation for some gates, including X, H, and C_{not}. For most other gates, it is challenging to find a fault-tolerant implementation, and for some gates, including the Toffoli gate and $\pi/8$-gate, the only fault-tolerant implementations known require auxiliary qubits. For a fault-tolerant implementation of the logical $\tilde{X}$ operation, recall from section 11.3.3 that the logical qubit $|\tilde{0}\rangle$ is an evenly weighted superposition of all elements of C, and that $|\tilde{1}\rangle$ is an evenly weighted superposition of all elements of $C^{\perp} - C$. Recall further that the elements of C that are not in $C^{\perp}$ are those obtained from elements of C by adding 1111111. Thus, applying X to every qubit in the seven-qubit block performs the logical $\tilde{X}$ gate taking $|\tilde{0}\rangle$ to $|\tilde{1}\rangle$, and $|\tilde{1}\rangle$ to $|\tilde{0}\rangle$. Expanding on this reasoning using relations such as "adding any element of C to any element of $C^{\perp}$ results in an element of $C^{\perp}$" shows that the logical $\widetilde{C_{not}}$ may be implemented by applying C_{not} operators between the corresponding qubits of the two blocks, as shown in figure 12.7. Both of these implementations are fault-tolerant because a single error cannot create multiple errors either in its own block or in another. Unfortunately, the transversal strategy applied in these examples, where gates are applied only between corresponding qubits in the blocks, does not work in most cases.

When the transversal strategy does not work, it can be highly nontrivial to find a fault-tolerant implementation. The construction of fault-tolerant procedures depends on the code used. For some codes it is not known how to construct fault-tolerant implementations of some logical gates. Even for the Steane code, most single-qubit operations cannot be implemented transversally. Applying the phase gate $P_{\frac{\pi}{2}} = |0\rangle\langle 0| + \mathbf{i}|1\rangle\langle 1|$ of section 5.5 to all seven qubits of the Steane code results in the logical gate $|\tilde{0}\rangle\langle\tilde{0}| - \mathbf{i}|\tilde{1}\rangle\langle\tilde{1}|$, which isn't quite $\tilde{P}_{\frac{\pi}{2}}$. In this case, applying $|0\rangle\langle 0| - \mathbf{i}|1\rangle\langle 1|$ to

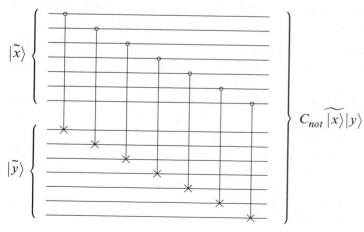

Figure 12.7
Fault-tolerant C_{not}.

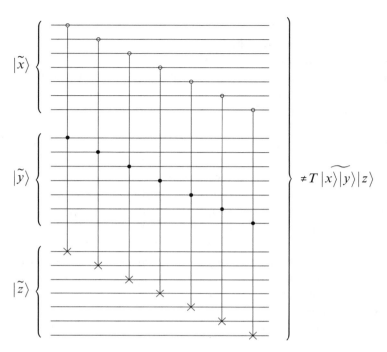

Figure 12.8
The transversal approach does not result in a logical Toffoli gate $\tilde{T}$.

each qubit results in a fault tolerant implementation of $\tilde{P}_{\frac{\pi}{2}}$, but in other cases, there is no easy fix. For example, not only does applying $P_{\frac{\pi}{4}} = |0\rangle\langle 0| + e^{i\frac{\pi}{4}}|1\rangle\langle 1|$ to each qubit not result in $\tilde{P}_{\frac{\pi}{4}}$, but it is not possible to implement $\tilde{P}_{\frac{\pi}{4}}$ in any transversal way. No transversal implementation of $\tilde{P}_{\frac{\pi}{4}}$ is known. For $\tilde{P}_{\frac{\pi}{4}}$, the only fault-tolerant implementations known require ancilla qubits.

The logical Toffoli gate $\tilde{T}$ cannot be implemented by the application of Toffoli gates to corresponding qubits in the three blocks as shown in figure 12.8 (see exercise 12.4). Like the $\tilde{P}_{\frac{\pi}{4}}$ gate, only nontransversal implementations of $\tilde{T}$ are possible for the Steane code.

In order to show that fault-tolerant computation can be done on data encoded using the Steane code, we need to show that all logical unitary operators can be approximated arbitrarily closely by the application of a sequence of fault-tolerant gates. We give fault-tolerant versions of the logical operations for the universally approximating set of gates described in section 5.5: the Hadamard gate H, the phase gate $P_{\frac{\pi}{2}}$, the controlled-NOT gate C_{not}, and the $\pi/8$-gate $P_{\frac{\pi}{4}}$. Fault-tolerant implementations for $\tilde{P}_{\frac{\pi}{2}}$ and $\widetilde{C}_{not}$ have already been described. The logical Hadamard gate $\tilde{H}$ can be implemented transversally by applying H to each of the qubits in the block. Finding a fault-tolerant implementation of $\tilde{P}_{\frac{\pi}{4}}$ is more work.

A number of fault-tolerant implementations use the same key idea: many transforms that can be implemented using a fault-tolerantly prepared ancilla state do not have a direct fault-tolerant

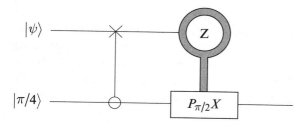

Figure 12.9
A circuit that forms the basis for a fault-tolerant implementation for $P_{\frac{\pi}{4}}$. Application of $P_{\pi/2}\,X$ is conditional on the outcome of the measurement with Z.

implementation. The trick is to use measurement. We illustrate these techniques by developing a fault-tolerant implementation of the $\pi/8$-gate. It is perhaps unsurprising that the state $|\pi/4\rangle = |0\rangle + e^{i\frac{\pi}{4}}|1\rangle$ can be used to implement the $\pi/8$-gate $P_{\frac{\pi}{4}}$. The circuit of figure 12.9 performs the $\pi/8$-gate $P_{\frac{\pi}{4}}$ on any input state $|\psi\rangle$. Since we already know fault-tolerant implementations of the C_{not}, $P_{\frac{\pi}{2}}$, and X, we must find a fault-tolerant preparation of the encoded state $|\widetilde{\pi/4}\rangle$ to realize this circuit. We first consider fault-tolerant measurement, which will be used as part of fault-tolerant state preparation.

12.2.4 Fault-Tolerant Measurement
Recall from section 11.4.3 that if M is both Hermitian and unitary, an indirect measurement may be performed with an additional ancilla qubit using the following circuit:

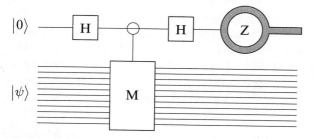

This construction is far from fault-tolerant, since a single error in the ancilla qubit could propagate to all n qubits. To make this construction fault-tolerant, we use a cat state as we did for the fault-tolerant syndrome measurement of section 12.2.2. For the present fault-tolerant construction we need an n-qubit cat state and, just as for fault-tolerant quantum error correction, we must perform checks on the cat state we construct and discard any states that fail those tests. Indirect measurement by M has a fault-tolerant implementation whenever a version of M controlled correctly by the cat state can be constructed. If M has a transversal implementation in terms of single-qubit operators, a controlled version is easy to obtain: control each single qubit operator with the corresponding qubit in the cat state, so that either all single-qubit operators or none at all are performed. The use of this construction is illustrated in the fault-tolerant preparation of the state $|\widetilde{\pi/4}\rangle$ described in the next section.

12.2.5 Fault-Tolerant State Preparation of $|\widetilde{\pi/4}\rangle$

To prepare a state $|\tilde{\phi}\rangle$ fault-tolerantly, it suffices to find an efficiently and fault-tolerantly imple-mentable measurement operator $\tilde{M}$ for which $|\tilde{\phi}\rangle$ is an eigenstate. Any fault-tolerantly prepared state that is not orthogonal to $|\tilde{\phi}\rangle$ yields $|\tilde{\phi}\rangle$ with positive probability when measured by $\tilde{M}$. When an incorrect eigenstate is obtained after such a measurement, the process can be repeated until the correct state is obtained or, in many cases, a fault-tolerant gate can be used to transform the obtained eigenstate into the desired one.

To obtain an efficiently and fault-tolerantly implementable $\tilde{M}$ for the $|\widetilde{\pi/4}\rangle$ state, we begin with a general observation about the operators

$$P_\theta = |0\rangle\langle 0| + e^{i\theta}|1\rangle\langle 1|$$

and the states $|\theta\rangle = \frac{1}{\sqrt{2}}(|0\rangle + e^{i\theta}|1\rangle)$. Since X has eigenvectors $|+\rangle$ and $|-\rangle$, with eigenvalues 1 and -1 respectively, $P_\theta X P_\theta^{-1}$ has eigenvectors $P_\theta|+\rangle = \frac{1}{\sqrt{2}}(|0\rangle + e^{i\theta}|1\rangle)$ and $P_\theta|-\rangle = \frac{1}{\sqrt{2}}(|0\rangle - e^{i\theta}|1\rangle)$ with eigenvalues 1 and -1 respectively. At first, this fact does not seem useful; yes, $|\pi/4\rangle$ is an eigenstate of $M = P_{\pi/4} X P_{\pi/4}^{-1}$, but it is $P_{\pi/4}$ we are trying to implement in the first place. However, the commutation relation $X P_\theta^{-1} = e^{-i\theta} P_\theta X$ implies that

$$M = P_{\pi/4} X P_{\pi/4}^{-1} = e^{-i\frac{\pi}{4}} P_{\pi/4} P_{\pi/4} X = e^{-i\frac{\pi}{4}} P_{\pi/2} X,$$

and we know how to implement $\tilde{P}_{\pi/2}$ and $\tilde{X}$ fault-tolerantly.

For the indirect measurement construction to work, we do not need to implement full controlled versions of these gates; instead, we need only to implement versions that are correctly controlled by the cat state used to fault tolerantly implement the measurement, a much easier task. To obtain the logical analog of indirect measurement by M, apply a controlled $e^{-i\frac{\pi}{4}}$ phase gate between the first qubit of the cat state and the first qubit of the ancilla followed by seven controlled $P_{\pi/2}X$ gates, between the seven pairs of corresponding qubits of the cat state and the ancilla, implements $\tilde{P}_{\pi/2}\tilde{X}$ (see figure 12.10). The cat state construction is then undone and the remaining qubit measured in the standard basis. If the measurement result is 0, the desired state $|\widetilde{\pi/4}\rangle$ is obtained. If the measurement result is 1, the resulting state is $\frac{1}{\sqrt{2}}(|\tilde{0}\rangle - e^{i\frac{\pi}{4}}|\tilde{1}\rangle)$ and the desired state can be obtained by applying $\tilde{Z}$.

To see that this circuit performs the measurement $\tilde{M}$, let us consider what happens at each stage. The Hadamard transformation together with the six C_{not} operations result in the state $|\phi_0\rangle|\tilde{\psi}\rangle$. The next eight gates perform $\tilde{M}$ on the computational qubits controlled by the cat state, which results in the state

$$\frac{1}{\sqrt{2}}(|0\rangle^{\otimes 7}|\tilde{\psi}\rangle + |1\rangle^{\otimes 7} \tilde{M}|\tilde{\psi}\rangle).$$

The six C_{not} result in the state

$$\frac{1}{\sqrt{2}}(|0\rangle^{\otimes 7}|\tilde{\psi}\rangle + |1\rangle|0\rangle^{\otimes 6} \tilde{M}|\tilde{\psi}\rangle).$$

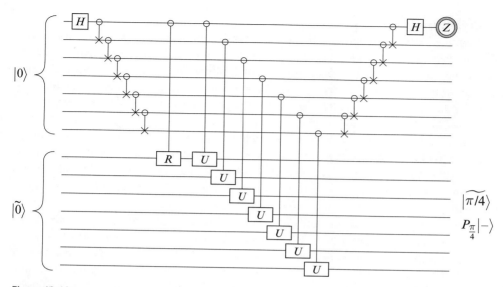

Figure 12.10

Fault-tolerant construction of $|\widetilde{\pi/4}\rangle$, where $R = e^{-i\frac{\pi}{4}}$ and $U = P_{\frac{\pi}{2}} X$. The first set constructs a cat state, the next set applies $\tilde{M}$ controlled by the cat state, then the cat state construction is undone and the first qubit of the cat state register is measured. Either $|\widetilde{\pi/4}\rangle$ or $P_{\frac{\pi}{2}}|-\rangle$ is obtained. In the latter case, a $\tilde{Z}$ operator could be applied to obtain $|\widetilde{\pi/4}\rangle$.

The final Hadamard transformation results in the state

$$\frac{1}{2}(|+\rangle|0\rangle^{\otimes 6}|\tilde{\psi}\rangle + |-\rangle|0\rangle^{\otimes 6}\tilde{M}|\tilde{\psi}\rangle),$$

which is equal to

$$\frac{1}{2\sqrt{2}}(|0\rangle|0\rangle^{\otimes 6}(|\tilde{\psi}\rangle + \tilde{M}|\tilde{\psi}\rangle) + |1\rangle|0\rangle^{\otimes 6}(|\tilde{\psi}\rangle - \tilde{M}|\tilde{\psi}\rangle)).$$

Just as in section 11.4.3, one or the other of the eigenstates of $\tilde{M}$ is obtained when the first qubit is measured in the standard basis.

12.3 Robust Quantum Computation

Section 12.3.1 describes concatenated coding that iteratively replaces a circuit with a larger and more robust one. That section also analyzes how many levels of concatenation are required to obtain a given accuracy to show that a polynomial increase in resources (qubits and gates) can achieve an exponential increase in accuracy. With these tools in hand, section 12.3.2 describes a threshold result.

12.3.1 Concatenated Coding

Let Q_0 be a time-partitioned circuit (section 12.1) for a computation we wish to make robust. Suppose we want the computation to succeed with probability at least $1 - \epsilon$. Let Q_{i+1} be the time-partitioned circuit obtained by encoding each of the qubits of circuit Q_i with Steane's code, replacing all of the basic gates used in Q_i with fault-tolerant logical equivalents, and performing fault-tolerant error correction after the completion of the logical equivalent of each of Q_i's time intervals. In other words, the circuit Q_i is obtained from Q_0 by i rounds of concatenated coding. Figure 12.11 schematically represents two levels of concatenated coding. In circuit Q_i there are i different levels of error correction: error correction on blocks of seven qubits are done most often, and error correction on blocks corresponding to the final logical qubits least often. This hierarchical application of error correction enables an exponential level of robustness to be achieved with only polynomially many resources, qubits and gates.

This paragraph provides a rough heuristic argument for how polynomially many resources suffice to obtain exponential accuracy. When encoding qubits using an error correcting code, we can think of it roughly as having replaced parts that fail under a single-qubit error with an ensemble of parts that fails only in the presence of two or more errors. If, within a given time period, the probability of a part failing is p, the ensemble fails with probability cp^2. Suppose a machine M_0 that is composed of N parts, each of which fails with probability p in a single time interval, runs for T time intervals. The chance that M operates without fault is $(1 - p)^{NT}$. Suppose a new machine, M_1, is created in which each of the N parts is replaced by K parts that together perform the operation of the original part, and that while each part still fails with probability p, the ensemble of K parts fails to perform the desired operation with probability only cp^2 for some constant $c < 1/p$. The basic parts of machine M_1 now can be replaced with ensembles of K parts. After continuing in this way i times, the hierarchical ensembles in machine M_i, making the equivalent of a single part in machine M_0, fail to perform the desired operation with probability only $c^{2^i-1}p^{2^i}$, so overall the machine M_i succeeds with probability $(1 - c^{2^i-1}p^{2^i})^{NT}$. The number of parts K^i in the hierarchical ensemble corresponding to a single part in machine M_0 increases only exponentially in i, while the accuracy increases doubly exponentially in i. Thus, for the ensemble to achieve a failure rate of no more than $(1/2)^r$, we need encode only $O(\log_2 r)$ times: For any $i > \log_2 \left(\frac{\log_2 c - r}{\log_2(cp)} \right)$, the failure rate is less that $(1/2)^r$ because

$$i > \log_2 \left(\frac{\log_2 c - r}{\log_2(cp)} \right)$$

implies that

$$2^i > \frac{r - \log_2 c}{-\log_2(cp)}.$$

The denominator $-\log_2(cp)$ is positive, since $cp < 1$, so

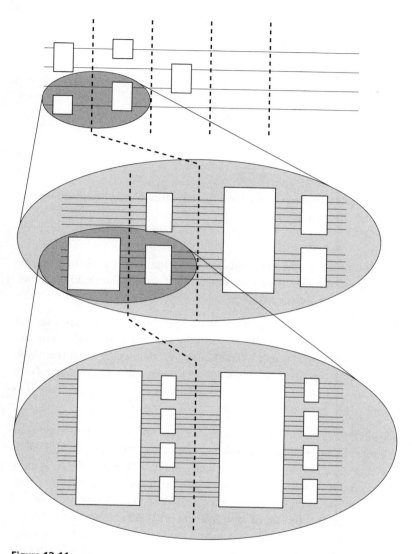

Figure 12.11
Schematic diagram with circuits Q_0, Q_1, and Q_2 showing two levels of concatenated coding. The number of qubits is only suggestive: for the Steane code, the circuit Q_2 would use forty-nine qubits for each qubit of Q_0.

$$-2^i \log_2(cp) > r - \log_2 c,$$

which implies

$$-r > 2^i \log_2(cp) - \log_2 c = \log_2 \left(\frac{(cp)^{2^i}}{c} \right).$$

Thus,

$$2^{-r} > \frac{(cp)^{2^i}}{c},$$

where $\frac{(cp)^{2^i}}{c} = c^{2^i - 1} p^{2^i}$ is the failure rate we computed for an ensemble in machine M_i that replaces a single part in the original machine M_0.

12.3.2 A Threshold Theorem

This section begins by stating a threshold theorem and explaining the meaning of the concepts used in the statement of the theorem. It then briefly describes more general threshold theorems, and numerical estimates for thresholds obtained so far.

A Threshold Theorem For any $[[n, 1, 2t + 1]]$ quantum error correcting code that has a full set of fault-tolerant procedures, there exists a threshold p_T with the following properties. For any $\epsilon > 0$, and any ideal circuit C, there exists a fault-tolerant circuit C' that, under local stochastic noise of error rate $p < p_T$, produces output that is within ϵ, in the statistical distance metric, from the output of C, and fewer than $a|C|$ qubits and time steps are used in C', where the factor a is polylogarithmic in $\frac{|C|}{\epsilon}$, where $|C|$ is the number of locations in C.

An error correcting code has a *full set of fault-tolerant procedures* if it has fault-tolerant procedures for a set of universal logical gates, error correction steps, state preparation, and measurement. Suppose a circuit C has been divided into time steps in which each qubit is subjected to at most one preparation, gate, or measurement. A *location* in C is a gate, a preparation, a measurement, or a wait (the identity transformation storing the qubit for the next step). A *fault-tolerant protocol* based on an error correcting code with a full set of fault-tolerant procedures replaces each location with a fault-tolerant procedure followed by an accompanying fault-tolerant error correcting procedure. The circuit C' in the threshold theorem is obtained by applying the fault-tolerant protocol iteratively, in each stage replacing the gates, preparations, and measurements contained in the fault-tolerant procedures making up the circuit obtained in the previous round, as explained in section 12.3.1. The number of iterations that need to be carried out depends on the accuracy desired.

In a *local stochastic error model*, the probability of errors at all locations in a set of locations during one time step decreases exponentially with the size of the set. More precisely, for every subset S of locations in a circuit C, the total probability of having faults at every location in S (and possibly outside S as well) is at most $\prod_i p_i$, where p_i is the fixed probability of having a fault at location L_i. If the error probability p_i for all locations is less than p, the *error rate* is said to be less than p. The type of error that occurs at these locations is not specified; for analysis, it is helpful to imagine that the error is chosen by an adversary who is trying to disrupt the computations as much as possible. Stochastic means that the locations of the faults are chosen randomly.

The *statistical distance* ϵ between two probability distributions P and Q, with probabilities p_i and q_i respectively for the N outcomes i, is the L^1-distance

$$\epsilon = ||P - Q||_1 = \sum_{i=1}^{N} |p_i - q_i|.$$

In our case, let P be the probability distribution over the measurement outcomes obtained if the ideal circuit were executed perfectly and the resulting state were measured in the standard basis. Let Q be the probability distribution obtained by applying C' under local stochastic noise of rate p and then measuring the logical qubits. Prior to measurement, the output state of C in the ideal case, and the output state of C' in the noisy case, can be written as density operators ρ and σ respectively. The statistical distance between the distributions obtained by measuring ρ and σ in the standard basis is the same as the *trace distance* between the Hermitian operators ρ and σ. The *trace distance* or *trace metric* comes from the trace norm for Hermitian operators: the trace norm $||A||_{Tr}$ of A is defined as $||A||_{Tr} = \mathbf{tr}|A|$, where $|A| = \sqrt{A^\dagger A}$ is the positive square root of the operator $A^\dagger A$. Let ρ and ρ' be two density operators. The trace metric $d_{Tr}(\rho, \rho')$ on density operators is defined to be

$$\epsilon = d_{Tr}(\rho, \rho') = ||\rho - \rho'||_{Tr} = \mathbf{tr}|\rho - \rho'|.$$

Threshold theorems have been obtained for other error models, including more general models. For example, threshold theorems exist for error models in which each basic gate is replaced with one that interacts with the environment but remains close to the original gate. More precisely, each basic gate, acting perfectly, is modeled as $U \otimes I$, where U is the basic gate acting on the computational system and I is the identity acting on the environment. Each perfect gate has a noisy counterpart modeled by a unitary operator V acting on the computational system, together with the environment where V is constrained to be within η_T of $U \otimes I$,

$$||V - U \otimes I||_{Tr} < \eta_T$$

for some threshold value η_T. This noise model is quite general. In particular, it subsumes the local stochastic error model.

Estimates for threshold values such as p_T or η_T have been obtained for a variety of codes, fault-tolerant procedures, and error models. Early results had thresholds on the order of $\eta' = 10^{-7}$, and these have been improved to $\eta' = 10^{-3}$. Further improvements are needed to reach $\eta' = 10^{-2}$, a value that begins to be in reach of implementation experiments. Better understanding of realistic noise models, the development of more advanced error codes, and improved fault-tolerant techniques and analyses will improve these values.

12.4   References

The first paper on fault tolerance was by Shor [252]. Preskill [233] surveys issues and results related to fault tolerance. Both Aliferis [18] and Gottesman [138] give detailed, rigorous, but nevertheless highly readable accounts of fault tolerance, including threshold theorem proofs.

   Early threshold results were proved by Aharonov and Ben-Or [9, 10], Kitaev [173], and Knill, Laflamme, and Zurek [180, 181]. Improved error thresholds on the order of 10^{-3} have been found; see, for example, Steane [263], Knill [178], and Aliferis, Gottesman, and Preskill [19]. Threshold results have been found for a variety of noise models, including non-Markovian noise [268]. Threshold results have also been found for alternative models of computation [219, 220] such as cluster state quantum computing, which will be discussed in section 13.4. Steane [260] estimates realistic decoherence times.

   Steane [261] overviews a number of different universally approximating finite sets of gates from the point of view of fault-tolerant quantum computing. Eastin and Knill [108] showed that no code admits a universal transversal gate set. Cross, DiVincenzo, and Terhal [92] provide a comparison of many quantum error correcting codes from the point of view of fault tolerance.

12.5   Exercises

Exercise 12.1.   Show that a single-qubit gate followed by a single-qubit error is equivalent to a (possibly different) single-qubit error followed by the same gate.

Exercise 12.2.   Why do we consider the preparation of the cat state in figure 12.6 to be fault-tolerant, even though it includes two C_{not} gates from the qubit on which the Z measurement is made?

Exercise 12.3.   What effect does applying $P_{\frac{\pi}{4}}$ to each of the qubits in the Steane seven-qubit encoding have?

Exercise 12.4.   Show that the transversal circuit shown in figure 12.8 does not implement the Toffoli gate. Consider the effect of the circuit on $|\tilde{1}\rangle \otimes |\tilde{0}\rangle \otimes |\tilde{0}\rangle$.

Exercise 12.5.   Design a fault-tolerant version of the Toffoli gate for the Steane code.

13 Further Topics in Quantum Information Processing

This chapter gives brief overviews of topics that we were not able to discuss fully. Section 13.1 surveys more recent results in quantum algorithms. Known limitations of quantum computation are discussed in section 13.2. Other approaches to robust quantum computation, as well as a few of the many advances in quantum error correction, are described in section 13.3. Section 13.4 briefly describes alternative models of quantum computation, including cluster state quantum computation, adiabatic quantum computation, holonomic quantum computation, and topological quantum computation, and their implications for quantum algorithms, robustness, and approaches to building quantum computers. Section 13.5 makes a quick tour of the extensive area of quantum cryptography, and touches upon quantum games, quantum interactive protocols, and quantum information theory. Insights from quantum information processing that led to breakthroughs in classical computer sciences are discussed in section 13.6.

Section 13.7 briefly surveys approaches to building quantum computers, starting with criteria for scalable quantum computers. This discussion leads into the consideration of simulations of quantum systems in section 13.8. Section 13.9 discusses the still poorly understood question of where the power of quantum computation comes from, with an emphasis on the status of entanglement. Finally, section 13.10 discusses computation in theoretical variants of quantum theory.

This overview is not meant to be complete. In an area advancing as quickly as this one, there are new results every day. Exploring the quantum physics section of the e-print archive (http://arXiv.org/archive/quant-ph) is an excellent way to discover additional topics and to keep up with the latest developments in the field (but be aware that the papers there are not refereed).

13.1 Further Quantum Algorithms

After Grover's algorithm, there was a hiatus of more than five years before a significantly different algorithm was found. The field advanced during this time, with researchers finding variants on the techniques of Shor and Grover to provide algorithms for a wider range of problems, but no algorithmic breakthroughs occurred. Grover and others extended his techniques to provide small speedups for a number of problems, as mentioned in section 9.6. Shor's algorithms were extended to provide solutions to the hidden subgroup problem over a variety of non-Abelian groups that are close to being Abelian [244, 162, 161, 29], including a solution for the hidden subgroup problem for normal subgroups of arbitrary finite groups [147, 141] and groups that are *almost Abelian* in the sense that the intersection of the normalizers for all subgroups is large [141]. On the negative side, Grigni et al. [141] showed in 2001 that for most non-Abelian groups and their subgroups, the standard Fourier sampling method used by Shor and successors could yield only exponentially little information about the hidden subgroup. On the other hand, Ettinger et al. [114] showed in 2004 that there is no information theoretic barrier to solving this problem; they showed that the query complexity of the general non-Abelian hidden subgroup problem is polynomial.

Most researchers expect that quantum computers cannot solve NP-complete problems in polynomial time. There is no proof (a proof would imply $\mathbf{P} \neq \mathbf{NP}$). As section 13.10 discusses in more detail, Aaronson goes so far as to suggest that this limit on computational power be viewed as a principle governing any reasonable physical theory capable of describing our universe. A lot of focus has been given to candidate *NP-intermediate problems*, problems that are in $\mathbf{NP}$, not in $\mathbf{P}$, and are not $\mathbf{NP}$ complete. Ladner's theorem says that if $\mathbf{P} \neq \mathbf{NP}$, then there exist *NP intermediate problems*. Factoring and the discrete logarithm problem are both candidate NP-intermediate problems. Other candidate problems include graph isomorphism, the gap shortest lattice vector problem, and many hidden subgroup problems [254, 13]. While polynomial time quantum algorithms have been found for a few hidden subgroup problems, particularly cases that are close to Abelian, these problems remain some of the most important open questions in the field of quantum computation.

Two special cases of the hidden subgroup problem have received the most attention: the symmetric group S_n, the full permutation group of n elements, and the dihedral group D_n, the group of symmetries of a regular n-sided polygon. An early result of Beals [34] provided a quantum Fourier transform for the symmetric group, but a solution to the hidden subgroup problem for the symmetric group continues to elude researchers. This problem is of particular interest since a solution would yield a solution to the graph isomorphism problem. The hidden subgroup problem for the dihedral group attracted even more attention when Regev [237] showed in 2002 that any efficient algorithm to the dihedral hidden subgroup problem that uses Fourier sampling, a generalization of Shor's technique, would enable the construction of an efficient algorithm for the gap shortest vector problem, a problem of cryptographic interest. In 2003, Kuperberg found a subexponential (but still superpolynomial) algorithm for the dihedral group [189], which Regev improved upon by reducing the space requirements to polynomial while retaining the subexponential time complexity [239]. Alagic et al. have extended these techniques to a solution of Simon's problem for general non-Abelian groups [17]. Lomont surveys hidden subgroup results and techniques in [191].

In 2002, Hallgren found an efficient quantum algorithm for solving Pell's equation [146]. Solving Pell's equation is believed to be harder than factoring or the discrete log problem. The security of the Buchmann-Williams classical key exchange and the Buchmann-Williams public key cryptosystem is based on the difficulty of solving Pell's equation. So even the Buchmann-Williams public key cryptosystem, which was believed to have a stronger security guarantee than standard public key encryption algorithms, is now known to be insecure in a world with quantum computers. In 2003, van Dam, Hallgren, and Ip found an efficient quantum algorithm for the shifted Legendre symbol problem [272]. The shifted Legendre symbol problem is the basis for the security of some algebraically homomorphic cryptosystems that are used, for example, in certain cryptographic-grade random number generators. The existence of van Dam et al.'s algorithm means that quantum computers can predict these random number generators, thus rendering them insecure. In 2007, Farhi, Goldstone, and Gutmann [115] found a quantum algorithm for evaluating NAND trees in $O(\sqrt{N})$, a problem that had puzzled quantum computing researchers for many years.

In the past five years, a new family of quantum algorithms has been discovered that uses techniques of quantum random walks to solve a variety of problems. Childs et al. [80] solve a black box graph traversal problem in polynomial time that cannot be solved in subexponential time classically. Magniez et al. [201] prove a Grover-type speedup result for a different graph problem using a quantum random walk approach. Magniez and Nayak [200] apply quantum random walks to the problem of testing commutativity of a group, Buhrman and Špalek [75] to matrix product verification, and Ambainis [23] to element distinctness. Krovi and Brun [186] study hitting times of quantum walks on quotient graphs. Both Ambainis [22] and Kempe [169] provide overviews of quantum walks and quantum walk–based algorithms.

Quantum learning theory [70, 246, 132, 160, 27] provides a conceptual framework that unites Shor's algorithm and Grover's algorithm. Quantum learning is part of computational learning theory that is concerned with concept learning. A concept is modeled by a membership function, a Boolean function $c : \{0, 1\}^n \rightarrow \{0, 1\}$. Let $C = \{c_i\}$ be a class of concepts. Generally, a quantum learning problem involves querying an oracle O_c for one of the concepts c in C, and the job is to discover the concept c. The types of oracles vary. A common one is a membership oracle, which upon input of x outputs $c(x)$. Common models include exact learning and probably approximately correct (PAC) learning. In the quantum case, oracles output a superposition upon input of a superposition of inputs. Servedio and Gortler [132] establish a negative result, that the number of classical and quantum queries required for any concept class does not differ by more than a polynomial in either the exact or the PAC model. On the positive side, the same paper shows that for computational efficiency, rather than query complexity, the story is quite different. In the exact model, the existence of *any* classical one-way function guarantees the existence of a concept class that is polynomial-time learnable in the quantum case but not in the classical. For the PAC model, a slightly weaker result is known in terms of a particular one-way function.

13.2 Limitations of Quantum Computing

Beals and colleagues [35] proved that, for a broad class of problems, quantum computation can provide at most a small polynomial speedup. Their proof established lower bounds on the number of time steps any quantum algorithm must use to solve these problems. Their methods were used by others to provide *lower bounds* for other types of problems. Ambainis [21] found another powerful method for establishing lower bounds.

In 2002, Aaronson answered negatively the question of whether there could be efficient quantum algorithms for the collision problem [1]. His results were generalized by Shi and himself [248, 6].

This result was of great interest because it showed that there could not exist a generic quantum attack on all cryptographic hash functions. Aaronson's result says that any attack must use specific properties of the hash function under consideration. Shor's algorithms break some cryptographic hash functions, and quantum attacks on others may yet be discovered.

Section 9.3 showed that Grover's search algorithm is optimal. In 1999, Ambainis [20], building on work of Buhrman and de Wolf [73] and Farhi, Goldstone, Gutmann, and Sipser [117], showed that for searching an ordered list, quantum computation can give no more than a constant factor improvement over the best possible classical algorithms. Childs and colleagues [81, 82] improved estimates for this constant. Aaronson [5] provides a high-level overview of the limits of quantum computation.

13.3 Further Techniques for Robust Quantum Computation

While quantum error correction is one of the most advanced areas of quantum information processing, many open questions remain. As more quantum information processing devices are built, finding quantum codes or other robustness methods optimized for the particular errors to which the devices are most vulnerable will remain a rich area of research.

For transmitting quantum information, either as part of quantum communication protocols or to move information around inside a quantum computer, not only are efficient error detection and the trade-off between data expansion and strength of the code important, but the decoding efficiency is as well. One longtime frustration has been the difficulty of using certain classical codes with efficient decoding properties, such as low-density parity check (LDPC) codes, as the basis for constructing quantum codes with similarly efficient decoding. The duality constraint in the CSS code construction was too much of a barrier for these codes, and no one knew what else to do. In 2006, Brun, Devetak, and Hsieh realized that by using a side resource of entanglement between the sender and the receiver, quantum versions of many more classical codes, including LDPC codes, could be obtained [68, 137]. This construction may also be useful beyond quantum communication.

Instead of encoding the states so that we can detect and correct common errors, we may be able to place the states in subspaces unaffected by these errors. Such approaches, complementary to the error correcting codes we have seen, go under the various headings of *error avoiding codes*, *noiseless quantum computation*, or, most commonly, *decoherence-free subspaces*. Under certain conditions, we expect a system to be subject to systematic errors affecting all the qubits of the system. The quantum codes we have seen, while effective on errors involving small numbers of qubits, are not effective on systematic errors affecting all qubits. Lidar and Whaley provide a detailed review of decoherence-free subspaces in [195]. Operator error correction [184, 185] provides a framework that unifies quantum error correcting codes and decoherence-free subspaces. Quantum computers built according the topological model of quantum computation (described in section 13.4.4) would have robustness built in from the start.

Here we give a few simple examples to illustrate the general approach of decoherence-free subspaces.

Example 13.3.1 *Systematic bit-flip errors.* Suppose a system tends to be subject to errors that perform a quantum bit flip on all qubits of the system. Bit-flip errors have no effect on the states $|++\rangle$ and $|--\rangle$ (or any linear combination of them). For example, a bit-flip error takes

$$|--\rangle = \frac{1}{2}(|0\rangle - |1\rangle)(|0\rangle - |1\rangle)$$

to

$$\frac{1}{2}(|1\rangle - |0\rangle)(|1\rangle - |0\rangle) = |--\rangle.$$

If we encode every $|0\rangle$ and $|1\rangle$ as two qubit states $|++\rangle$ and $|--\rangle$ respectively, we will have succeeded in protecting our computational states from all systematic bit-flip errors by embedding them in states of a $2n$-qubit system that are immune from these errors.

Example 13.3.2 *Systematic phase errors.* Suppose a system tends to be subject to errors that perform the same relative phase shift $E = |0\rangle\langle 0| + e^{i\phi}|1\rangle\langle 1|$ on all qubits of the system. If we encode each single-qubit state $|0\rangle$ and $|1\rangle$ as the two qubit states

$$|\psi_0\rangle = \frac{1}{\sqrt{2}}(|01\rangle + |10\rangle)$$

and

$$|\psi_1\rangle = \frac{1}{\sqrt{2}}(|01\rangle - |10\rangle),$$

the error becomes a physically irrelevant global phase, so the computational states are entirely protected from these errors:

$$(E \otimes E)\frac{1}{\sqrt{2}}(|01\rangle \pm |10\rangle) = \frac{1}{\sqrt{2}}(|0\rangle \otimes e^{i\phi}|1\rangle \pm e^{i\phi}|1\rangle \otimes |0\rangle)$$

$$= e^{i\phi}\frac{1}{\sqrt{2}}(|01\rangle \pm |10\rangle)$$

$$\sim \frac{1}{\sqrt{2}}(|01\rangle \pm |10\rangle).$$

Thus, the two-dimensional space spanned by $\{|\psi_0\rangle, |\psi_1\rangle\}$ can be used as a binary quantum system that is error-free within an environment that produces only systematic relative phase errors.

We would like to combine these approaches to obtain an encoding that protects against all systematic qubit errors. A subspace that is immune to both systematic X and Z errors is certainly immune to $Y = ZX$ errors, and therefore to any systematic single-qubit error, since it is immune to any linear combination of these errors. The following example is due to Zanardi and Rasetti [292]. This example was designed for the error environment of qubits encoded in photon polarization as affected by a quartz crystal. This decoherence-free subspace method has been experimentally verified by Kwiat et al. [190].

Example 13.3.3 *Systematic single-qubit errors.* The reader can check that all quantum states represented by the elements of the two-dimensional space spanned by the vectors

$$|\varphi_0\rangle = \tfrac{1}{2}(|1001\rangle - |0101\rangle + |0110\rangle - |1010\rangle)$$

$$|\varphi_1\rangle = \tfrac{1}{2}(|1001\rangle - |0011\rangle + |0110\rangle - |1100\rangle)$$

are left invariant by systematic X and Z errors. Since $|\varphi_0\rangle$ and $|\varphi_1\rangle$ are not orthogonal, we cannot encode $|0\rangle$ and $|1\rangle$ as these two vectors. By using the Gram-Schmidt process, we can find orthonormal vectors: we can replace $|\varphi_1\rangle$ with $|\varphi_1'\rangle$, the normalized component of $|\varphi_1\rangle$ perpendicular to $|\varphi_0\rangle$, by taking $|\varphi_2\rangle = |\varphi_1\rangle - \langle\varphi_0|\varphi_1\rangle|\varphi_0\rangle$, and then normalizing to obtain

$$|\varphi_1'\rangle = \frac{1}{\sqrt{\langle\varphi_2|\varphi_2\rangle}}|\varphi_2\rangle,$$

which by construction is perpendicular to $|\varphi_0\rangle$. By encoding all $|0\rangle$ and $|1\rangle$ as $|\varphi_0\rangle$ and $|\varphi_1'\rangle$, we can protect against all systematic X and Z errors and therefore against all systematic single-qubit errors. Thus, by embedding the states of an n-qubit system in the states of a $4n$-qubit system, we have obtained a computational subspace immune to all systematic single-qubit errors.

Decoherence-free subspace approaches have been developed for a variety of complex situations. See [195] for a survey.

13.4 Alternatives to the Circuit Model of Quantum Computation

The circuit model of quantum computing of section 5.6 is well designed for comparisons between quantum algorithms and classical algorithms. We have seen its use in comparing the efficiency of quantum algorithms to classical algorithms and for showing that any classical computation can be done on a quantum computer in comparable time. Other models rival the circuit model for inspiring the discovery of new quantum algorithms or for giving insight into the limitations of quantum computation. Furthermore, other models better support certain promising approaches toward ways of physically realizing quantum computers and understanding the robustness of these implementations.

Two significant alternatives to the circuit model have been developed so far: cluster state quantum computing and adiabatic quantum computing. The next four subsections briefly describe these two models and their applications, holonomic quantum computation, a hybrid of adiabatic quantum computing and the standard circuit model, and topological quantum computing, which is related to holonomic quantum computation.

13.4.1 Measurement-Based Cluster State Quantum Computation

The elegant *cluster state model of quantum computation* [235, 218] makes exceptionally clear use of quantum entanglement, quantum measurement, and classical processing. In contrast to the standard circuit model, cluster state quantum computation makes no use of unitary operations in its processing of information; all computations are accomplished by measurement of qubits in a cluster state, the maximally connected, highly persistent entangled states of section 10.2.4. In a cluster state algorithm, the order in which the qubits are measured is set; only the basis in which each of the qubits is measured is determined by the results of previous measurements.

The initial cluster state is independent of the algorithm to be performed; it depends only on the size of the problem to be solved. All of the processing, including input and output, takes place entirely by a series of single-qubit measurements, so the entanglement between the qubits can only decrease in the course of the algorithm. For this reason, cluster state quantum computation is sometimes called *one-way* quantum computation. In cluster state quantum computation, the entanglement creation and the computational stages of a quantum computation are cleanly separated.

Cluster state quantum computation has been shown to be computationally equivalent to the standard circuit model of quantum computation. Cluster states, therefore, provide a universal entanglement resource for quantum computation. The proof of computational equivalence relies on a mapping of the time sequence of quantum gates in a quantum circuit to a spatial dimension of the 2-D lattice in which the cluster state lives. The processing proceeds from left to right, with the input placed in states on the far left of the cluster, and the output appearing in the states on the far right of the cluster once the algorithm is complete. A single qubit in the quantum circuit model is mapped to a row of qubits in the cluster state; thus, the single qubits of the cluster state are distinct from the logical qubits being processed by the computation. Many qubits in the cluster are not associated with any qubit in the circuit model. These qubits connect the qubit rows and, together with measurement, enable quantum gates to be carried out as the qubits of the cluster are measured from left to right. The measurements of qubits in a single column can be carried out in parallel.

General cluster state computations use more general structures than those arising as analogs of quantum circuits. For example, the measurements do not necessarily proceed from left to right, and rows in the cluster state may have no obvious meaning. There is no reason for them to represent a *logical* qubit, a concept from the circuit model of quantum computation that does not have an analog in general cluster state quantum computation. Any computation in the cluster state model partitions the cluster into sets of qubits $Q_1, Q_2, \ldots, Q_L$. The qubits within a set can be measured in any order; in particular, they may be measured in parallel. All qubits in the set Q_i must be measured before any qubit of Q_{i+1} is measured. How a qubit in Q_{i+1} is measured may depend on the results of measurements of qubits in $Q_1, Q_2, \ldots, Q_L$. The interpretation of the final result may depend on measurements results obtained at earlier states. Raussendorf, Browne, and Briegel [235] define the logical depth of a computation to be the minimum number of sets Q_i needed to carry out the computation. Both the interpretation of the final results and the decision of what basis to use for a measurement given the previous results require classical computation that must be taken into account in terms of the efficiency of the algorithm.

For some computations, the logical depth is surprisingly low. For example, take any quantum circuit consisting entirely of elements of the Clifford group, the group generated by the C_{not}, Hadamard, and $\pi/2$-phase shift gates. While the corresponding computation in the cluster state model proceeds by measuring columns of qubits from right to left, it turns out that for all cluster computations corresponding to Clifford group circuits, one can simply measure all the qubits at once. Thus, the logical depth of computations using only Clifford gates is 1; there are no dependencies between the measurements needed to accomplish the computation. This result implies that the only computation going on is the classical interpretation of the results and determination of intermediate measurements. Thus, a quantum circuit consisting of entirely of Clifford gates has a classical analog of equivalent efficiency. This result, known as the Gottesman-Knill theorem [133], is not trivial in that, for example, the Walsh-Hadamard transformation is contained in the Clifford group. The cluster state model provides a particularly simple proof of this theorem.

The cluster state model is of great theoretical interest since it clarifies the role of entanglement in quantum computation and provides means of analyzing quantum computation. It has also had substantial impact on approaches to building quantum computers, particularly optical quantum computers. It will be discussed again in section 13.7 in that context. Furthermore, as will be discussed in section 13.9, it has clarified the role of entanglement in quantum competition in surprising ways.

13.4.2 Adiabatic Quantum Computation

To describe *adiabatic quantum computation*, we must first describe the Hamiltonian framework for quantum mechanics on which it rests. Quantum systems evolve by unitary operators, so the state of any system, initially in state $|\Psi_0\rangle$, as it evolves over time t can be described by $|\Psi_t\rangle = U_t|\Psi_0\rangle$, where U_t is a unitary operator for each t. Furthermore, the evolution must be continuous and additive: $U_{t_1+t_2} = U_{t_2}U_{t_1}$ for all times t_1 and t_2. Any unitary operator U can be written $U = e^{-iH}$ for some Hermitian H. Any continuous and additive family of unitary operators can be written as $U_t = e^{-itH}$ for some Hermitian operator H called the Hamiltonian for the system. Schrödinger's equation provides an equivalent formulation: the Hamiltonian H must satisfy

$$i\frac{d}{dt}|\Psi(t)\rangle = H|\Psi(t)\rangle$$

using units in which Planck's constant $\hbar = 1$. Let λ_0 be the smallest eigenvalue of H. Any λ_0-eigenstate of H is called a *ground state* of H. The Hamiltonian framework and Schrödinger's equation can be found in any quantum mechanics book.

To solve a problem using adiabatic quantum computation, an appropriate Hamiltonian H_1 must be found, one for which a solution to the problem can be represented as the ground state of the Hamiltonian. An adiabatic algorithm begins with the system in the ground state of a known and easily implementable Hamiltonian H_0. A path H_t is chosen between the initial Hamiltonian and the final Hamiltonian $H = H_1$, and the Hamiltonian is gradually perturbed to follow this path. The theory of adiabatic quantum computation rests on the adiabatic theorem [210], which says that as long as the path is traversed slowly enough the system will remain in the ground state, and thus at the end of computation it will be in the solutions state, the ground state of H_1. How slowly the path must be traversed depends on the eigengap, the difference between the two lowest eigenvalues. In general, it is hard to obtain bounds on this gap, so the art of designing an adiabatic algorithm is first in finding a mapping of the problem to an appropriate Hamiltonian, and then in finding a short path for which one can show that the eigengap never becomes too narrow.

Adiabatic quantum computation was introduced by Farhi, Goldstone, Gutmann, and Sipser [118]. Childs, Farhi, and Preskill [79] show that adiabatic quantum computation has some inherent protection against decoherence, which means that it may be a particularly good model both for designing robust implementations of quantum computers and robust algorithms [79]. Roland and Cerf [243] show how to recapture Grover's algorithm, and the optimality proof, within the adiabatic context.

Aharonov et al. [16] develop a model for adiabatic quantum computation and prove that it is computationally equivalent to universal quantum computation in the circuit model. Other models of adiabatic computation exist. Some are equivalent in power only to classical computation [63], while for others, the extent of their power in not yet understood. This situation complicates not only discussions of adiabatic quantum computing but also implementation efforts. For example, some small adiabatic devices have been built for which it has not been possible to determine whether they perform universal quantum computation or not.

Aharonov and Ta-Shma's wide-ranging paper [15], after developing tools for adiabatic quantum computation, investigates the use of adiabatic models for understanding which states, particularly superpositions of states drawn from probability distributions, can be efficiently generated. Initial interest centered on the possibility of using adiabatic methods to develop a quantum algorithm to solve NP-complete problems [116, 78, 153], because adiabatic algorithms were not subject to the lower bound results proven for other approaches. Vazirani and van Dam [273] and Reichardt [240] rule out a variety of adiabatic approaches to solving NP-complete problems in polynomial time.

13.4.3 Holonomic Quantum Computation

Holonomic, or *geometric, quantum computation* [293, 76] is a hybrid between adiabatic quantum computation and the standard circuit model in which the quantum gates are implemented via adiabatic processes. Holonomic quantum computation makes use of non-Abelian geometric phases that arise from perturbing a Hamiltonian adiabatically along a loop in its parameter space. The phases depend only on topological properties of the loop, and so are insensitive to perturbations.

This property means that holonomic quantum computation has good robustness with respect to errors in the control driving the Hamiltonian's evolution. Early experimental efforts have been carried out using a variety of underlying hardware.

13.4.4 Topological Quantum Computation

In 1997, prior to the development of holonomic quantum computation, Kitaev proposed *topological quantum computing*, a more speculative approach to quantum computing that also has excellent robustness properties [174, 125, 233, 87]. Kitaev recognized that topological properties are totally unaffected by small perturbations, so encoding quantum information in topological properties would give intrinsic robustness. The type of topological quantum computing Kitaev proposed makes use of the Aharonov-Bohm effect, in which a particle that travels around a solenoid acquires a phase that depends only on how many times it has encircled the solenoid. This topological property is highly insensitive to even large disturbances in the particle's path.

Kitaev defined quantum computation in this model and showed that, by using non-Abelian Aharonov-Bohm effects, such a quantum computer would be universal in the sense of being able to simulate computations in the quantum circuit model without a significant loss of efficiency. However, only a few non-Abelian Aharonov-Bohm effects have been found in nature, and all of these are unsuitable for quantum computation. Researchers are working to engineer such effects, but even the most basic building blocks of topological quantum computation have yet to be realized experimentally in the laboratory. In the long term, the robustness properties of topological quantum computing may enable it to win out over other approaches. In the meantime, it is of significant theoretical interest. For example, it led to a novel type of quantum algorithm that provides a polynomial time approximation of the Jones polynomial [11].

13.5 Quantum Protocols

The most famous quantum protocols are quantum key distribution schemes, such as those of sections 2.4 and 3.4. Quantum key distribution was the first example of a quantum cryptographic protocol. Since then, quantum approaches to a wide variety of cryptographic and communication tasks have been developed.

Some quantum cryptographic protocols, such as the quantum key distribution schemes we described, use quantum means to secure classical information. Others secure quantum information. Many are *unconditionally* secure in that their security is based entirely on properties of quantum mechanics. Others are only quantum computationally secure in that their security depends on a problem being computationally intractable for quantum computers. For example, *unconditionally* secure bit commitment is known to be impossible to achieve through either classical or quantum means [205, 197, 93]. Weaker forms of bit commitment exist. In particular, quantum computationally secure bit commitments schemes exist as long as there exist quantum one-way functions [8, 106]. Kashefi and Kerenidis discuss the status of quantum one-way functions [168].

Closely related to quantum key distribution schemes are protocols for *unclonable encryption* [136]. Uncloneable encryption is a symmetric key encryption scheme that guarantees that an eavesdropper cannot even copy an encrypted message, say for later attempts at decryption, without being detected. In addition to providing a stronger security guarantee than most symmetric key encryption systems, the keys can be reused as long as eavesdropping is not detected. Uncloneable encryption has strong ties with quantum authentication [33]. One type of authentication is digital signatures. Shor's algorithms break all standard digital signature schemes. Quantum digital signature schemes have been developed [139], but the keys involved can be used only a limited number of times. In this respect they resemble classical schemes such as Merkle's one-time digital signature scheme [207].

Some quantum secret sharing protocols protect classical information in the presence of eavesdroppers [151]. Others protect a quantum secret. Cleve et al. [86] provide quantum protocols for (k, n) threshold quantum secrets. Gottesman et al. [134] provide protocols for more general quantum secret sharing. There is a strong tie between quantum secret sharing and CSS quantum error correcting codes. Quantum multiparty function evaluation schemes exist [91].

Fingerprinting is a mechanism for identifying strings such that equality of two strings can be determined with high probability by comparing their respective fingerprints. It has been shown that classical fingerprints for bit strings of length n need to be of at least length $O(\sqrt{n})$. Buhrman et al. [72] show that a quantum fingerprint of classical data can be exponentially smaller; they can be constructed with only $O(\log(n))$ qubits.

In 2005, Watrous [280] was able to show that many classical zero-knowledge interactive protocols are zero knowledge against a quantum adversary. A significant part of the challenge was to find a reasonable and sufficiently general definition of quantum zero knowledge. The problems on which statistical zero-knowledge protocols are generally based are candidate NP-intermediate problems such as graph isomorphism, so for this reason also zero-knowledge protocols are of interest for quantum computation. Aharonov and Ta-Shma [15] detail intriguing connections between statistical zero-knowledge and adiabatic state generation.

There is a close connection between quantum interactive protocols and quantum games. An introduction to this field is provided by [192]. Early work in this area includes a discussion of a quantum version of the prisoner's dilemma [110]. See Meyer [212] for a lively discussion of other quantum games. Gutoski and Watrous [145] tie quantum games to quantum interactive proofs.

13.6 Insight into Classical Computation

A number of classical algorithmic results have been obtained by taking a quantum information processing viewpoint. Kerenidis and de Wolf [170] and Wehner et al. [282] use quantum arguments to prove lower bounds for locally decodable codes, Aaronson [2] for local search, Popescu et al. [230] for the number of gates needed for a classical reversible circuit, and de Wolf [98] for matrix rigidity. Aharonov and Regev [14] "dequantize" a quantum complexity result for a lattice problems to obtain a related classical result. The usefulness of the complex perspective for evaluating real valued integrals is sometimes used as an analogy to explain this phenomenon. Drucker and de Wolf survey these and other results in [105]. We know of two additional examples that were not included in their survey. One is the Gentry result [127] discussed at the end of the next paragraph. Another is an early example due to Kuperberg, his proof of Johansson's theorem [187]. We describe a couple of examples in greater detail.

Cryptographic protocols usually rely on the empirical hardness of a problem for their security; it is rare to be able to prove complete, information theoretic security. When a cryptographic protocol is designed based on a new problem, the difficulty of the problem must be established before the security of the protocol can be understood. Empirical testing of a problem takes a long time. Instead, whenever possible, *reduction* proofs are given that show that if the new problem were solved it would imply a solution to a known hard problem; the proofs show that the solution to the known problem can be reduced to a solution of the new problem. Regev [238] designed a novel, purely classical cryptographic system based on a certain problem. He was able to reduce a known hard problem to this problem, but only by using a quantum step as part of the reduction proof. Thus, he has shown that if the new problem is efficiently solvable in any way, there is an efficient quantum algorithm for the old problem. But it says nothing about whether there would be a classical algorithm. This result is of practical importance; his new cryptographic algorithm is a more efficient lattice-based public key encryption system. Lattice-based systems are currently the leading candidate for public key systems secure against quantum attacks. Four years after Regev's original result, Peikert provided a completely classical reduction [224]. At the same conference, however, Gentry presented his spectacular result, a fully homomorphic encryption system [128], answering a thirty-year open question. As part of his work, he uses a related, but different, quantum reduction argument for an otherwise completely classical result [127].

In another spectacular, if less practical, result, Aaronson found a new solution to a notorious conjecture about a purely classical complexity class **PP** [4]. From 1972 until 1995, this question remained open. Aaronson defines a new quantum complexity class **PostBQP**, an extension of the standard quantum complexity class **BQP**, motivated by the use of postselection in certain quantum arguments. It takes him a page to show that **PostBQP=PP**, and then only three lines to prove the conjecture. The original 1995 proof, while entirely classical, was significantly more complicated. Thus, it seems, for this question at least, the right way to view the classical class **PP** is through the eyes of quantum information processing.

13.7 Building Quantum Computers

DiVincenzo developed widely used requirements for the building of a quantum computer. Obtaining n qubits does not suffice, just like n bits, say n light switches, does not make a classical computer; the bits or qubits must interact in a controllable fashion. It is relatively easy to obtain n qubits, but it is hard to get them to interact with each other and with control devices, while preventing them from interacting with anything else. DiVincenzo's criteria [104] are, roughly:

- Scalable physical system with well-characterized qubits,
- Ability to initialize the qubits in a simple state,
- Robustness to environmental noise: long decoherence times, much longer than the gate operation time,
- Ability to realize high fidelity universal quantum gates,
- High-efficiency, qubit-specific measurements.

Two other criteria were added later in recognition of the need for *flying qubits* used to transmit information between different parts of a quantum computer:

- Ability to interconvert stationary and flying qubits,
- Faithful transmission of flying qubits between specified locations.

DiVincenzo's criteria are rooted in the standard circuit model of quantum computation. Pérez-Delgado and Kok [227] give more general criteria, including formal operational definitions of a quantum computer, that are meant to encompass alternative models of quantum computation.

There are daunting technical difficulties in actually building such a machine. Research teams around the world are actively studying ways to build practical quantum computers. The field is changing rapidly. It is impossible even for experts to predict which of the many approaches are likely to succeed. Both [295] and [157] contain detailed evaluations of the various approaches. No one has yet made a detailed proposal that meets all of the DiVincenzo criteria, let alone realize it in a laboratory. A breakthrough will be needed to go beyond tens of qubits to hundreds of qubits.

The earliest small quantum computers [176] used liquid NMR [129]. NMR technology was already highly advanced due to its use in medicine. The NMR approach uses the nuclear spin state of atoms. Many copies of one molecule are contained in a macroscopic amount of liquid. A quantum bit is encoded in the average spin state of a large number of nuclei. Each qubit corresponds to a particular atom of the molecule, so the atoms for one qubit can be distinguished from those of other qubits by their nuclei's characteristic frequency. The spin states can be manipulated by magnetic fields, and the average spin state can be measured with NMR techniques. NMR quantum computers work at room temperature. However, liquid NMR has severe scaling problems—the measured signal scales as $1/2^n$ with the number of qubits n—so liquid NMR appears unlikely to lead implementation efforts much longer, let alone achieve a scalable quantum computer.

As an example of how hard it is to predict which approaches are most likely to lead to a scalable quantum computer, in 2000 optical approaches were considered unpromising. Optical methods were recognized as the unrivaled approach for quantum communications applications such as quantum key distribution, and also as *flying qubits* sending information between different parts of a quantum computer, because photons do not interact much with other things and so have long decoherence times. This same trait, however, means that it is difficult to get photons to interact with each other, which made them appear unsuitable as the fundamental qubits on which computation would be done. While *nonlinear* optical materials induce some photon-photon interactions, no known material has a strong enough nonlinearity to act as a C_{not} gate, and scientists doubt that such a material will ever be found. Knill, Laflamme, and Milburn's 2001 paper [179] showed how, by clever use of measurement, C_{not} gates could be achieved, avoiding the issue of nonlinear optical elements altogether. While this result, known as the KLM approach, was a huge breakthrough for the field of optical quantum computing, major difficulties remained. The overhead required by these methods was enormous. In 2004, Nielsen showed how this overhead could be greatly reduced by combining the KLM approach with cluster state quantum computing. O'Brien [222] gives a brief but insightful overview of optical approaches to quantum computers, now viewed as one of the more promising approaches in spite of the many hurdles that remain.

Ion trap approaches are currently the most advanced approach that appear possibly scalable. The field has made steady progress. In an ion trap quantum computer [84, 258], individual ions, each representing a qubit, are confined by electric fields. Lasers are directed at individual ions to perform single-qubit quantum gates and two-qubit operations between adjacent ions. All operations necessary for quantum computation have been demonstrated in the laboratory for small numbers of ions. To scale this technology, proposed architectures include quantum memory and processing elements where qubits are moved back and forth either through physical movement of the ions [171] or by using photons to transfer their state [262]. More recently, architectural designs for quantum computers have begun to be studied. Van Meter and Oskin [211] survey architectural issues and approaches for quantum computers.

Many other approaches exist, including cavity QED, neutral atom, and various solid state approaches. See [295] and [157] for descriptions of these approaches, their experimental status at the time the reports were written, and their perceived strengths and weaknesses. Hybrid approaches are also being pursued. Of particular interest are interfaces between optical qubits and qubits in some of these other forms.

Once a quantum information processing device is built, it must be tested to determine if it works as expected and to learn what sorts of errors occur. Finding good, efficient methods of testing is a far from trivial task, given the exponentially large state space, and that measurement affects the state. *Quantum state tomography* studies methods for experimentally characterizing a quantum state by examining multiple copies of the state. *Quantum process tomography* aims to characterize experimentally sequences of operations performed by a device. Early work includes Poyatos et al. [231, 232] and Chuang and Nielsen [83]. D'Ariano et al. provide a review of quantum tomography [94]. While a full characterization of an n-qubit system requires exponentially many probes of the system, some features can be determined with less. Of particular interest is determining the decoherence to which a process is subjected. A recent breakthrough by Emerson et al. provides a symmetrization process that reduces the number of probes needed to characterize the decoherence to only polynomially many [113, 28].

The efforts and success in creating highly entangled states for use in quantum information processing devices have found a number of other applications, and they have enabled deeper experimental exploration of quantum mechanics [157, 295]. Highly entangled states, and the improvements in quantum control, have been used in quantum microlithography to affect matter at scales below the wavelength limit and in quantum metrology to achieve extremely accurate sensors. Applications include clock accuracy beyond that of current atomic clocks, which are limited by the quantum noise of atoms, optical resolution beyond the wavelength limit, ultrahigh resolution spectroscopy, and ultraweak absorption spectroscopy.

13.8 Simulating Quantum Systems

A major application of quantum computers is to the simulation of quantum systems. Long before we have quantum computers capable of simulating any quantum system, special-purpose quantum devices capable of simulating small quantum systems will be built. The simulations run on these special purpose devices will have applications in fields ranging from chemistry to biology to material science. They will also support the design and implementation of yet larger special purpose devices, a process that ideally leads all the way to the building of scalable general-purpose quantum computers.

Early work on quantum simulation of quantum systems includes [285, 196, 289]. Somma et al.'s overview [257] discusses what types of physical problems simulation on quantum computers could solve. Clearly, a simulation cannot efficiently output the amplitudes of the state, as expressed in the standard basis, at all times, since even at just one point in time this information can be exponential in the size of the system. What is meant by a full simulation of a quantum system by a quantum computer is an algorithm that gives a measurement outcome with the same probability as an analogous measurement on the actual system no matter when or what measurement is performed. Even on a universal quantum computer, there are limits to what information can be gained from a simulation. For some quantities of interest, it is not obvious how to extract efficiently that information from a simulation; for some quantities there may be an information theoretic barrier, for others algorithmic advances are needed.

Many quantum systems can be efficiently simulated classically. After all, we live in a quantum world but nevertheless have been able to use classical methods to simulate a wide variety of natural phenomena effectively. Some entangled quantum systems can be efficiently simulated classically [278]. The question of which quantum systems can be efficiently simulated classically remains open. New approaches to classical simulation of quantum systems continue to be developed, many benefiting from the quantum information processing viewpoint [249, 204]. The quantum information processing viewpoint has also lead to improvements in a commonly used classical approach to simulating quantum systems, the DMRG approach [276].

While universal quantum computers will be able to simulate a wide variety of quantum systems, they cannot efficiently simulate some theoretical quantum systems, systems that satisfy Schrödinger's equation but have not been found in nature. They cannot simulate efficiently, even approximately, most quantum systems in the theoretical sense, abstract systems whose dynamics are described by e^{-itH} for some Hamiltonian H. The proof of this fact follows directly from the fact most unitary operators are not efficiently implementable. It is conjectured [99], but not known, that all physically realizable quantum systems are efficiently simulatable on a quantum computer. If it turns out that this conjecture is wrong and a natural phenomenon is discovered that is not efficiently simulatable on quantum computers as we have defined them, then we will have to revise our notion of a quantum computer to incorporate this phenomenon. But we would also have discovered an additional, potentially powerful, computational resource.

13.9 Where Does the Power of Quantum Computation Come From?

Entanglement is the most common answer given as to where the power of quantum computation comes from. Other common answers include quantum parallelism, the exponential size of the state space, and quantum Fourier transforms. Section 7.6 discussed the inadequacy of quantum parallelism and the size of the state space as answers. Quantum Fourier transforms, while central to most quantum algorithms, cannot be the answer in light of the result, mentioned in the reference section of chapter 7, that quantum Fourier transforms can be efficiently simulated classically. The rest of this section is devoted to explaining why the answer *entanglement* is also unsatisfactory, followed by a challenge to our readers to contribute to ongoing efforts to understand what Vlatko Vedral [274] terms "the elusive source of quantum effectiveness."

One reason entanglement is so often cited as the source of quantum computing's power is Jozsa and Linden's result [167] that any pure state quantum algorithm achieving an exponential speedup over classical algorithms must make use of entanglement between a number of qubits that increases with the size of the input to the algorithm. In the same paper, however, Jozsa and Linden speculate that, in spite of this result, entanglement should not be viewed as the key resource for quantum computation. They suggest that similar results can be proved for other properties quite different from entanglement. For example, the Gottesman-Knill theorem, discussed in section 13.4.1, implies that states that do not have polynomially sized stabilizer descriptions are also essential for quantum computation. This property is distinct from entanglement. Since the Clifford group contains the C_{not}, this set of states includes certain entangled states.

An analog of Jozsa and Linden's result does not hold for less dramatic improvements over the classical case. In fact, improvements can be obtained with no entanglement whatsoever; Meyer [213] shows that in the course of the Bernstein-Vazirani algorithm, which achieves an n to 1 reduction in the number of queries required, no qubits become entangled. More obviously, there exist other applications of quantum information processing that require no entanglement. For example, the BB84 quantum key distribution protocol makes no use of entanglement. Looking at the question from the opposite side, many entangled systems have been shown to be classically simulatable [278, 204].

The cluster state model of quantum computation, on the other hand, suggests the centrality of entanglement to quantum computation. Other closely related models with other types of highly entangled initial states have been shown to be universal for quantum computation. While it was known that these states are, in some measures of entanglement, far from maximally entangled, many researchers conjectured that in theory most classes of sufficiently entangled quantum states could be used as the basis of universal one-way quantum computation, but that finding measurement strategies for many of these classes might be prohibitively difficult. This conjecture, however, turns out to be false.

Two groups of researchers [142, 64] showed that most quantum states are too entangled to be useful as a substrate for universal one-way quantum computation. For a few months, it was thought that perhaps these results would not apply to efficiently constructable quantum states, but Low [199] quickly exhibited classes of efficiently constructable quantum states that were too entangled to be useful as the basis for one-way quantum computation. Most of these states, however, are useful for quantum information processing applications such as quantum teleportation.

These observations prompt two questions: what types of entanglement are useful, and for what. As mentioned in chapter 10, multipartite entanglement remains only poorly understood. Another intriguing challenge is to find a view of quantum information processing that makes obvious its limitations. For example, is there a vantage point from which the $\Omega(\sqrt{N})$ lower bound on quantum algorithms for exhaustive search, proved in section 9.3, becomes a one-line observation? The route toward understanding what aspects of quantum mechanics are responsible for the power of quantum information processing is even less obvious. We hope readers of this book will contribute toward an improved understanding of these fundamental questions.

13.10 What if Quantum Mechanics Is Not Quite Correct?

Quantum mechanics may be wrong. Physicists have not yet understood how to reconcile quantum mechanics with general relativity. A complete physical theory would need to make modifications to one of general relativity or quantum mechanics, possibly both. Any modifications to quantum mechanics would have to be subtle, however; quantum mechanics is one of the most tested theories of all time, and its predictions hold to great accuracy. Most of the predictions of quantum mechanics will continue to hold, at least approximately, once a more complete theory is found. Since no one knows how to reconcile the two theories, no one knows what, if any, modifications would be necessary. Once the new physical theory is known, its computational power can be analyzed. In the meantime, theorists have looked at what computational power would be possible if certain changes in quantum mechanics were made.

So far these changes imply greater computational power rather than less; computers built on those principles could do everything a quantum computer could do and substantially more. For example, Abrams and Lloyd [7] showed that if quantum mechanics were nonlinear, even slightly, computation using that nonlinearity could solve all problems in the class **#P**, a class that contains all **NP** problems and substantially more, in polynomial time. Aaronson [4] showed that if a certain exponent in the axioms of quantum mechanics were anything other than 2, all **PP** problems, another class substantially larger than **NP**, would be solvable in polynomial time. These results mean that modification to quantum mechanics would not necessarily destroy the power obtained by computers making use of these physical principles; in fact, in many cases it would increase the power. With these results in mind, Aaronson [5] suggests that limits on computational power should be considered a fundamental principle guiding our search for physical theories of the universe, much as is done for the laws of thermodynamics.

Many intriguing questions as to the extent and source of the power of quantum computation remain, and they are likely to remain for many years while we humans struggle to understand what Nature allows us to compute efficiently and why.

APPENDIXES

A   Some Relations Between Quantum Mechanics and Probability Theory

The inherently probabilistic nature of quantum mechanics is well known, but the close relationship between the formal structures underlying quantum mechanics and probability theory is surprisingly neglected. This appendix describes standard probability theory in a somewhat nonstandard way, in a language closer to the standard way of describing quantum mechanics. This rephrasing illuminates the parallels and differences between the two theories. Probability theory helps in understanding quantum mechanics, not only by placing structures such as tensor products in a more familiar context, but also because the mathematical formalisms underlying quantum theory can be precisely and usefully viewed as an extension of probability theory. This view clarifies relationships between quantum theory and probability theory, including differences between entanglement and classical correlation.

A.1   Tensor Products in Probability Theory

Tensor products are rarely mentioned in probability textbooks, but the tensor product is as much a part of probability theory as of quantum mechanics. The tensor product structure inherent in probability theory should be stressed more often; one of the sources of mistaken intuition about probabilities is a tendency to try to impose the more familiar direct product structure on what is actually a tensor product structure.

Let A be a finite set of n elements. A probability distribution μ on A is a function

$$\mu : A \to [0, 1]$$

such that $\sum_{a \in A} \mu(a) = 1$. The space $\mathcal{P}^A$ of all probability distributions over A has dimension $n - 1$. We can view $\mathcal{P}^A$ as the $(n-1)$-dimensional simplex $\sigma_{n-1} = \{x \in \mathbf{R}^n | x_i \geq 0, x_1 + x_2 + \cdots + x_n = 1\}$, which is contained in the n-dimensional space $\mathbf{R}^A$, the space of all functions from A to $\mathbf{R}$,

$$\mathbf{R}^A = \{f : A \to \mathbf{R}\}$$

(see figure A.1). For $n = 2$, the simplex σ_{n-1} is the line segment from $(1, 0)$ to $(0, 1)$. Each vertex of the simplex corresponds to an element $a \in A$ in that it represents the probability distribution

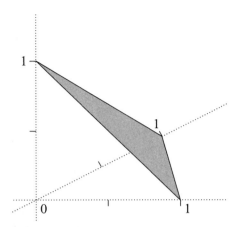

Figure A.1
Simplex σ_2, which corresponds to the set of all probability distributions over a set A of three elements.

that is 1 on a and 0 for all other elements of A. An arbitrary probability distribution μ maps to the point in the simplex $x = (\mu(a_1), \mu(a_2), \ldots, \mu(a_n))$.

Let B be a finite set of m elements. Let $A \times B$ be the Cartesian product $A \times B = \{(a, b)|a \in A, b \in B\}$. What is the relation between $\mathcal{P}^{A \times B}$, the space of all probability distributions over $A \times B$, and the spaces $\mathcal{P}^A$ and $\mathcal{P}^B$? The tempting guess is not correct: $\mathcal{P}^{A \times B} \neq \mathcal{P}^A \times \mathcal{P}^B$. The following dimension check shows that this relationship cannot hold. First, consider the relationship between $\mathbf{R}^{A \times B}$ and $\mathbf{R}^A$ and $\mathbf{R}^B$. Since $A \times B$ has cardinality $|A \times B| = |A||B| = nm$, $\mathbf{R}^{A \times B}$ has dimension nm, which is not equal to $n + m$, the dimension of $\mathbf{R}^A \times \mathbf{R}^B$. Since $dim(\mathcal{P}^A) = dim(\mathbf{R}^A) - 1$, $dim(\mathcal{P}^{A \times B}) = nm - 1$, which is not equal to $n + m - 2$, the dimension of $\mathcal{P}^A \times \mathcal{P}^B$. Thus,

$$\mathcal{P}^{A \times B} \neq \mathcal{P}^A \times \mathcal{P}^B.$$

Instead, $\mathbf{R}^{A \times B}$ is the tensor product $\mathbf{R}^A \otimes \mathbf{R}^B$ of $\mathbf{R}^A$ and $\mathbf{R}^B$, and $\mathcal{P}^{A \times B} \subset \mathbf{R}^A \otimes \mathbf{R}^B$. Before showing that this relationship holds, we give an example to help build intuition.

Example A.1.1 Let $A_0 = \{0_0, 1_0\}$, $A_1 = \{0_1, 1_1\}$, and $A_2 = \{0_2, 1_2\}$. Let 1_0 and 0_0 correspond to whether or not the next person you meet is interested in quantum mechanics, A_1 to whether she knows the solution to the Monty Hall problem, and A_2 to whether she is at least $5'6''$ tall. So $1_0 1_1 0_2$ corresponds to someone under $5'6''$ who is interested in quantum mechanics and knows the solution to the Monty Hall problem. We often write 110 instead of $1_0 1_1 0_2$; the subscripts are implied by the position. A probability distribution over the set of eight possibilities, $A_0 \times A_1 \times A_2$, has form

$$\vec{p} = (p_{000}, p_{001}, p_{010}, p_{011}, p_{100}, p_{101}, p_{110}, p_{111}).$$

More generally, a probability distribution over $A_0 \times A_1 \times \cdots \times A_k$, where the A_i are all 2 element sets, is a vector of length 2^k. We always order the entries so that the binary subscripts increase. Thus, the dimension of the space of probability distributions over the Cartesian product of n two-element sets increases exponentially with n.

This paragraph shows that $\mathbf{R}^{A \times B} = \mathbf{R}^A \otimes \mathbf{R}^B$. Given functions $f : A \to \mathbf{R}$ and $g : B \to \mathbf{R}$, define the tensor product $f \otimes g : A \times B \to \mathbf{R}$ by $(a, b) \mapsto f(a)g(b)$. The reader should check that this definition satisfies the axioms for a tensor product. Furthermore, the linear combination of functions in $\mathbf{R}^{A \times B}$ is a function in $\mathbf{R}^{A \times B}$. Thus $\mathbf{R}^A \otimes \mathbf{R}^B \subseteq \mathbf{R}^{A \times B}$. Conversely, we must show that any function $h \in \mathbf{R}^{A \times B}$ can be written as a linear combination of functions $f_i \otimes g_i$ where $f_i \in \mathbf{R}^A$ and $g_i \in \mathbf{R}^B$. Define a family of functions $f_b^A \in \mathbf{R}^A$, one for each $b \in B$, by

$$f_b^A : A \to \mathbf{R}$$
$$a \mapsto h(a, b).$$

Similarly, for each $a \in A$, define

$$g_a^B : B \to \mathbf{R}$$
$$b \mapsto h(a, b).$$

Furthermore, define the probability distributions

$$\delta_a^A : A \to \mathbf{R}$$
$$a' \mapsto \begin{cases} 1 & \text{if } a = a' \\ 0 & \text{otherwise} \end{cases}$$

and

$$\delta_b^B : B \to \mathbf{R}$$
$$b' \mapsto \begin{cases} 1 & \text{if } b = b' \\ 0 & \text{otherwise.} \end{cases}$$

Then $h(a, b) = \sum_{a' \in A} \delta_{a'}^A(a) g_{a'}^B(b)$, so $h = \sum_{a' \in A} \delta_{a'}^A \otimes g_{a'}^B$. Therefore, $h \in \mathbf{R}^A \otimes \mathbf{R}^B$. For completeness, we mention that by symmetry $h = \sum_{b' \in B} f_{b'}^A \otimes \delta_{b'}^B$.

Now let us restrict our attention to probability distributions. If μ and ν are probability distributions, then so is $\mu \otimes \nu$:

$$\sum_{(a,b) \in A \times B} (\mu \otimes \nu)(a, b) = \sum_{(a,b) \in A \times B} \mu(a)\nu(b)$$

$$= \sum_a \sum_b \mu(a)\nu(b)$$

$$= \left(\sum_a \mu(a) \right) \left(\sum_b \nu(b) \right) = 1.$$

Furthermore, the linear combination of probability distributions is a probability distribution as long as the linear factors sum to 1. Conversely, we show that any probability distribution $\eta \in \mathcal{P}^{A \times B}$ is the linear combination of tensor products of probability distributions in $\mathcal{P}^A$ and $\mathcal{P}^B$ with linear factors summing to 1. Define a family of probability distributions, one for each $a \in A$,

$$h_a^B : B \to \mathbf{R}$$

$$b \mapsto \frac{\eta(a, b)}{\sum_{b' \in B} \eta(a, b')}.$$

Let $c_a = \sum_{b \in B} \eta(a, b)$ and $c_b = \sum_{a \in A} \eta(a, b)$. Observe that $\sum_{a \in A} c_a = 1$. Then

$$\eta(a, b) = \sum_{a' \in A} c_{a'} h_{a'}^B \delta_{a'}^A$$

$$= \sum_{a' \in A} c_{a'} \delta_{a'}^A \otimes h_{a'}^B.$$

Since $\delta_{a'}^A$ is a probability distribution in $\mathcal{P}^A$, every probability distribution over $A \times B$ is in $\mathcal{P}^A \otimes \mathcal{P}^B$.

A joint distribution $\mu \in \mathcal{P}^{A \times B}$ is *independent* or *uncorrelated* with respect to the decomposition $\mathcal{P}^A \otimes \mathcal{P}^B$ if it can be written as a tensor product $\mu_A \otimes \mu_B$ of distributions $\mu_A \in \mathcal{P}^A$ and $\mu_B \in \mathcal{P}^B$. The vast majority of joint distributions do not have this form, in which case they are *correlated*. For any joint distribution $\mu \in \mathcal{P}^{A \times B}$, define a *marginal* distribution $\mu_A \in \mathcal{P}^A$ by

$$\mu_A : a \mapsto \sum_{b \in B} \mu(a, b).$$

An uncorrelated distribution is the tensor product of its marginals. Other distributions cannot be reconstructed from their marginals; information has been lost. One of the sources of mistaken intuition about probabilities is a tendency to try to impose the more familiar direct product structure, which does support reconstruction, on what is actually a tensor product structure; the relationship between a distribution and its marginals properly understood only within a tensor product structure.

A distribution $\mu : A \to R$ that is concentrated entirely at one element is said to be *pure*; on a set A of n elements there are exactly n pure distributions $\mu_a : A \to [0, 1]$, one for each element of A, where

$$\mu_a : a' \mapsto \begin{cases} 1 & \text{if } a' = a \\ 0 & \text{otherwise.} \end{cases}$$

These are exactly the distributions that correspond to the vertices of the simplex. All other distributions are said to be *mixed*.

When an observation is made, the probability distribution is updated accordingly. All states incompatible with the observation are ruled out, and the remaining probabilities are normalized to sum to 1. Much noise is made about the *collapse* of the state due to quantum measurement. But this *collapse* occurs in classical probability; it is known as *updating* a probability distribution in light of new information.

Example A.1.2 Suppose your friend is about to toss two fair coins. The probability distribution for the four outcomes HH, HT, TH, and TT is $\vec{p}_I = (1/4, 1/4, 1/4, 1/4)$. After she tosses the two coins, she tells you that the two coins agreed. To compute the new probability distributions, the possibilities compatible with your friend's observation, HT and TH, are ruled out, and the remaining possibilities are normalized to sum to 1, resulting in the probability distribution $\vec{p}_F = (1/2, 0, 0, 1/2)$.

Example A.1.3 Let us return to the example of the traits for the next person you meet. Unless you know all of these traits, the distribution $\vec{p}_I = (p_{000}, \ldots, p_{111})$ is a mixed distribution. When you meet the person you can observe her traits. Once you have made these observations, the distribution *collapses* to a pure distribution. For example, if the person is interested in quantum mechanics, does not know the solution to the Monty Hall problem, and is 5'8", the *collapsed* distribution is $\vec{p}_F = (0, 0, 0, 0, 0, 1, 0, 0)$.

The true surprise in quantum mechanics is that quantum states cannot generally be modeled by probability distributions — the content of Bell's theorem. Overly simplified versions of the EPR paradox, in which only one basis is considered, reduce to an unsurprising classical result that instant, faster-than-light knowledge of a faraway state may be possible upon the observation of a local state.

Example A.1.4 Suppose someone prepares two sealed envelopes with identical pieces of paper and sends them to opposite sides of the universe. Half the time, both envelopes contain 0; half the time, 1. The initial distribution is $\vec{p}_I = (1/2, 0, 0, 1/2)$. If someone then opens one of the envelopes and observes a 0, the state of the contents of the other envelope is immediately known — known faster than light can travel between the envelopes — and the distribution after the observation is $\vec{p}_F = (1, 0, 0, 0)$.

To understand fully the relationship between quantum mechanics and probability theory, it is useful to view probability distributions as operators. Consider the set of linear operators $\mathcal{M}^A = \{M : \mathbf{R}^A \to \mathbf{R}^A\}$. To every function $f : A \to \mathbf{R}$, there is an associated operator $M_f : \mathbf{R}^A \to \mathbf{R}^A$ given by $M_f : g \mapsto fg$. In particular, for any probability distribution μ on A, there is an associated operator $M_\mu : \mathbf{R}^A \to \mathbf{R}^A$. An operator M is said to be a projector if $M^2 = M$. The set of probability distributions μ whose corresponding operators M_μ are projectors is exactly the set of pure distributions. The matrix for the operator corresponding to a function is always diagonal. For a probability distribution, this matrix is trace 1 as well as diagonal. For example, the operator corresponding to the probability distribution $\vec{p}_I = (1/2, 0, 0, 1/2)$ has matrix

$$\begin{pmatrix} 1/2 & 0 & 0 & 0 \\ 0 & 0 & 0 & 0 \\ 0 & 0 & 0 & 0 \\ 0 & 0 & 0 & 1/2 \end{pmatrix}.$$

Updating the probability distribution with information from an observation involves setting some of the matrix entries to zero and renormalizing the diagonal to sum to 1.

Example A.1.5 The matrix for the initial probability distribution in example A.1.2 is

$$\begin{pmatrix} 1/4 & 0 & 0 & 0 \\ 0 & 1/4 & 0 & 0 \\ 0 & 0 & 1/4 & 0 \\ 0 & 0 & 0 & 1/4 \end{pmatrix}.$$

The matrix for the updated probability distribution after the measurement involves setting the probabilities of HT and TH to 0 and renormalizing the matrix to obtain a trace 1 matrix:

$$\begin{pmatrix} 1/2 & 0 & 0 & 0 \\ 0 & 0 & 0 & 0 \\ 0 & 0 & 0 & 0 \\ 0 & 0 & 0 & 1/2 \end{pmatrix}.$$

Example A.1.6 The matrix for the initial probability distribution in example A.1.3 is

$$\begin{pmatrix} 1/2 & 0 & 0 & 0 \\ 0 & 0 & 0 & 0 \\ 0 & 0 & 0 & 0 \\ 0 & 0 & 0 & 1/2 \end{pmatrix}.$$

The matrix for the updated probability distribution after the envelope has been opened involves setting the probability of both envelopes containing 1 to 0 and renormalizing the matrix to obtain a trace 1:

$$\begin{pmatrix} 1 & 0 & 0 & 0 \\ 0 & 0 & 0 & 0 \\ 0 & 0 & 0 & 0 \\ 0 & 0 & 0 & 0 \end{pmatrix}.$$

A.2 Quantum Mechanics as a Generalization of Probability Theory

The remainder of this appendix relies on the density operator formalism and the notions of pure and mixed quantum from section 10.1. This section describes how pure and mixed quantum states generalize the classical notion of pure and mixed probability distributions. This viewpoint helps clarify the distinction between quantum entanglement and classical correlations in mixed quantum states.

Let ρ be a density operator. Section 10.1.2 showed that every density operator ρ can be written as a probability distribution over pure quantum states $\sum_i p_i |\psi_i\rangle\langle\psi_i|$, where the $|\psi_i\rangle$ are mutually orthogonal eigenvectors of ρ, and the p_i are the eigenvalues, with $p_i \in [0, 1]$ and $\sum p_i = 1$. Conversely, any probability distribution μ over a set of orthogonal quantum states $|\psi_1\rangle, |\psi_2\rangle, \ldots, |\psi_L\rangle$ with $\mu : |\psi_i\rangle \to p_i$ has a corresponding density operator $\rho_\mu = \sum_i p_i |\psi_i\rangle\langle\psi_i|$. In the basis $\{|\psi_i\rangle\}$, the density operator ρ_μ is diagonal:

$$\begin{pmatrix} p_1 & & & \\ & p_2 & & \\ & & \ddots & \\ & & & p_L \end{pmatrix}.$$

Thus, a probability distribution over a set of orthonormal quantum states $\{|\psi_i\rangle\}$ can be viewed as a trace 1 diagonal matrix acting on $\mathbf{R}^L$. Under the isomorphism between $\mathbf{R}^L$ and the subspace of V generated by $|\psi_1\rangle, |\psi_2\rangle, \ldots, |\psi_L\rangle$, the density operator ρ_μ realizes the operator M_μ of section A.1; a probability distribution over a set of orthonormal quantum states $\{|\psi_i\rangle\}$ can be viewed as a trace 1 diagonal matrix acting on $\mathbf{R}^L$. In this way, density operators are a direct generalization of probability distributions.

Although every density operator can be viewed as a probability distribution over a set of orthogonal quantum states, this representation is not unique in general. More importantly, for most pairs of density operators ρ_1 and ρ_2, there is no basis over which both ρ_1 and ρ_2 are diagonal. Thus, although each density operator of dimension N can be viewed as a probability distribution over N states, the space of all density operators is much larger than the space of probability distributions over N states; the space of all density operators contains many different overlapping copies of the space of probability distributions over N states, one for each orthonormal basis.

Let $\rho : V \to V$ be a density operator. By exercise 10.4 a density operator ρ corresponds to a pure state if and only if it is a projector. This statement is analogous to that for probability distributions; the pure states correspond exactly to rank 1 density operators, and mixed states have rank greater than 1. As explained in section 10.3, density operators are also used to model probability distributions over pure states, particularly probability distributions over the possible outcomes of a measurement yet to be performed. This use is analogous to the classical use of probability distributions to model the probabilities of possible traits before they can be observed.

A pure quantum state $|\psi\rangle$ is *entangled* with respect to the tensor decomposition into single qubits if it cannot be written as the tensor product of single-qubit states. For a mixed quantum state, it is important to determine if all of its correlation comes from being a mixture in the classical sense or if it is also correlated in a quantum fashion. A mixed quantum state $\rho : V \otimes W \to V \otimes W$ is said to be *uncorrelated* with respect to the decomposition $V \otimes W$ if $\rho = \rho_V \otimes \rho_W$ for some density operators $\rho_V : V \to V$ and $\rho_W : W \to W$. Otherwise ρ is said to be *correlated*. A mixed quantum state ρ is said to be *separable* if it can be written $\rho = \sum_{j=1}^{L} p_j |\psi_j^V\rangle\langle\psi_j^V| \otimes |\phi_j^W\rangle\langle\phi_j^W|$ where $|\psi_j^V\rangle \in V$ and $|\phi_j^W\rangle \in W$. In other words, ρ is separable if all the correlation comes from its being a classical mixture of uncorrelated quantum states. If a mixed state ρ is not separable, it is *entangled*. For example, the mixed state $\rho_{cc} = \frac{1}{2}(|00\rangle\langle00| + (|11\rangle\langle11|)$ is classically correlated but not entangled, whereas the Bell state $|\Phi^+\rangle\langle\Phi^+| = \frac{1}{2}(|00\rangle + |11\rangle)(\langle00| + \langle11|)$ is entangled. The marginals of a pure distribution are always pure, but the analogous statement is not true for quantum states; all of the partial traces of a pure state are pure only if the original pure state was not entangled. The partial traces of the Bell state $|\Phi^+\rangle$, a pure state, are not pure. Most pure quantum states are entangled, exhibiting quantum correlations with no classical analog. All pure probability distributions are completely uncorrelated.

Classical and quantum analogs:

Classical probability	Quantum mechanics
probability distribution μ viewed as operator M_μ	density operator ρ
pure distribution: M_μ is a projector	pure state: ρ is a projector
simplex: $\sigma_{n-1} = \{x \in \mathbf{R}^n \mid x_i \geq 0, x_1 + \cdots + x_n = 1\}$	Bloch region: set of trace 1 positive Hermitian operators
marginal distribution	partial trace
A distribution is *uncorrelated* if it is the tensor product of its marginals	A state is *uncorrelated* if it is the tensor product of its partial traces

Key difference:

Classical	Quantum
pure distributions are always uncorrelated	pure states contain no classical correlation but can be entangled
A marginal of a pure distribution is a pure distribution	The partial trace of a pure state may be a mixed state

A.3 References

The view of quantum mechanics as an extension of probability theory is discussed in many quantum mechanics references, particularly those concerned with the deeper mathematical aspects of the theory. Aaronson gives a playful account [3]. Rieffel treats this subject in [241]. In their chapter on quantum probability, Kitaev et al. outline parallels between quantum mechanics and probability theory [175]. Kuperberg's *A Concise Introduction to Quantum Probability, Quantum Mechanics, and Quantum Computation* also serves as an excellent reference [188]. Sudbery [267] gives a brief account in his section of *Statistical Formulations of Classical and Quantum Mechanics*. An early account of some of these ideas can be found in Mackey's *Mathematical Foundations of Quantum Mechanics*. Strocchi's *An Introduction to the Mathematical Structure of Quantum Mechanics* gives a detailed and readable account [266]. A number of papers by Summers, including [236], address relations and distinctions between quantum mechanics and probability theory.

A.4 Exercises

Exercise A.1. Show that an independent joint distribution is the tensor product of its marginals.

Exercise A.2. Show that a general distribution cannot be reconstructed from its marginals. Exhibit three distinct distributions with the same marginals.

Exercise A.3.

a. Show that the tensor product of a pure distribution is pure.

b. Show that any distribution is a linear combination of pure distributions. Conclude that the set of distributions on a finite set A is convex.

c. Show that any pure distribution on a joint system $A \times B$ is uncorrelated.

d. A distribution is said to be extremal if it cannot be written as a linear combination of other distributions. Show that the extremal distributions are exactly the pure distributions.

Exercise A.4. Show that the probability distributions μ whose corresponding operators M_μ are projectors are exactly the pure distributions.

Exercise A.5. For each of the states $|0\rangle$, $|-\rangle$, and $|i\rangle = \frac{1}{\sqrt{2}}(|0\rangle + i|1\rangle)$, give the matrix for the corresponding density operator in the standard basis, and write each of these states as a probability distribution over pure states. For which of these states is this distribution unique?

Exercise A.6.

a. Give an example of three density operators no two of which can be simultaneously diagonalized in that there does not exist a basis with respect to which both are diagonal.

b. Show that if a set of density operators commute, then they can be simultaneously diagonalized.

Exercise A.7.   Show that the binary operator $f \otimes g : (a, b) \mapsto f(a)g(b)$ for $f \in \mathbf{R}^A$ and $g \in \mathbf{R}^B$ satisfies the relations defining a tensor product structure on $\mathbf{R}^{A \times B}$ given in section 3.1.2.

Exercise A.8.   Show that a separable pure state must be uncorrelated.

Exercise A.9.   Show that if a density operator $\rho \in V \otimes W$ is uncorrelated with respect to the tensor decomposition $V \otimes W$, then it is the tensor product of its partial traces with respect to V and W.

B Solving the Abelian Hidden Subgroup Problem

This appendix covers the solution to the Abelian hidden subgroup problem using a generalization of Shor's factoring algorithm. Recall from box 8.4 that any finite Abelian group can be written as the product of cyclic groups.

Finite Abelian Hidden Subgroup Problem Let G be a finite Abelian group with cyclic decomposition $G \cong \mathbf{Z}_{n_0} \times \cdots \times \mathbf{Z}_{n_L}$. Suppose G contains a subgroup $H < G$ that is implicitly defined by a function f on G in that f is constant and distinct on every coset of H. Find a set of generators for H.

This appendix shows that, for finite Abelian groups, if

$$U_f : |g\rangle|0\rangle \rightarrow |g\rangle|f(g)\rangle$$

can be computed in poly-log time, then generators for H can be computed in poly-log time.

This appendix makes use of deeper aspects of group theory, such as group representations, than the rest of the book. Basic elements of group theory were reviewed in the boxes accompanying section 8.6. Section B.1 reviews group representations of finite Abelian groups, including Schur's lemma. Section B.2 defines quantum Fourier transforms over finite Abelian groups. Section B.3 explains how these quantum Fourier transforms enable the solution of the Abelian hidden subgroup problem. Section B.4 looks at Simon's problem and Shor's factoring algorithm as instances of this general solution to the Abelian hidden subgroup problem. The appendix concludes in section B.5 with a few remarks on the non-Abelian hidden subgroup problem.

B.1 Representations of Finite Abelian Groups

A *representation* of an Abelian group G is a group homomorphism χ from G to the multiplicative group of complex numbers $\mathbf{C}$:

$$\chi : G \rightarrow \mathbf{C}.$$

More generally, representations of groups are group homomorphisms into the space of linear operators on a vector space. However, in the Abelian case it suffices to consider only *characters*, the representations into the multiplicative group of complex numbers.

For the additive group $\mathbf{Z}_n$, the homomorphism condition implies that any representation χ of $\mathbf{Z}_n$ must send $0 \mapsto 1$, and the generator 1 of $\mathbf{Z}_n$ must map to one of the n roots of unity since

$$\sum_{i=1}^{n} 1 = 0$$

implies

$$\prod_{i=1}^{n} \chi(1) = \chi(0) = 1.$$

Since $\chi(1)$ determines the image of all other elements in $\mathbf{Z}_n$ there can be at most n representations. Any nth root of unity works, so the n representations χ_j

$$\chi_j : x \mapsto \exp\left(\frac{2\pi \mathbf{i}}{n} jx\right)$$

for all $j \in \mathbf{Z}_n$ form the complete set of representations of $\mathbf{Z}_n$. Many of the representations are not one-to-one: for example the trivial representation that we have labeled by $0 \in \mathbf{Z}_n$ sends all group elements to 1. We have labeled the representations by group elements $j \in \mathbf{Z}_n$ in one way. We use this labeling as our standard labeling throughout this appendix. Other labelings by group elements are possible.

More generally, for any Abelian group, the homomorphism condition $\chi(gh) = \chi(g)\chi(h)$ implies that $\chi(e) = 1$, $\chi(g^{-1}) = \overline{\chi(g)}$, and that every $\chi(g)$ is a kth root of unity, where k is the order of g. An Abelian group of order $|G|$ has exactly $|G|$ distinct representations χ_i.

Example B.1.1   The two representations for $\mathbf{Z}_2$ are $\chi_i(j) = -1^{ij}$ or

$$\chi_0(x) = 1$$

$$\chi_1(x) = \begin{cases} 1 & \text{if } x = 0 \\ -1 & \text{if } x = 1. \end{cases}$$

Example B.1.2   The four representations $\chi_i(j) = \exp(2\pi \mathbf{i}\frac{ij}{4})$ of $\mathbf{Z}_4$ are given in the following table:

	0	1	2	3
χ_0	1	1	1	1
χ_1	1	i	-1	$-i$
χ_2	1	-1	1	-1
χ_3	1	$-i$	-1	i

The representations of a product $\mathbf{Z}_n \times \mathbf{Z}_m$ can be defined in terms of the representation of each of its factors. Let χ_i be the n different representations of $\mathbf{Z}_n$ and χ'_j be the m different representations of $\mathbf{Z}_m$. Then

$$\hat{\chi}_{ij}((g, h)) = \chi_i(g)\chi'_j(h)$$

are all nm distinct representations of $\mathbf{Z}_n \times \mathbf{Z}_m$. We have labeled these representations by group elements $(i, j) \in \mathbf{Z}_n \times \mathbf{Z}_m$.

Example B.1.3 The 2^n representations of $\mathbf{Z}_2^n$ have a particularly nice form. If we write each element b of $\mathbf{Z}_2^n$ as $b = (b_0, b_1, \ldots, b_{n-1})$, where each b_i is a binary variable, then the group representation χ_b is the n-way product of the two representations χ_0 and χ_1 for $\mathbf{Z}_2$,

$$\chi_b(a) = \chi_{b_0}(a_0), \ldots, \chi_{b_{n-1}}(a_{n-1}) = (-1)^{a \cdot b},$$

where $a \cdot b$ is the standard dot product of the vectors a and b.

Since any finite Abelian group is isomorphic to a finite product $\mathbf{Z}_{n_0} \times \cdots \times \mathbf{Z}_{n_k}$ of cyclic groups, the definition of χ_i, together with the result about representations for product groups, provides an effective way to construct all of the representations for any finite Abelian group. These representation may be labeled by group elements as before.

For Abelian groups, the set of representations itself forms a group denoted by $\hat{G}$ where

- the representation $\chi(g) = 1$ for all $g \in G$ is the identity,
- the product $\chi = \chi_i \circ \chi_j$ of two representations χ_i and χ_j defined by $\chi(g) = \chi_i(g)\chi_j(g)$ for all $g \in G$ is itself a representation, and
- the inverse of any representation χ is defined by

$$\chi^{-1}(g) = 1/\chi(g) = \overline{\chi(g)}$$

for all $g \in G$.

For a subgroup $H < G$, let $H^\perp = \{g \in G | \chi_g(h) = 1, \forall h \in H\}$. Since G is Abelian, the set of cosets of H in G forms a group G/H, the quotient group of G modulo H, of order $[G : H] = |G|/|H|$. The $[G : H]$ representations of G/H are in one-to-one correspondence with representations of G that map all elements of H to 1. Thus, there are exactly $[G : H]$ representations in $H^\perp$. The set $H^\perp$ forms a group that has representations in its own right. Since $H^\perp$ has size $[G : H]$, there are exactly $[G : H]$ distinct representations of the group $H^\perp$. An element $g' \in G$ acts as a representation $\chi_{g'}$ of $H^\perp$ in the following way:

$$g' : H^\perp \to \mathbf{C}$$
$$\chi_g \mapsto \chi_g(g').$$

Not all of these representations are distinct, however. All $h \in H$ act as the trivial representation on $H^\perp$:

$$h : H^\perp \to \mathbf{C}$$
$$\chi_g \mapsto \chi_g(h) = 1.$$

The group $H^{\perp\perp} = \{g' \in G | \chi_{g'}(g) = 1, \forall g \in H^\perp\}$ has size $|G|/[G : H] = |H|$. By definition of $H^\perp$ and $\chi_{g'}$,

$$H^{\perp\perp} = \{g' \in G | g'(\chi_g) = 1, \forall g \in H^\perp\}$$
$$= \{g' \in G | \chi_g(g') = 1, \forall g \in H^\perp\}.$$

Thus, all elements of H are contained in $H^{\perp\perp}$. Since $|H^{\perp\perp}| = |H|$,

$$H^{\perp\perp} = H.$$

Chapter 11 discusses groups C that are classical error correcting codes. The dual group $C^\perp$ to a classical code C is defined in the way we just discussed. Classical codes and their duals form the basis for the construction of the quantum CSS codes discussed in section 11.3.

Example B.1.4 Any subgroup H of $G = \mathbf{Z}_2^n$ is isomorphic to $\mathbf{Z}_2^k$ for some k. Since there are $[G : H] = 2^{n-k}$ elements of $H^\perp < G$, $H^\perp$ is isomorphic to $\mathbf{Z}_2^{n-k}$. Using the expression for the representations of $\mathbf{Z}_2^n$ of example B.1.3, the elements of $H^\perp$ are the elements b such that $\chi_b(a) = (-1)^{a \cdot b} = 1$ for all $a \in H$. Thus, $H^\perp = \{b | a \cdot b = 0 \bmod 2, \forall a \in H\}$.

To define the quantum Fourier transform for a general Abelian group, we need a technical result, Schur's lemma, that is a generalization of identity 11.7 for the Walsh-Hadamard transformation.

B.1.1 Schur's Lemma

Schur's lemma   Let χ_i and χ_j be representations of an Abelian group G. Then,

$$\sum_{g \in G} \chi_i(g)\overline{\chi_i(g)} = |G|,$$

and

$$\sum_{g \in G} \chi_i(g)\overline{\chi_j(g)} = 0 \text{ for } \chi_i \neq \chi_j.$$

The first case follows by observing that $\omega\overline{\omega} = 1$ for any root of unity. For $i \neq j$,

$$\chi_i(h) \sum_{g \in G} \chi_i(g)\overline{\chi_j(g)} = \sum_{g \in G} \chi_i(h)\chi_i(g)\overline{\chi_j(g)}$$

$$= \sum_{g \in G} \chi_i(hg)\overline{\chi_j(h^{-1}hg)}$$

$$= \sum_{g \in G} \chi_i(g)\overline{\chi_j(h^{-1}g)}$$

$$= \sum_{g \in G} \chi_i(g)\chi_j(h)\overline{\chi_j(g)}$$

$$= \chi_j(h) \sum_{g \in G} \chi_i(g)\overline{\chi_j(g)}.$$

Since $\chi_i(h) \neq \chi_j(h)$ for some h, it follows that $\sum_{g \in G} \chi_i(g)\overline{\chi_j(g)} = 0$.

If we think of χ_i as a complex vector of n elements $(\chi_i(g_0), \ldots \chi_i(g_{n-1}))$, then Schur's lemma says that χ_i has length $|G|$ and any two different vectors χ_i and χ_j are orthogonal.

Schur's lemma for subgroups A simple corollary of Schur's lemma holds for representations χ of G restricted to subgroups:

$$\sum_{h \in H} \chi(h) = \begin{cases} |H| & \text{if } \chi(h) = 1, \forall h \in H \\ 0 & \text{otherwise.} \end{cases}$$

Since any representation χ of G is a representation of H when restricted to H, we can apply Schur's lemma directly to χ viewed as a representation of H to obtain this equality.

B.2 Quantum Fourier Transforms for Finite Abelian Groups

This section defines quantum Fourier transforms over finite Abelian groups. Section B.2.1 defines the Fourier basis for an Abelian group. This basis is used in the definition of the quantum Fourier transform over an Abelian group given in section B.2.2.

B.2.1 The Fourier Basis of an Abelian Group

To an Abelian group G with $|G| = n$, we associate an n-dimensional complex vector space V by labeling a basis for the vector space with the n elements of the group $\{|g_0\rangle, \ldots |g_{n-1}\rangle\}$. The Fourier transform of section 7.8 takes elements of this basis to another, the Fourier basis. As the first step to generalizing the Fourier transform to general Abelian groups, this section defines the Fourier basis for V corresponding to the basis $\{|g_0\rangle, \ldots |g_{n-1}\rangle\}$. The Fourier basis is defined in terms of the set of group representations χ_g of G.

A group G acts in a natural way upon itself: for every group element $g \in G$, there is a map from G to G that sends a to ga for all elements $a \in G$. This map can be viewed as a unitary transform T_g acting on V that takes

$$T_g : |a\rangle \rightarrow |ga\rangle.$$

The transformation T_g is unitary for any g because it is reversible, $T_{g^{-1}} T_g = I$, and maps basis states to basis states.

The Fourier basis of G with respect to a particular labeling χ_g of the representations of G consists of all $\{|e_k\rangle | k \in G\}$, where

$$|e_k\rangle = \frac{1}{\sqrt{|G|}} \sum_{g \in G} \overline{\chi_k(g)} |g\rangle.$$

From Schur's lemma and the fact that $\langle g'|g\rangle = 0$ for $g \neq g'$, it is easy to see that this set forms a basis, since

$$\langle e_j | e_k \rangle = \frac{1}{|G|} \left(\sum_{g' \in G} \chi_j(g') \langle g'| \right) \left(\sum_{g \in G} \overline{\chi_k(g)} |g\rangle \right)$$

$$= \frac{1}{|G|} \sum_{g' \in G} \sum_{g \in G} \chi_j(g') \overline{\chi_k(g)} \langle g'|g\rangle$$

$$= \frac{1}{|G|} \sum_{g \in G} \chi_j(g) \overline{\chi_k(g)}$$

$$= \delta_{jk}.$$

For each $k \in G$, the vector $|e_k\rangle$ is an eigenvector of $T_j = |h\rangle \mapsto |jh\rangle$ with eigenvalue $\chi_k(j)$:

$$T_j |e_k\rangle = \frac{1}{\sqrt{|G|}} \sum_{g \in G} \overline{\chi_k(g)} T_j |g\rangle$$

$$= \frac{1}{\sqrt{|G|}} \sum_{g \in G} \overline{\chi_k(g)} |jg\rangle$$

$$= \frac{1}{\sqrt{|G|}} \sum_{g \in G} \overline{\chi_k(j^{-1}) \chi_k(jg)} |jg\rangle$$

$$= \chi_k(j) \frac{1}{\sqrt{|G|}} \sum_{h \in G} \overline{\chi_k(h)} |h\rangle$$

$$= \chi_k(j) |e_k\rangle.$$

Example B.2.1 The Fourier basis for $\mathbf{Z}_2$ is

$$|e_0\rangle = \tfrac{1}{\sqrt{2}}(\overline{\chi_0(0)}|0\rangle + \overline{\chi_0(1)}|1\rangle) = \tfrac{1}{\sqrt{2}}(|0\rangle + |1\rangle)$$
$$|e_1\rangle = \tfrac{1}{\sqrt{2}}(\overline{\chi_1(0)}|0\rangle + \overline{\chi_1(1)}|1\rangle) = \tfrac{1}{\sqrt{2}}(|0\rangle - |1\rangle).$$

Similarly, for $\mathbf{Z}_4$ we get

$$|e_0\rangle = \tfrac{1}{2}\sum_{i=0}^{3}\overline{\chi_0(i)}|i\rangle = \tfrac{1}{2}(|0\rangle + |1\rangle + |2\rangle + |3\rangle)$$
$$|e_1\rangle = \tfrac{1}{2}\sum_{i=0}^{3}\overline{\chi_1(i)}|i\rangle = \tfrac{1}{2}(|0\rangle - \mathbf{i}|1\rangle - |2\rangle + \mathbf{i}|3\rangle)$$
$$|e_2\rangle = \tfrac{1}{2}\sum_{i=0}^{3}\overline{\chi_2(i)}|i\rangle = \tfrac{1}{2}(|0\rangle - |1\rangle + |2\rangle - |3\rangle)$$
$$|e_3\rangle = \tfrac{1}{2}\sum_{i=0}^{3}\overline{\chi_3(i)}|i\rangle = \tfrac{1}{2}(|0\rangle + \mathbf{i}|1\rangle - |2\rangle - \mathbf{i}|3\rangle).$$

B.2.2 The Quantum Fourier Transform Over a Finite Abelian Group

The quantum Fourier transform for an Abelian group G is the transformation $\mathcal{F}$ that maps $|e_g\rangle$ to $|g\rangle$,

$$\mathcal{F} = \sum_{g\in G}|g\rangle\langle e_g|.$$

Consider the effect of $\mathcal{F}$ on a group element $|h\rangle$. With $\langle e_k| = \frac{1}{\sqrt{|G|}}\sum_{g\in G}\chi_k(g)\langle g|$ we get

$$\langle e_k|h\rangle = \frac{1}{\sqrt{|G|}}\sum_{g\in G}\chi_k(g)\langle g|h\rangle = \frac{1}{\sqrt{|G|}}\chi_k(h)$$

and thus,

$$\mathcal{F}|h\rangle = \sum_{g\in G}|g\rangle\langle e_g|h\rangle = \frac{1}{\sqrt{|G|}}\sum_{g\in G}\chi_g(h)|g\rangle.$$

It follows that the matrix for $\mathcal{F}$ in the standard basis has entries

$$\mathcal{F}_{gh} = \langle g|\mathcal{F}|h\rangle = \frac{\chi_g(h)}{\sqrt{|G|}}.$$

The inverse Fourier transform is

$$\mathcal{F}^{-1} = \sum_{g\in G}|e_g\rangle\langle g|.$$

With $\mathcal{F}^{-1}|h\rangle = |e_h\rangle = \frac{1}{\sqrt{|G|}}\sum_{g\in G}\overline{\chi_h(g)}|g\rangle$, the matrix for $\mathcal{F}^{-1}$ in the standard basis has entries

$$\mathcal{F}_{gh}^{-1} = \frac{\overline{\chi_h(g)}}{\sqrt{|G|}}.$$

Suppose that $\mathcal{F}_G$ and $\mathcal{F}_H$ are Fourier transforms for G and H, respectively. If the elements $(g, h) \in G \times H$ are encoded as $|g\rangle|h\rangle$, then $\mathcal{F}_{G \times H} = \mathcal{F}_G \otimes \mathcal{F}_H$ is a Fourier transform for $G \times H$.

Example B.2.2 The Hadamard transformation H is the Fourier transform for $\mathbf{Z}_2$:

$$\mathcal{F}_2 = \frac{1}{\sqrt{2}} \begin{pmatrix} \chi_0(0) & \chi_0(1) \\ \chi_1(0) & \chi_1(1) \end{pmatrix} = \frac{1}{\sqrt{2}} \begin{pmatrix} 1 & 1 \\ 1 & -1 \end{pmatrix} = H.$$

The k-bit Walsh-Hadamard transform W is the Fourier transform for $\mathbf{Z}_2^k$. In standard labeling, the representations for $\mathbf{Z}_2^k$ are of the form $\chi_i(j) = (-1)^{i \cdot j}$. For instance, $\mathcal{F}_{2 \times 2}$, for Fourier transform for $\mathbf{Z}_2 \times \mathbf{Z}_2$ is

$$\mathcal{F}_{2 \times 2} = H \otimes H = \frac{1}{2} \begin{pmatrix} 1 & 1 & 1 & 1 \\ 1 & -1 & 1 & -1 \\ 1 & 1 & -1 & -1 \\ 1 & -1 & -1 & 1 \end{pmatrix}.$$

By comparison, $\mathcal{F}_4$, the Fourier transform for $\mathbf{Z}_4$ is

$$\mathcal{F}_4 = \frac{1}{2} \begin{pmatrix} \mathbf{i}^0 & \mathbf{i}^0 & \mathbf{i}^0 & \mathbf{i}^0 \\ \mathbf{i}^0 & \mathbf{i}^1 & \mathbf{i}^2 & \mathbf{i}^3 \\ \mathbf{i}^0 & \mathbf{i}^2 & \mathbf{i}^4 & \mathbf{i}^6 \\ \mathbf{i}^0 & \mathbf{i}^3 & \mathbf{i}^6 & \mathbf{i}^9 \end{pmatrix} = \frac{1}{2} \begin{pmatrix} 1 & 1 & 1 & 1 \\ 1 & \mathbf{i} & -1 & -\mathbf{i} \\ 1 & -1 & 1 & -1 \\ 1 & -\mathbf{i} & -1 & \mathbf{i} \end{pmatrix}.$$

The quantum Fourier transform can be defined for non-Abelian groups as well. The definition is in terms of group representations, but the set of representations for non-Abelian groups is much more complicated than for the Abelian case. All of these quantum Fourier transforms have efficient implementations. Even in the Abelian case, some of the implementations are simpler than others. One useful property is that if U_1 and U_2 are two quantum algorithms implementing the quantum Fourier transforms for groups G_1 and G_2 respectively, then $U_1 \otimes U_2$ implements the quantum Fourier transform for $G_1 \times G_2$. Section 7.8 gave an $O(n^2)$ implementation for quantum Fourier transforms over the groups $\mathbf{Z}_{2^n}$. Section B.6 gives pointers to papers on efficient implementations for quantum Fourier transforms over other groups. We now turn to the use of quantum Fourier transforms in solving the hidden subgroup problem for Abelian groups.

B.3   General Solution to the Finite Abelian Hidden Subgroup Problem

This section explains how to solve the finite Abelian hidden subgroup problem. Suppose a group G, with cyclic decomposition $G \cong \mathbf{Z}_{n_0} \times \cdots \times \mathbf{Z}_{n_L}$, contains a subgroup $H < G$ that is implicitly defined by a function $f : G \to G$ in that f is constant and distinct on every coset of H. Suppose further that U_f can be computed in polylogarithmic time with respect to the size of the group G.

This section shows how, with high probability, generators for H can be found in polylogarithmic time.

A general procedure used to solve the Abelian hidden subgroup problem consists of four steps followed by a final measurement. This procedure is repeated a number of times that depends on the desired level of certainty $1 - \epsilon$.

$$\text{initialization:} \quad \frac{1}{\sqrt{|G|}} \sum_{g \in G} |0\rangle$$

$$U_f: \quad \frac{1}{\sqrt{|G|}} \sum_{g \in G} |g\rangle |f(g)\rangle$$

$$\text{measurement:} \quad \frac{1}{\sqrt{|H|}} \sum_{h \in H} |\tilde{g}h\rangle$$

$$\mathcal{F}_G: \quad \frac{1}{\sqrt{|G||H|}} \sum_{g \in G} \chi_g(\tilde{g}) \left(\sum_{h \in H} \chi_g(h) \right) |g\rangle.$$

A measurement of this state returns with equal probability a $g \in H^{\perp}$ such that $\chi_g(h) = 1$ for all $h \in H$.

We now go through these steps in more detail. After computing U_f on the superposition of all group elements,

$$U_f \left(\frac{1}{\sqrt{|G|}} \sum_{g \in G} |g\rangle |0\rangle \right) = \frac{1}{\sqrt{|G|}} \sum_{g \in G} |g\rangle |f(g)\rangle,$$

a measurement of the second register randomly yields a single $f(\tilde{g})$ for some $\tilde{g} \in G$. Since $f(\tilde{g}) = f(\tilde{g}h)$ for all $h \in H$, and by assumption f is different on every coset, $f(\tilde{g})$ is the value of f on all elements of the coset $\tilde{g}H$ and on no others. After this measurement, we have the state

$$|\psi\rangle = \frac{1}{\sqrt{|H|}} \sum_{h \in H} |\tilde{g}h\rangle,$$

a superposition over only elements of the coset $\tilde{g}H$. Each coset is equally likely to be the result of this measurement, so measuring $|\psi\rangle$ at this point yields a random element $g \in G$ with equal probability. The key insight is that the Fourier transform of the state $|\psi\rangle$ eliminates the constant $\tilde{g}$ and allows us to extract information about H.

The state $|\psi\rangle$ is the image of the state $\frac{1}{\sqrt{|H|}} \sum_{h \in H} |h\rangle$ under the transformation

$$T_{\tilde{g}} : G \to G$$

$$T_{\tilde{g}} : |g\rangle \mapsto |\tilde{g}g\rangle.$$

Apply the quantum Fourier transform to $|\psi\rangle$:

$$\mathcal{F} \frac{1}{\sqrt{|H|}} \sum_{h \in H} |\tilde{g}h\rangle = \frac{1}{\sqrt{|G||H|}} \sum_{h \in H} \sum_{g \in G} \chi_g(\tilde{g}h)|g\rangle$$

$$= \frac{1}{\sqrt{|G||H|}} \sum_{h \in H} \sum_{g \in G} \chi_g(\tilde{g})\chi_g(h)|g\rangle$$

$$= \frac{1}{\sqrt{|G||H|}} \sum_{g \in G} \chi_g(\tilde{g}) \left(\sum_{h \in H} \chi_g(h) \right) |g\rangle.$$

Schur's lemma for subgroups says $\sum_{h \in H} \chi_g(h) \neq 0$ if and only if $\chi_g(h) = 1$ for all $h \in H$. It follows that measuring this state returns the index g of some representation χ_g that is constant (1) on H. For product groups $G = G_0 \times \cdots \times G_L$, the element g of $H^\perp$ returned is in the form $g = (g_0, g_1, \ldots, g_L)$, where g_i is an element of G_i.

To obtain a complete set of generators for $H^\perp$, we repeat the preceding algorithm a number of times that depends on our desired level of certainty $1 - \epsilon$. If the first $n - 1$ group elements returned do not yet generate all of $H^\perp$, the next run through the algorithm has at least a 50 percent chance of returning an element of $H^\perp$ not generated by the previous elements, since any proper subgroup has index at least 2 in the whole group. Thus, by repeating this procedure the appropriate number of times, we can obtain any desired level of certainty $1 - \epsilon$.

We have now completed the quantum part of the solution. From a sufficient number of elements of $H^\perp$, classical methods can efficiently find a full set of generators for H.

B.4 Instances of the Abelian Hidden Subgroup Problem

B.4.1 Simon's Problem
Simon's problem works with the group $G = \mathbf{Z}_2^n$ that has representations $\chi_x(y) = (-1)^{x \cdot y}$. The function

$$f(g \oplus a) = f(g)$$

defines a subgroup

$$A = \{0, a\}.$$

The measurement at the end of one run through the four-step procedure for solving Abelian hidden subgroup problems returns an element

$$x_j \in A^\perp = \{x | (-1)^{x \cdot y} = 1 \text{ for all } y \in A\}.$$

The element x_j must satisfy $x_j \cdot y = 0 \bmod 2$ for all $y \in A$. With sufficiently many values x_j, we can solve for a. In this problem, we know that we have found a solution when there is a unique non-zero solution for a.

B.4.2 Shor's Algorithm: Finding the Period of a Function

For simplicity, assume that r divides n (see section 8.2.1 for the general case) and work with the group

$$G = \mathbf{Z}_n.$$

The periodic function f has the property

$$f(x+r) = f(x),$$

which defines the subgroup

$$H = \{kr \mid k \in [0, \ldots n/r)\}.$$

The problem is to find the generator r of the subgroup. In the standard labeling of representations for $\mathbf{Z}_n$,

$$\chi_g(h) = \exp\left(2\pi i \frac{gh}{n}\right)$$

and

$$H^\perp = \{x \mid \exp\left(2\pi i \frac{xh}{n}\right) = 1 \text{ for all } h \in H\}$$

$$= \{x \mid xkr = 0 \bmod n \text{ for all } k \in [0, \ldots n/r)\}.$$

Measurement after one round of the four-step procedure yields $x \in H^\perp$. The element x satisfies $xkr = 0 \bmod n$ for all $k \in [0, \ldots n/r)$. In particular, $xr = 0 \bmod n$, so x is a multiple of n/r. The period r can now be computed as in section 8.2.1.

B.5 Comments on the Non-Abelian Hidden Subgroup Problem

No one knows how to solve the general hidden subgroup problem. Quantum Fourier transforms can be defined over non-Abelian groups. In fact, efficient implementations of quantum Fourier transforms over all finite groups are known. It is not known, however, how to use the quantum Fourier transformation to extract information about the generators of hidden subgroups for most non-Abelian groups. Worse still, researchers have proved that Fourier sampling, a general technique based on Shor's technique, cannot be used to solve the general hidden subgroup problem. Section 13.1 briefly describes more recent progress in understanding quantum approaches to the non-Abelian hidden subgroup problem.

B.6 References

Kitaev [172] presents a solution for the Abelian stabilizer problem and relates it to factoring and discrete logarithms. The general hidden subgroup problem as presented in this appendix and its solution were introduced by Mosca and Ekert [214]. Ekert and Jozsa [112] and Jozsa [165] analyze the quantum Fourier transform in the context of the hidden subgroup problem. Hallgren [148] studies extensions to the non-Abelian case. Grigni et al. [141] showed in 2001 that for most non-Abelian groups, Fourier sampling yields only exponentially little information about the hidden subgroup.

B.7   Exercises

Exercise B.1.   Let G and H be finite graphs. A map $f : G \rightarrow H$ is a *graph isomorphism* if it is one-to-one and $f(g_1)$ and $f(g_2)$ have an edge between them if and only if g_1 and g_2 do. An *automorphism* of G is a graph isomorphism from G to itself, $f : G \rightarrow G$. A graph automorphism of G is a permutation of its vertices. The *graph isomorphism problem* is to find an efficient algorithm for determining whether there is an isomorphism between two graphs or not.

a. Show that the set $\text{Aut}(G)$ of automorphisms of a graph G forms a group, a subgroup of the permutation group S_n, where $n = |G|$.

b. Two graphs G_1 and G_2 are isomorphic if there exists at least one automorphism in $\text{Aut}(G_1 \cup G_2) < S^{2n}$ that maps nodes of G_1 to G_2 and vice versa. Show that if G_1 and G_2 are nonisomorphic connected graphs, then $\text{Aut}(G_1 \cup G_2) = \text{Aut}(G_1) \times \text{Aut}(G_2)$.

c. Show that if $\text{Aut}(G_1 \cup G_2)$ is strictly bigger than $\text{Aut}(G_1) \times \text{Aut}(G_2)$, then there must be an element of $\text{Aut}(G_1 \cup G_2)$ that swaps G_1 and G_2.

d. Express the graph isomorphism problem as a hidden subgroup problem.

Exercise B.2.   Write out the algorithm that solves Simon's problem using the hidden subgroup framework of section B.3.

Exercise B.3.   Write out an algorithm that finds the period of a function using the hidden subgroup framework of section B.3.

Exercise B.4.   Find an efficient algorithm that solves the discrete logarithm problem.

Bibliography

[1] Scott Aaronson. Quantum lower bounds for the collision problem. In *Proceedings of STOC '02*, pages 635–642, 2002.

[2] Scott Aaronson. Lower bounds for local search by quantum arguments. In *Proceedings of STOC '04*, pages 465–474, 2004.

[3] Scott Aaronson. Are quantum states exponentially long vectors? arXiv:quant-ph/0507242, 2005.

[4] Scott Aaronson. Quantum computing, postselection, and probabilistic polynomial-time. *Proceedings of the Royal Society A*, 461:3473–3482, 2005.

[5] Scott Aaronson. The limits of quantum computers. *Scientific American*, 298(3):62–69, March 2008.

[6] Scott Aaronson and Yaoyun Shi. Quantum lower bounds for the collision and the element distinctness problems. *Journal of the ACM*, 51(4):595–605, 2004.

[7] Daniel S. Abrams and Seth Lloyd. Nonlinear quantum mechanics implies polynomial-time solution for NP-complete and #P problems. *Physical Review Letters*, 81:3992–3995, 1998.

[8] Mark Adcock and Richard Cleve. A quantum Goldreich-Levin theorem with cryptographic applications. In *Proceedings of STACS '02*, pages 323–334, 2002.

[9] Dorit Aharonov and Michael Ben-Or. Fault-tolerant quantum computation with constant error. In *Proceedings of STOC '97*, pages 176–188, 1997.

[10] Dorit Aharonov and Michael Ben-Or. Fault-tolerant quantum computation with constant error rate. arXiv:quant-ph/9906129, 1999.

[11] Dorit Aharonov, Vaughan Jones, and Zeph Landau. A polynomial quantum algorithm for approximating the Jones polynomial. In *Proceedings of STOC '06*, pages 427–436, 2006.

[12] Dorit Aharonov, Zeph Landau, and Johann Makowsky. The quantum FFT can be classically simulated. Los Alamos Physics Preprint Archive, http://xxx.lanl.gov/abs/quant-ph/0611156, 2006.

[13] Dorit Aharonov and Oded Regev. A lattice problem in quantum NP. In *Proceedings of FOCS '03*, pages 210–219, 2003.

[14] Dorit Aharonov and Oded Regev. Lattice problems in NP ∩ coNP. *Journal of the ACM*, 52(5):749–765, 2005.

[15] Dorit Aharonov and Amnon Ta-Shma. Adiabatic quantum state generation and statistical zero knowledge. In *Proceedings of STOC '03*, pages 20–29, 2003.

[16] Dorit Aharonov, Wim van Dam, Julia Kempe, Zeph Landau, Seth Lloyd, and Oded Regev. Adiabatic quantum computation is equivalent to standard quantum computation. *SIAM Journal on Computing*, 37(1):166–194, 2007.

[17] Gorjan Alagic, Cristopher Moore, and Alexander Russell. Quantum algorithms for Simon's problem over general groups. In *Proceedings of SODA '07*, pages 1217–1224, 2007.

[18] Panos Aliferis. Level reduction and the quantum threshold theorem. Ph.D. thesis, Caltech, 2007.

[19] Panos Aliferis, Daniel Gottesman, and John Preskill. Quantum accuracy threshold for concatenated distance-3 codes. *Quantum Information and Computation*, 6(2):97–165, 2006.

[20] Andris Ambainis. A better lower bound for quantum algorithms searching an ordered list. In *Proceedings of FOCS '99*, pages 352–357, 1999.

[21] Andris Ambainis. Quantum lower bounds by quantum arguments. In *Proceedings of STOC '00*, pages 636–643, 2000.

[22] Andris Ambainis. Quantum walks and their algorithmic applications. *International Journal of Quantum Information*, 1:507–518, 2003.

[23] Andris Ambainis. Quantum walk algorithm for element distinctness. In *Proceedings of FOCS'02*, pages 22–31, 2004.

[24] Alain Aspect, Jean Dalibard, and Gérard Roger. Experimental test of Bell's inequalities using time-varying analyzers. *Physical Review Letters*, 49:1804–1808, 1982.

[25] Alain Aspect, Philippe Grangier, and Gérard Roger. Experimental tests of realistic local theories via Bell's theorem. *Physical Review Letters*, 47:460–463, 1981.

[26] Alain Aspect, Philippe Grangier, and Gérard Roger. Experimental realization of Einstein-Podolsky-Rosen-Bohm gedanken experiment: A new violation of Bell's inequalities. *Physical Review Letters*, 49:91–94, 1982.

[27] Alp Atici and Rocco Servedio. Improved bounds on quantum learning algorithms. *Quantum Information Processing*, 4(5):355–386, 2005.

[28] Dave Bacon. Does our universe allow for robust quantum computation? *Science*, 317(5846):1876, 2007.

[29] Dave Bacon, Andrew Childs, and Wim van Dam. From optimal measurement to efficient quantum algorithms for the hidden subgroup problem over semidirect product groups. In *Proceedings of FOCS '05*, 2005.

[30] Paul Bamberg and Shlomo Sternberg. *A Course in Mathematics for Students of Physics*, volume 2. Cambridge University Press, 1990.

[31] Adriano Barenco, Charles H. Bennett, Richard Cleve, David P. DiVincenzo, Norman H. Margolus, Peter W. Shor, Tycho Sleator, John A. Smolin, and Harald Weinfurter. Elementary gates for quantum computation. *Physical Review A*, 52(5):3457–3467, 1995.

[32] Adriano Barenco, Artur K. Ekert, Kalle-Antti Suominen, and Päivi Törmä. Approximate quantum Fourier transform and decoherence. *Physical Review A*, 54(1):139–146, July 1996.

[33] Howard Barnum, Claude Crépeau, Daniel Gottesman, Adam Smith, and Alain Tapp. Authentication of quantum messages. In *Proceedings of FOCS '02*, pages 449–458, 2002.

[34] Robert Beals. Quantum computation of Fourier transforms over the symmetric group. In *Proceedings of STOC '97*, pages 48–53, 1997.

[35] Robert Beals, Harry Buhrman, Richard Cleve, Michele Mosca, and Ronald de Wolf. Quantum lower bounds by polynomials. *Journal of the ACM*, 48(4):778–797, 2001.

[36] John S. Bell. On the Einstein-Podolsky-Rosen paradox. *Physics*, 1:195–200, 1964.

[37] C. H. Bennett, F. Bessette, G. Brassard, L. Salvail, and J. Smolin. Experimental quantum cryptography. *Journal of Cryptology*, 5(1):3–28, 1992.

[38] C. H. Bennett and P. W. Shor. Quantum information theory. *IEEE Transactions on Information Theory*, 44(6):2724–2742, 1998.

[39] Charles H. Bennett. Logical reversibility of computation. *IBM Journal of Research and Development*, 17:525–532, 1973.

[40] Charles H. Bennett. Time/space trade-offs for reversible computation. *SIAM Journal on Computing*, 18(4):766–776, 1989.

[41] Charles H. Bennett, Ethan Bernstein, Gilles Brassard, and Umesh V. Vazirani. Strengths and weaknesses of quantum computing. *SIAM Journal on Computing*, 26(5):1510–1523, 1997.

[42] Charles H. Bennett and Gilles Brassard. Quantum cryptography: Public key distribution and coin tossing. In *Proceedings of IEEE International Conference on Computers, Systems, and Signal Processing*, pages 175–179, 1984.

[43] Charles H. Bennett and Gilles Brassard. Quantum public key distribution reinvented. *SIGACT News*, 18, 1987.

[44] Charles H. Bennett, Gilles Brassard, Claude Crépeau, Richard Jozsa, A. Peres, and William K. Wootters. Teleporting an unknown quantum state via dual classical and Einstein-Podolsky-Rosen channels. *Physical Review Letters*, 70:1895–1899, 1993.

[45] Charles H. Bennett, Gilles Brassard, and Artur K. Ekert. Quantum cryptography. *Scientific American*, 267(4):50, October 1992.

[46] Charles H. Bennett and Stephen J. Wiesner. Communication via one- and two-particle operators on Einstein-Podolsky-Rosen states. *Physical Review Letters*, 69:2881–2884, 1992.

[47] Daniel Bernstein, Johannes Buchmann, and Erik Dahmen. *Post-Quantum Cryptography*. Springer Verlag, 2009.

[48] Ethan Bernstein and Umesh V. Vazirani. Quantum complexity theory. In *Proceedings of STOC '93*, pages 11–20, 1993.

[49] Ethan Bernstein and Umesh V. Vazirani. Quantum complexity theory. *SIAM Journal on Computing*, 26(5):1411–1473, 1997.

[50] André Berthiaume and Gilles Brassard. The quantum challenge to structural complexity theory. In *Proceedings of the Seventh Annual Structure in Complexity Theory Conference*, pages 132–137, 1992.

[51] J. Bienfang, A. J. Gross, A. Mink, B. J. Hershman, A. Nakassis, X. Tang, R. Lu, D. H. Su, C. W. Clark, D. J. Williams, E. W. Hagley, and J. Wen. Quantum key distribution with 1.25 gbps clock synchronization. *Optics Express*, 12:2011–2016, 2004.

[52] David Biron, Ofer Biham, Eli Biham, Markus Grassel, and David A. Lidar. Generalized Grover search algorithm for arbitrary initial amplitude distribution. In *Selected Papers from QCQC '98*, pages 140–147, 1998.

[53] Arno Bohm. *Quantum Mechanics: Foundations and Applications*. 3rd ed. Springer Verlag, 1979.

[54] David Bohm. The paradox of Einstein, Rosen, and Podolsky. *Quantum Theory*, pages 611–623, 1951.

[55] Ravi B. Boppana and Michael Sipser. The complexity of finite functions. In J. van Leeuwen, editor, *Handbook of Theoretical Computer Science*, volume A, pages 757–804. Elsevier, 1990.

[56] D. Boschi, S. Branca, F. De Martini, L. Hardy, and S. Popescu. Experimental realization of teleporting an unknown pure quantum state via dual classical and Einstein-Podolski-Rosen channels. *Physical Review Letters*, 80:1121–1125, 1998.

[57] Dirk Bouwmeester, Jian-Wei Pan, Klaus Mattle, Manfred Eibl, Harald Weinfurter, and Anton Zeilinger. Experimental quantum teleportation. *Nature*, 390:575, 1997.

[58] Michel Boyer, Gilles Brassard, Peter Høyer, and Alain Tapp. Tight bounds on quantum search. In *Proceedings of PhysComp '96*, pages 36–43, 1996.

[59] Gilles Brassard. Quantum communication complexity (a survey). arXiv:quant-ph/0101005, 2001.

[60] Gilles Brassard, Richard Cleve, and Alain Tapp. The cost of exactly simulating quantum entanglement with classical communication. *Physical Review Letters*, 83:1874–1877, 1999.

[61] Gilles Brassard, Peter Høyer, and Alain Tapp. Quantum algorithm for the collision problem. *SIGACT News*, 28:14–19, 1997.

[62] Gilles Brassard, Peter Høyer, and Alain Tapp. Quantum counting. *Lecture Notes in Computer Science*, 1443:820–831, 1998.

[63] Sergey Bravyi and Barbara Terhal. Complexity of stoquastic frustration-free Hamiltonians. arXiv:0806.1746, 2008.

[64] Michael J. Bremner, Caterina Mora, and Andreas Winter. Are random pure states useful for quantum computation? *Physical Review Letters*, 102:190502, 2009.

[65] Hans Briegel and Robert Raussendorf. Persistent entanglement in arrays of interacting particles. *Physical Review Letters*, 86(5):910–913, 2001.

[66] E. Oran Brigham. *The Fast Fourier Transform*. Prentice-Hall, 1974.

[67] D. E. Browne. Efficient classical simulation of the quantum Fourier transform. *New Journal of Physics*, 9:146, 2007.

[68] Todd A. Brun, Igor Devetak, and Min-Hsiu Hsieh. Correcting quantum errors with entanglement. *Science*, 314(5798):436–439, 2006.

[69] Dagmar Bruss. Characterizing entanglement. *Journal of Mathematical Physics*, 43(9):4237–4250, 2002.

[70] Nader H. Bshouty and Jeffrey C. Jackson. Learning DNF over the uniform distribution using a quantum example oracle. *SIAM Journal on Computing*, 28:1136–1142, 1999.

[71] Jeffrey Bub. *Interpreting the Quantum World*. Cambridge University Press, 1997.

[72] Harry Buhrman, Richard Cleve, John Watrous, and Ronald de Wolf. Quantum fingerprinting. *Physical Review Letters*, 87(16), 2001.

[73] Harry Buhrman and Ronald de Wolf. A lower bound for quantum search of an ordered list. *Information Processing Letters*, 70(5):205–209, 1999.

[74] Harry Buhrman and Ronald de Wolf. Communication complexity lower bounds by polynomials. In *Proceedings of CCC '01*, pages 120–130, 2001.

[75] Harry Buhrman and Robert Špalek. Quantum verification of matrix products. In *Proceedings of SODA '06*, pages 880–889, 2006.

[76] Angelo C. M. Carollo and Vlatko Vedral. Holonomic quantum computation. arXiv:quant-ph/0504205, 2005.

[77] Nicolas J. Cerf, Lov K. Grover, and Colin P. Williams. Nested quantum search and structured problems. *Physical Review A*, 61(3):032303, 2000.

[78] Andrew Childs, Edward Farhi, Jeffrey Goldstone, and Sam Gutmann. Finding cliques by quantum adiabatic evolution. *Quantum Information and Computation*, 2(181):181–191, 2002.

[79] Andrew Childs, Edward Farhi, and John Preskill. Robustness of adiabatic quantum computation. *Physical Review A*, 65:012322, 2001.

[80] Andrew M. Childs, Richard Cleve, Enrico Deotto, Edward Farhi, Sam Gutmann, and Daniel A. Spielman. Exponential algorithmic speedup by a quantum walk. In *Proceedings of STOC '03*, pages 59–68, 2003.

[81] Andrew M. Childs, Andrew J. Landahl, and Bablo A. Parrilo. Improved quantum algorithms for the ordered search problem via semidefinite programming. *Physical Review A*, 75:032335, 2007.

[82] Andrew M. Childs and Troy Lee. Optimal quantum adversary lower bounds for ordered search. *Lecture Notes in Computer Science*, 5125:869–880, 2008.

[83] Isaac L. Chuang and Michael Nielsen. Prescription for experimental determination of the dynamics of a quantum black box. *Journal of Modern Optics*, 44:2567–2573, 1997.

[84] J. Ignacio Cirac and Peter Zoller. Quantum computations with cold trapped ions. *Physical Review Letters*, 74:4091–4094, 1995.

[85] Richard Cleve. An introduction to quantum complexity theory. arXiv:quant-ph/9906111v1, 1999.

[86] Richard Cleve, Daniel Gottesman, and Hoi-Kwong Lo. How to share a quantum secret. *Physical Review Letters*, 83(3):648–651, 1999.

[87] Graham P. Collins. Computing with quantum knots. *Scientific American*, 294(4):56–63, April 2006.

[88] James W. Cooley and John W. Tukey. An algorithm for the machine calculation of complex Fourier series. *Mathematics of Computation*, 19(90):297–301, 1965.

[89] Don Coppersmith. An approximate Fourier transform useful in quantum factoring. Research Report RC 19642, IBM, 1994.

[90] Thomas H. Cormen, Charles E. Leiserson, Ronald L. Rivest, and Clifford Stein. *Introduction to Algorithms*. MIT Press, 2001.

[91] Claude Crépeau, Daniel Gottesman, and Adam Smith. Secure multi-party quantum computation. In *Proceedings of STOC '02*, pages 643–652, 2002.

[92] Andrew Cross, David P. DiVincenzo, and Barbara Terhal. A comparative code study for quantum fault tolerance. arXiv:quant-ph/0711.1556v1, 2007.

[93] G. M. D'Ariano, D. Kretschmann, D. Schlingemann, and R. F. Werner. Reexamination of quantum bit commitment: The possible and the impossible. *Physical Review A*, 76(3):032328, 2007.

[94] G. Mauro D'Ariano, Matteo G. A. Paris, and Massimiliano F. Sacchi. Quantum tomography. *Advances in Imaging and Electron Physics*, 128:205–308, 2003.

[95] Christopher M. Dawson and Michael Nielsen. The Solovay-Kitaev algorithm. *Quantum Information and Computation*, 6:81–95, 2006.

[96] Ronald de Wolf. Characterization of non-deterministic quantum query and quantum communication complexity. In *Proceedings of CCC '00*, pages 271–278, 2000.

[97] Ronald de Wolf. Quantum communication and complexity. *Theoretical Computer Science*, 287(1):337–353, 2002.

[98] Ronald de Wolf. Lower bounds on matrix rigidity via a quantum argument. *Lecture Notes in Computer Science*, 4051:299–310, 2006.

[99] David Deutsch. Quantum theory, the Church-Turing principle and the universal quantum computer. *Proceedings of the Royal Society of London Ser. A*, A400:97–117, 1985.

[100] David Deutsch. Quantum computational networks. *Proceedings of the Royal Society of London Ser. A*, A425:73–90, 1989.

[101] David Deutsch, Adriano Barenco, and Artur K. Ekert. Universality in quantum computation. *Proceedings of the Royal Society of London Ser. A*, 449:669–677, 1995.

[102] David Deutsch and Richard Jozsa. Rapid solution of problems by quantum computation. *Proceedings of the Royal Society of London Ser. A*, A439:553–558, 1992.

[103] P. A. M. Dirac. *The Principles of Quantum Mechanics*. 4th ed. Oxford University Press, 1958.

[104] David P. DiVincenzo. The physical implementation of quantum computation. *Fortschritte der Physik*, 48:771–784, 2000.

[105] Andrew Drucker and Ronald de Wolf. Quantum proofs for classical theorems. arXiv:0910.3376, 2009.

[106] Paul Dumais, Dominic Mayers, and Louis Salvail. Perfectly concealing quantum bit commitment from any quantum one-way permutation. *Lecture Notes in Computer Science*, 1807:300–315, 2000.

[107] Wolfgang Dür, Guifre Vidal, and J. Ignacio Cirac. Three qubits can be entangled in two inequivalent ways. *Physical Review A*, 62:062314, 2000.

[108] Bryan Eastin and Emanuel Knill. Restrictions on transversal encoded quantum gate sets. *Physical Review Letters*, 102(11)110502, 2009.

[109] Albert Einstein, Boris Podolsky, and Nathan Rosen. Can quantum-mechanical description of physical reality be considered complete? *Physical Review*, 47:777–780, 1935.

[110] Jens Eisert, Martin Wilkens, and Maciej Lewenstein. Quantum games and quantum strategies. *Physical Review Letters*, 83, 1999.

[111] Artur K. Ekert. Quantum cryptography based on Bell's theorem. *Physical Review Letters*, 67(6):661–663, August 1991.

[112] Artur K. Ekert and Richard Jozsa. Quantum algorithms: Entanglement enhanced information processing. In *Proceedings of Royal Society Discussion Meeting "Quantum Computation: Theory and Experiment."* Philosophical Transactions of the Royal Society of London Ser. A, 1998.

[113] J. Emerson, M. Silva, O. Moussa, C. Ryan, M. Laforest, J. Baugh, D. Cory, and R. Laflamme. Symmetrized characterization of noisy quantum processes. *Science*, 317(5846):1893–1896, 2007.

[114] M. Ettinger, P. Høyer, and E. Knill. The quantum query complexity of the hidden subgroup problem is polynomial. *Information Processing Letters*, 91(1):43–48, 2004.

[115] E. Farhi, J. Goldstone, and S. Gutmann. A quantum algorithm for the Hamiltonian NAND tree. arXiv:quant-ph/0702144, 2007.

[116] Edward Farhi, Jeffrey Goldstone, Sam Gutmann, Joshua Lapan, Andrew Lundgren, and Daniel Preda. A quantum adiabatic evolution algorithm applied to instances of an NP-complete problem. *Science*, 292:5516, 2001.

[117] Edward Farhi, Jeffrey Goldstone, Sam Gutmann, and Michael Sipser. A limit on the speed of quantum computation for insertion into an ordered list. arXiv:quant-ph/9812057, 1998.

[118] Edward Farhi, Jeffrey Goldstone, Sam Gutmann, and Michael Sipser. Quantum computation by adiabatic evolution. arXiv:quant-ph/0001106, January 2000.

[119] Richard Feynman. Simulating physics with computers. *International Journal of Theoretical Physics*, 21(6–7):467–488, 1982.

[120] Richard Feynman. Quantum mechanical computers. *Optics News*, 11, 1985.

[121] Richard Feynman. *Feynman Lectures on Computation*. Addison-Wesley, 1996.

[122] Richard P. Feynman, Robert B. Leighton, and Matthew Sands. *Lectures on Physics, Vol. III*. Addison-Wesley, 1965.

[123] Joseph Fourier. *Théorie analytique de la chaleur*. Firmin Didot, 1822.

[124] Edward Fredkin and Tommaso Toffoli. Conservative logic. *International Journal of Theoretical Physics*, 21:219–253, 1982.

[125] Michael H. Freedman, Alexei Kitaev, Michael J. Larsen, and Zhenghan Wang. Topological quantum computation. *Bulletin of the American Mathematical Society*, 40(1):31–38, 2001.

[126] Murray Gell-Mann. Questions for the future. In *The Nature of Matter; Wolfson College Lectures 1980*. Clarendon Press, 1981.

[127] Craig Gentry. A fully homomorphic encryption scheme. Ph.D. thesis, Stanford University, 2009.

[128] Craig Gentry. Fully homomorphic encryption using ideal lattices. In *Proceedings of STOC '09*, pages 169–178, 2009.

[129] Neil A. Gershenfeld and Isaac L. Chuang. Bulk spin resonance quantum computing. *Science*, 275:350–356, 1997.

[130] Nicolas Gisin, Gregoire Ribordy, Wolfgang Tittel, and Hugo Zbinden. Quantum cryptography. *Reviews of Modern Physics*, 74(1):145–195, January 2002.

[131] Oded Goldreich. *Computational Complexity*. Cambridge University Press, 2008.

[132] Steven Gortler and Rocco Servedio. Equivalences and separations between quantum and classical learnability. *SIAM Journal on Computing*, 33(5):1067–1092, 2004.

[133] Daniel Gottesman. The Heisenberg representation of quantum computers. arXiv:quant-ph/9807006, July 1998.

[134] Daniel Gottesman. On the theory of quantum secret sharing. *Physical Review A*, 61, 2000.

[135] Daniel Gottesman. Stabilizer codes and quantum error correction. Ph.D. thesis, Caltech, May 2000.

[136] Daniel Gottesman. Uncloneable encryption. *Quantum Information and Computation*, 3:581–602, 2003.

[137] Daniel Gottesman. Jump-starting quantum error correction with entanglement. *Science*, 314:427, 2006.

[138] Daniel Gottesman. An introduction to quantum error correction and fault-tolerant quantum computation. arXiv:0904.2557, 2009.

[139] Daniel Gottesman and Isaac L. Chuang. Quantum digital signatures. arXiv:quant-ph/0105032, November 2001.

[140] George Greenstein and Arthur G. Zajonc. *The Quantum Challenge*. Jones and Bartlett, 1997.

[141] Michelangelo Grigni, Leonard Schulman, Monica Vazirani, and Umesh V. Vazirani. Quantum mechanical algorithms for the nonabelian hidden subgroup problem. In *Proceedings of STOC '01*, pages 68–74, 2001.

[142] D. Gross, S. T. Flammia, and J. Eisert. Most quantum states are too entangled to be useful as computational resources. *Physical Review Letters*, 102:190501, 2009.

[143] Lov K. Grover. Quantum computers can search arbitrarily large databases by a single query. *Physical Review Letters*, 79(23):4709–4712, 1997.

[144] Lov K. Grover. A framework for fast quantum mechanical algorithms. In *Proceedings of STOC '98*, pages 53–62, 1998.

[145] Gus Gutoski and John Watrous. Toward a general theory of quantum games. In *Proceedings of STOC '07*, pages 565–574, 2007.

[146] Sean Hallgren. Polynomial-time quantum algorithms for Pell's equation and the principal ideal problem. In *Proceedings of STOC '02*, pages 653–658, 2002.

[147] Sean Hallgren, Alexander Russell, and Amnon Ta-Shma. Normal subgroup reconstruction and quantum computing using group representations. In *Proceedings of STOC '00*, pages 627–635, 2000.

[148] Sean Hallgren, Alexander Russell, and Amnon Ta-Shma. The hidden subgroup problem and quantum computation using group representations. *SIAM Journal on Computing*, 32(4):916–934, 2003.

[149] G. H. Hardy and E. M. Wright. *An Introduction to the Theory of Numbers*. Oxford University Press, 1979.

[150] Anthony J. G. Hey. *Feynman and Computation*. Perseus Books, 1999.

[151] Mark Hillery, Vladimir Buzek, and Andre Berthiaume. Quantum secret sharing. *Physical Review A*, 59:1829–1834, 1999.

[152] Kenneth M. Hoffman and Ray Kunze. *Linear Algebra*. 2nd ed. Prentice Hall, 1971.

[153] Tad Hogg. Adiabatic quantum computing for random satisfiability problems. *Physical Review A*, 67:022314, 2003.

[154] Tad Hogg, Carlos Mochon, Wolfgang Polak, and Eleanor Rieffel. Tools for quantum algorithms. *International Journal of Modern Physics*, C10:1347–1362, 1999.

[155] Peter Høyer and Ronald de Wolf. Improved quantum communication complexity bounds for disjointness and equality. In *Proceedings of STACS '02*, pages 299–310, 2002.

[156] R. J. Hughes, J. E. Nordholt, D. Derkacs, and C. G. Peterson. Practical free-space quantum key distribution over 10km in daylight and at night. *New Journal of Physics*, 4:43.1–43.14, 2002.

[157] Richard Hughes et al. Quantum cryptography roadmap, version 1.1. http://qist.lanl.gov, July 2004.

[158] Thomas A. Hungerford. *Algebra*. Springer Verlag, 1997.

[159] Thomas W. Hungerford. *Abstract Algebra: An Introduction*. Saunders College Publishing, 1997.

[160] Markus Hunziker, David A. Meyer, Jihun Park, James Pommersheim, and Mitch Rothstein. The geometry of quantum learning. arXiv:quant-ph/0309059, 2003.

[161] Yoshifumi Inui and François LeGall. An efficient quantum algorithm the non-Abelian hidden subgroup problem over a class of semi-direct product groups. *Quantum Information and Computation*, 7(5):559–570, 2007.

[162] Gabor Ivanyos, Frederic Magniez, and Miklos Santha. Efficient quantum algorithms for some instances of the non-Abelian hidden subgroup problem. *International Journal of Foundations of Computer Science*, 14(5):723–740, 2003.

[163] Thomas Jennewein, Christoph Simon, Gregor Weihs, Harald Weinfurter, and Anton Zeilinger. Quantum cryptography with entangled photons. *Physical Review Letters*, 84:4729–4732, 2000.

[164] David S. Johnson. A catalog of complexity classes. In J. van Leeuwen, editor, *Handbook of Theoretical Computer Science*, volume A, pages 67–162. Elsevier, 1990.

[165] Richard Jozsa. Quantum algorithms and the Fourier transform. *Proceedings of the Royal Society of London Ser. A*, pages 323–337, 1998.

[166] Richard Jozsa. Searching in Grover's algorithm. arXiv:quant-ph/9901021, 1999.

[167] Richard Jozsa and Noah Linden. On the role of entanglement in quantum computational speed-up. *Proceedings of the Royal Society of London Ser. A*, 459:2011–2032, 2003.

[168] Elham Kashefi and Iordanis Kerenidis. Statistical zero knowledge and quantum one-way functions. *Theoretical Computer Science*, 378(1):101–116, 2007.

[169] Julia Kempe. Quantum random walks—an introductory overview. *Contemporary Physics*, 44(4):307–327, 2003.

[170] Iordanis Kerenidis and Ronald de Wolf. Exponential lower bound for 2-query locally decodable codes via a quantum argument. In *Proceedings of STOC '03*, pages 516–525, 2003.

[171] David Kielpinski, Christopher R. Monroe, and David J. Wineland. Architecture for a large-scale ion-trap quantum computer. *Nature*, 417:709–711, 2002.

[172] Alexei Kitaev. Quantum measurements and the Abelian stabilizer problem. arXiv: quant-ph/9511026, 1995.

[173] Alexei Kitaev. Quantum computations: Algorithms and error correction. *Russian Mathematical Surveys*, 52(6):1191–1249, 1997.

[174] Alexei Kitaev. Fault-tolerant quantum computation by anyons. *Annals of Physics*, 303:2, 2003.

[175] Alexei Kitaev, Alexander Shen, and Mikhail N. Vyalyi. *Classical and Quantum Computation*. American Mathematical Society, 2002.

[176] E. Knill, R. Laflamme, R. Martinez, and C.-H. Tseng. A cat-state benchmark on a seven bit quantum computer. arXiv:quant-ph/9908051, 1999.

[177] Emanuel Knill. Approximation by quantum circuits. arXiv:quant-ph/9508006, 1995.

[178] Emanuel Knill. Quantum computing with realistically noisy devices. *Nature*, 434:39–44, 2005.

[179] Emanuel Knill, Raymond Laflamme, and Gerard Milburn. A scheme for efficient quantum computation with linear optics. *Nature*, 409:46–52, 2001.

[180] Emanuel Knill, Raymond Laflamme, and Wojciech H. Zurek. Resilient quantum computation. *Science*, 279:342–345, 1998.

[181] Emanuel Knill, Raymond Laflamme, and Wojciech H. Zurek. Resilient quantum computation: Error models and thresholds. *Proceedings of the Royal Society London A*, 454:365–384, 1998.

[182] Donald E. Knuth. *The Art of Computer Programming, volume 2: Seminumerical Algorithms*. Addison-Wesley, 2nd edition, 1981.

[183] Neal Koblitz and Alfred Menezes. A survey of public-key cryptosystems. *SIAM Review*, 46:599–634, 2004.

[184] David W. Kribs, Raymond Laflamme, and David Poulin. A unified and generalized approach to quantum error correction. *Physical Review Letters*, 94:199–218, 2005.

[185] David W. Kribs, Raymond Laflamme, David Poulin, and Maia Lesosky. Operator quantum error correction. *Quantum Information and Computation*, 6:382–399, 2006.

[186] Hari Krovi and Todd A. Brun. Quantum walks on quotient graphs. arXiv:quant-ph/0701173, 2007.

[187] Greg Kuperberg. Random words, quantum statistics, central limits, random matrices. *Methods and Applications of Analysis*, 9(1):101–119, 2002.

[188] Greg Kuperberg. A concise introduction to quantum probability, quantum mechanics, and quantum computation. Unpublished notes, available at www.math.ucdavis.edu/greg/intro.pdf, 2005.

[189] Greg Kuperberg. A subexponential-time quantum algorithm for the dihedral hidden subgroup problem. *SIAM Journal on Computing*, 35(1):170–188, 2005.

[190] Paul C. Kwiat, Andrew J. Berglund, Joseph B. Altepeter, and Andrew G. White. Experimental verification of decoherence-free subspaces. *Science*, 290:498–501, 2000.

[191] Chris Lomont. The hidden subgroup problem: Review and open problems. arXiv:quant-ph/0411037, 2004.

[192] Steven E. Landsburg. Quantum game theory. *Notices of the American Mathematical Society*, 51(4):394–399, 2004.

[193] Arjen Lenstra and Hendrik Lenstra, editors. *The Development of the Number Field Sieve*, volume 1554 of *Lecture Notes in Mathematics*. Springer Verlag, 1993.

[194] Richard L. Liboff. *Introductory Quantum Mechanics*. 3rd ed. Addison-Wesley, 1997.

[195] Daniel A. Lidar and K. Birgitta Whaley. Decoherence-free subspaces and subsystems. In *Irreversible Quantum Dynamics*, volume 622, pages 83–120, 2003.

[196] Seth Lloyd. Universal quantum simulators. *Science*, 273:1073–1078, 1996.

[197] Hoi-Kwong Lo and H. F. Chau. Why quantum bit commitment and ideal quantum coin tossing are impossible. *Physics D*, 120(1–2):177–187, 1998.

[198] Hoi-Kwong Lo and H. F. Chau. Unconditional security of quantum key distribution over arbitrarily long distances. *Science*, 283:2050–2056, 1999.

[199] Richard A. Low. Large deviation bounds for k-designs. arXiv:0903.5236, 2009.

[200] Frederic Magniez and Ashwin Nayak. Quantum complexity of testing group commutativity. *Algorithmica*, 48(3):221–232, 2007.

[201] Frederic Magniez, Miklos Santha, and Mario Szegedy. Quantum algorithms for the triangle problem. *SIAM Journal on Computing*, 37(2):413–424, 2007.

[202] Yuri I. Manin. Computable and uncomputable. Sovetskoye Radio, Moscow (in Russian), 1980.

[203] Yuri I. Manin. *Mathematics as Metaphor: Selected Essays of Yuri I. Manin*. American Mathematical Society, 2007.

[204] Igor Markov and Yaoyun Shi. Simulating quantum computation by contracting tensor networks. arXiv:quant-ph/0511069, 2005.

[205] Dominic Mayers. Unconditionally secure quantum bit commitment is impossible. *Physical Review Letters*, 78(17):3414–3417, 1997.

[206] Dominic Mayers. Unconditional security in quantum cryptography. *Journal of the ACM*, 48:351–406, 2001.

[207] Ralph C. Merkle. A certified digital signature. In *CRYPTO '89: Proceedings on Advances in Cryptology*, pages 218–238, 1989.

[208] N. David Mermin. Hidden variables and the two theorems of John Bell. *Reviews of Modern Physics*, 65:803–815, 1993.

[209] N. David Mermin. Copenhagen computation: How I learned to stop worrying and love Bohr. *IBM Journal of Research and Development*, 48:53, 2004.

[210] Albert Messiah. *Quantum Mechanics, Vol. II*. Wiley, 1976.

[211] Rodney Van Meter and Mark Oskin. Architectural implications of quantum computing technologies. *Journal on Emerging Technologies in Computing Systems*, 2(1):31–63, 2006.

[212] David A. Meyer. Quantum strategies. *Physical Review Letters*, 82:1052–1055, 1999.

[213] David A. Meyer. Sophisticated quantum search without entanglement. *Physical Review Letters*, 85:2014–2017, 2000.

[214] Michele Mosca and Artur Ekert. The hidden subgroup problem and eigenvalue estimation on a quantum computer. *Lecture Notes in Computer Science*, 1509:174–188, 1999.

[215] Geir Ove Myhr. Measures of entanglement in quantum mechanics. arXiv:quant-ph/0408094, August 2004.

[216] Ashwin Nayak and Felix Wu. The quantum query complexity of approximating the median and related statistics. In *Proceedings of STOC '99*, pages 384–393, 1999.

[217] Michael Nielsen. Conditions for a class of entanglement transformations. *Physics Review Letters*, 83(2):436–439, 1999.

[218] Michael Nielsen. Cluster-state quantum computation. arXiv:quant-ph/0504097, 2005.

[219] Michael Nielsen and Christopher M. Dawson. Fault-tolerant quantum computation with cluster states. *Physical Review A*, 71:042323, 2004.

[220] Michael Nielsen, Henry Haselgrove, and Christopher M. Dawson. Noise thresholds for optical quantum computers. *Physical Review A*, 96:020501, 2006.

[221] Michael Nielsen, Emanuel Knill, and Raymond Laflamme. Complete quantum teleportation using nuclear magnetic resonance. *Nature*, 396:52–55, 1998.

[222] Jeremy L. O'Brien. Optical quantum computing. *Science*, 318(5856):1567–1570, 2008.

[223] Christos H. Papadimitriou. *Computational Complexity*. Addison-Wesley, 1995.

[224] Chris Peikert. Public-key cryptosystems from the worst-case shortest vector problem: Extended abstract. In *Proceedings of STOC '09*, pages 333–342, 2009.

[225] Roger Penrose. *The Emperor's New Mind*. Penguin Books, 1989.

[226] Asher Peres. *Quantum Theory: Concepts and Methods*. Kluwer Academic, 1995.

[227] Pérez-Delgado and Pieter Kok. What is a quantum computer, and how do we build one? arXiv:0906.4344, 2009.

[228] Ray A. Perlner and David A. Cooper. Quantum resistant public key cryptography: A survey. In *IDtrust '09: Proceedings of the 8th Symposium on Identity and Trust on the Internet*, pages 85–93, 2009.

[229] Nicholas Pippenger and Michael J. Fischer. Relations among complexity measures. *Journal of the ACM*, 26(2):361–381, 1979.

[230] Sandu Popescu, Berry Groisman, and Serge Massar. Lower bound on the number of Toffoli gates in a classical reversible circuit through quantum information concepts. *Physical Review Letters*, 95:120503, 2005.

[231] J. F. Poyatos, R. Walser, J. I. Cirac, P. Zoller, and R. Blatt. Motion tomography of a single trapped ion. *Physical Review A*, 53(4):1966–1969, 1996.

[232] Juan Poyatos, J. Ignacio Cirac, and Peter Zoller. Complete characterization of a quantum process: The two-bit quantum gate. *Physical Review Letters*, 78(2):390–393, 1997.

[233] John Preskill. Fault-tolerant quantum computation. In H.-K. Lo, S. Popescu, and T. P. Spiller, editors, *Introduction to Quantum Computation and Information*, pages 213–269. World Scientific, 1998.

[234] H. Ramesh and V. Vinay. On string matching in $\tilde{o}(\sqrt{n} + \sqrt{m})$ quantum time. *Journal of Discrete Algorithms*, 1(1):103–110, 2001.

[235] Robert Raussendorf, Daniel Browne, and Hans Briegel. Measurement-based quantum computation with cluster states. *Physical Review A*, 68:022312, 2003.

[236] Miklos Redei and Stephen J. Summers. Quantum probability theory. arXiv:hep-th/0601158, 2006.

[237] Oded Regev. Quantum computation and lattice problems. In *Proceedings of FOCS '02*, pages 520–529, 2002.

[238] Oded Regev. On lattices, learning with errors, random linear codes, and cryptography. In *Proceedings of STOC '05*, pages 84–93, 2005.

[239] Oded Regev. A subexponential-time quantum algorithm for the dihedral hidden subgroup problem with polynomial space. arXiv:quant-ph/0406151, 2005.

[240] Ben W. Reichardt. The quantum adiabatic optimization algorithm and local minima. In *Proceedings of STOC '04*, pages 502–510, 2004.

[241] Eleanor Rieffel. Certainty and uncertainty in quantum information processing. In *Proceedings of the AAAI Spring Symposium 2007*, pages 134–141, 2007.

[242] Eleanor Rieffel. Quantum computing. In *The Handbook of Technology Management*, pages 384–392. Wiley, 2009.

[243] Jeremie Roland and Nicholas Cerf. Quantum search by local adiabatic evolution. arXiv:quant-ph/0107015, 2001.

[244] Markus Rötteler and Thomas Beth. Polynomial-time solution to the hidden subgroup problem for a class of non-Abelian groups. arXiv:quant-ph/9812070, 1998.

[245] Arnold Schönhage and Volker Strassen. Schnelle Multiplikation grosser Zahlen [Fast multiplication of large numbers]. *Computing*, 7(3–4):281–292, 1971.

[246] Rocco A. Servedio. Separating quantum and classical learning. *Lecture Notes in Computer Science*, 2076:1065–1080, 2001.

[247] Ramamurti Shankar. *Principles of Quantum Mechanics*. 2nd ed. Plenum Press, 1980.

[248] Yaoyun Shi. Quantum lower bounds for the collision and the element distinctness problems. In *Proceedings of FOCS '02*, pages 513–519, 2002.

[249] Yaoyun Shi, Luming Duan, and Guifre Vidal. Classical simulation of quantum many-body systems with a tree tensor network. *Physical Review A*, 74(2):022320, August 2006.

[250] Peter W. Shor. Algorithms for quantum computation: Discrete log and factoring. In *Proceedings of FOCS'94*, pages 124–134, 1994.

[251] Peter W. Shor. Scheme for reducing decoherence in quantum memory. *Physical Review A*, 52, 1995.

[252] Peter W. Shor. Fault-tolerant quantum computation. In *Proceedings of FOCS '96*, pages 56–65, 1996.

[253] Peter W. Shor. Polynomial-time algorithms for prime factorization and discrete logarithms on a quantum computer. *SIAM Journal on Computing*, 26(5):1484–1509, 1997.

[254] Peter W. Shor. Progress in quantum algorithms. *Quantum Information Processing*, 3(1–5):5–13, 2004.

[255] Peter W. Shor and John Preskill. Simple proof of security of the BB84 quantum key distribution protocol. *Physical Review Letters*, 85:441–444, 2000.

[256] David R. Simon. On the power of quantum computation. *SIAM Journal on Computing*, 26(5):1474–1483, 1997.

[257] Rolando D. Somma, Gerardo Ortiz, Emanuel Knill, and James Gubernatis. Quantum simulations of physics problems. In *Quantum Information and Computation*, volume 5105, pages 96–103, 2003.

[258] Andrew Steane. The ion trap quantum information processor. *Applied Physics B*, 64:623–642, 1996.

[259] Andrew Steane. Multiple particle interference and quantum error correction. *Proceedings of the Royal Society of London Ser. A*, 452: 2551–2573, 1996.

[260] Andrew Steane. Quantum computing. *Reports on Progress in Physics*, 61(2):117–173, 1998.

[261] Andrew Steane. Efficient fault-tolerant quantum computing. *Nature*, 399:124–126, 1999.

[262] Andrew Steane and David M. Lucas. Quantum computing with trapped ions, atoms and light. *Fortschritte der Physik*, 48:839–858, 2000.

[263] Andrew M. Steane. Overhead and noise threshold of fault-tolerant quantum error correction. *Physical Review A*, 68(4):042322, 2003.

[264] Gilbert Strang. *Introduction to Applied Mathematics*. Wellesley-Cambridge Press, 1986.

[265] Gilbert Strang. *Linear Algebra and its Applications*. Harcourt Brace Jovanovich, 1988.

[266] Franco Strocchi. *An Introduction to the Mathematical Structure of Quantum Mechanics*. World Scientific, 2005.

[267] Anthony Sudbery. *Quantum Mechanics and the Particles of Nature*. Cambridge University Press, 1986.

[268] Barbara Terhal and Guido Burkard. Fault-tolerant quantum computation for local non-Markovian noise. *Physical Review A*, 71:012336, 2005.

[269] Barbara M. Terhal and John A. Smolin. Single quantum querying of a database. *Physical Review A*, 58(3):1822–1826, 1998.

[270] Tommaso Toffoli. Reversible computing. In J. W. de Bakker and Jan van Leeuwen, editors, *Automata, Languages and Programming*, pages 632–644. Springer Verlag, 1980.

[271] Joseph F. Traub and Henryk Woźniakowski. Path integration on a quantum computer. *Quantum Information Processing*, 1(5):365–388, 2002.

[272] Wim van Dam, Sean Hallgren, and Lawrence Ip. Quantum algorithms for some hidden shift problems. In *Proceedings of SODA '03*, pages 489–498, 2003.

[273] Wim van Dam and Umesh V. Vazirani. Limits on quantum adiabatic optimization, 2003.

[274] Vlatko Vedral. The elusive source of quantum effectiveness. arXiv:0906.2656, 2009.

[275] Vlatko Vedral, Adriano Barenco, and Artur K. Ekert. Quantum networks for elementary arithmetic operations. *Physical Review A*, 54(1):147–153, 1996.

[276] Frank Verstraete, Diego Porras, and J. Ignacio Cirac. DMRG and periodic boundary conditions: A quantum information perspective. *Physical Review Letters*, 93:227205, 2004.

[277] George F. Viamontes, Igor L. Markov, and John P. Hayes. Is quantum search practical? *Computing in Science and Engineering*, 7(3):62–70, 2005.

[278] Guifre Vidal. Efficient classical simulation of slightly entangled quantum computations. *Physical Review Letters*, 91:147902, 2003.

[279] Heribert Vollmer. *Introduction to Circuit Complexity*. Springer, 1999.

[280] John Watrous. Zero-knowledge against quantum attacks. In *Proceedings of STOC '06*, pages 296–305, 2006.

[281] John Watrous. Quantum computational complexity. arXiv:0804.3401, 2008.

[282] Stephanie Wehner and Ronald de Wolf. Improved lower bounds for locally decodable codes and private information retrieval. *Lecture Notes in Computer Science*, 3580:1424–1436, 2005.

[283] Stephen B. Wicker. *Error Control Systems for Digital Communication and Storage*. Prentice-Hall, 1995.

[284] Stephen Wiesner. Conjugate coding. *SIGACT News*, 15:78–88, 1983.

[285] Stephen Wiesner. Simulations of many-body quantum systems by a quantum computer. arXiv:quant-ph/9603028, March 1996.

[286] William K. Wootters and Wojciech H. Zurek. A single quantum cannot be cloned. *Nature*, 299:802, 1982.

[287] Andrew Yao. Quantum circuit complexity. In *Proceedings of FOCS '93*, pages 352–361, 1993.

[288] Nadav Yoran and Anthony J. Short. Efficient classical simulation of the approximate quantum Fourier transform. *Physical Review A*, 76(4):060302, 2007.

[289] Christof Zalka. Simulating quantum systems on a quantum computer. *Royal Society of London Proceedings Series A*, 454:313–322, 1998.

[290] Christof Zalka. Grover's quantum searching algorithm is optimal. *Physical Review A*, 60(4):2746–2751, 1999.

[291] Christof Zalka. Using Grover's quantum algorithm for searching actual databases. *Physical Review A*, 62(5):052305, 2000.

[292] Paolo Zanardi and Mario Rasetti. Noiseless quantum codes. *Physical Review Letters*, 79(17):3306–3309, 1997.

[293] Paolo Zanardi and Mario Rasetti. Holonomic quantum computation. *Physical Review A*, 264:94–99, 1999.

[294] Anton Zeilinger. Quantum entangled bits step closer to IT. *Science*, 289:405–406, 2000.

[295] Peter Zoller et al. Quantum information processing and communication: Strategic report on current status, visions and goals for research in Europe. http://qist.ect.it, 2005.

Notation Index

Standard Notation

$\lvert x \rvert$	absolute value
$[x..y]$	closed interval
$\approx$	approximately equal
e	2.718281 ...
i	$\sqrt{-1}$
$\exp(x)$	e^x
log	logarithm base e
$\log_m$	logarithm base m
v	traditional vector notation
$\mathbf{v}^T$	transpose of a vector or matrix
a_{ij}	element i,j of matrix A
det A	determinant of A
λ	generic eigenvalue
U^{-1}	inverse of a unitary transformation, quantum algorithm
$U^{\dagger}$	conjugate transpose
C	the complex numbers
R	the real numbers
$\mathbf{R}^n$	n dimensional real space
Z	the natural numbers
$\mathbf{Z}_2$	the natural numbers modulo 2
$\mathbf{Z}_2^n$	group of n-bit strings under bitwise addition modulo 2
$\lvert G \rvert$	order of a group
$\hat{G}$	the group of representations of G
χ	group homomorphism
$H < G$	subgroup relation
o	generic group operation
$\cong$	isomorphism
$Z(S)$	the centralizer of subgroup S

General Concepts

Linear Algebra

Vectors

Matrices

Quantum States

Transformations, Operators

Index

Scientific and Engineering Computation

William Gropp and Ewing Lusk, editors; Janusz Kowalik, founding editor

Beowulf Cluster Computing with Linux, 2nd edition, William Gropp, Ewing Lusk, and Thomas Sterling, 2003

Beowulf Cluster Computing with Windows, Thomas Sterling, 2001

Scalable Input/Output: Achieving System Balance, Daniel A. Reed, 2003

Using OpenMP: Portable Shared Memory Parallel Programming, Barbara Chapman, Gabriele Jost, and Ruud van der Pas, 2008

Quantum Computing Without Magic, Zdzislaw Meglicki, 2008